M2M COMMUNICATIONS

M2M COMMUNICATIONS
A SYSTEMS APPROACH

Editors

David Boswarthick
ETSI, France

Omar Elloumi
Alcatel-Lucent, France

Olivier Hersent
Actility, France

WILEY
A John Wiley & Sons, Ltd., Publication

This edition first published 2012

Registered office
John Wiley & Sons Ltd, The Atrium, Southern Gate, Chichester, West Sussex, PO19 8SQ, United Kingdom

For details of our global editorial offices, for customer services and for information about how to apply for permission to reuse the copyright material in this book please see our website at www.wiley.com.

Library of Congress Cataloging-in-Publication Data

M2M communications : a systems approach / edited by David
Boswarthick, Omar Elloumi, Olivier Hersent.
p. cm.
Includes bibliographical references and index.
ISBN 978-1-119-99475-6 (cloth)
1. Machine-to-machine communications. I. Boswarthick, David.
II. Elloumi, Omar. III. Title: Machine-to-machine communications.
TK5105.67.M32 2012
621.39′8–dc23

2011044199

A catalogue record for this book is available from the British Library.

Print ISBN: 9781119994756

Typeset in 10/12 Times by Laserwords Private Limited, Chennai, India.

Contents

Foreword

It was with great pleasure that I accepted the invitation to write the foreword for this first of two books on M2M and Internet of Things, *M2M Communications: A Systems Approach*.

Although the market for Machine to Machine (M2M) devices and applications is still developing, we can already foresee that this technology will have a profound impact on our lives, as new application fields are explored. Numerous projections have been made for the growth of M2M: for example estimates include an increase from the current 6 billion cellular devices to eventually over 50 billion cellular-connected machines. Other estimates indicate the total market volume from M2M and Internet of Things reaching $11.5bn by 2012. Indeed the M2M market can already be segmented in numerous ways: distinguishing between hardware devices and software, between connection technologies, or according to specific industrial application segment.

Why are M2M markets taking off now, since many of the technologies used have existed for a number of years? A key factor in the growth of M2M today is the widespread availability of ubiquitous, low-cost connectivity. We have become used to cheap, high-speed home and business internet access. Now in many regions 3G and future LTE mobile networks offer similar access speeds at highly competitive prices. Suddenly a host of applications we have dreamt about and which require internet connectivity have become economically viable.

The large-scale deployment of IP-connected sensors, monitors and actuators, in the home and in industry, enables the development of new interconnected, interoperable services which hold the potential to transform our daily lives. M2M technologies offer a vision of mash-up applications founded in reality, utilising multiple new sources of information, in contrast to the virtual world of mash-up web services. This vision is sometimes referred to as 'The Internet of Things', but it's not the connected things which are important. Instead, what is important is the information which they provide us, and how we combine and present and use this information, and how we make decisions based upon it. The Internet of Things offers a technical viewpoint. We must look beyond that to see the societal impact, to understand how we will make use of this technology to change our lives for the better.

A key feature of this vision of the future is the variety and range of technologies, functionalities and requirements which we need to take into account. How can we develop a flexible architecture into which we can place today's and tomorrow's technologies? How can we enable interoperability? How can we preserve confidentiality and privacy of information while not restricting potentially beneficial new applications? How can we ensure the reliability of these systems we will build, as we grow increasingly dependent on them? The solutions to these challenges lie not with any one organization or individual. This requires cross-industry thinking, it requires collaboration between the different actors concerned and it requires co-ordination at an international level. Consensus-based international standards are essential to ensure the development of M2M technologies and markets, providing solutions to many of these challenges. ETSI's M2M Technical Committee is currently playing a leading role in driving international standards work in this domain.

Beyond the wider societal challenges, we also face many detailed technical issues. A rapid deployment and adoption of M2M technology will result in new demands being placed on our networks. M2M services often require high efficiency, low overhead, low power consumption and greater flexibility in networks. These requirements will compete with the demand for high speed, low latency and large capacity, which our networks are currently equipped to handle. We may need to re-think how we design and how we manage our networks, if M2M services become as widespread as forecast. We will need to consider new access technologies in order to enable new applications which are not well served by the current radio technologies on the market.

The growth of M2M applications will have a profound impact on the standards which define our telecoms and data networks. ETSI has many years of work ahead of it to develop the specifications and standards which will be needed. I am certain that this book and its accompanying volume will provide us with useful guidance in this task, helping us to better understand the issues we need to tackle in order to create an M2M-enabled world.

Luis Jorge Romero
ETSI Director General

List of Contributors

Samia Benrachi-Maassam
Network & Services Architect
Bouygues Telecom
299 Ter Avenue Division Leclerc
92290 Chatenay-Malabry, France

David Boswarthick
Technical Officer, TC M2M
ETSI
650, Route des Lucioles
06921 Sophia Antipolis, France

Ioannis Broustis
Member of Technical Staff
Alcatel-Lucent
600 Mountain Avenue
Murray Hill, NJ 07974, USA

Emmanuel Darmois
Vice President, Standards
Alcatel-Lucent
7–9 Avenue Morane Sauliner, BP 57
78141 Velizy, France

Omar Elloumi
Director Standardisation, M2M and Smart Technologies
Alcatel-Lucent
7–9 Avenue Morane Sauliner, BP 57
78141 Velizy, France

François Ennesser
Technical Marketing – Standardization & Technology
Gemalto S.A.
6 rue de Verrerie
92190 Meudon, France

Claudio Forlivesi
Research Engineer
Alcatel-Lucent
Copernicuslaan 50
2018 Antwerp, Belgium

Bruno Landais
Network Architect
Alcatel-Lucent
4 rue L. de Broglie, BP 50444
22304 Lannion, France

Ana Minaburo
Independant Consultant
Cesson Sevigne Cedex, France

Simon Mizikovsky
Technical Manager
Alcatel-Lucent
600 Mountain Ave.
Murray Hill, NJ 07974, USA

Toon Norp
Senior Business Consultant
TNO
Brassersplein 2,
NL-2612 Delft, Netherlands

Franck Scholler
E2E Network Solution Architect Manager
Alcatel-Lucent
7–9 Avenue Morane Sauliner, BP 57
78141 Velizy, France

Ganesh Sundaram
Distinguished Member of Technical Staff
Alcatel-Lucent
600 Mountain Ave.
Murray Hill, NJ 07974, USA

Laurent Toutain
Associate Professor
Telecom Bretagne
2 rue de la Chataigneraie, CS 17607
35576 Cesson Sevigne Cedex, France

Harish Viswanathan
CTO Advisor
Alcatel-Lucent
600 Mountain Ave.
Murray Hill, NJ 07974, USA

Gustav Vos
Director, Technology Standards
Sierra Wireless
13811 Wireless Way
Richmond, BC, V6V3A4, Canada

List of Acronyms

3GPP	Third-Generation Partnership Project
6LoWPAN	IPv6 Over Low Power and Lossy Networks
AAA	Authentication, Authorization and Accounting
ABRO	Authoritative Border Router Option
ADC	Analog-to-Digital Converter
AKA	Authentication and Key Agreement
ALG	Application Level Gateway
AMPS	American Mobile Phone Service
AODV	Ad hoc On-Demand Distance Vector
API	Application Programming Interface
APN	Access Point Name
ARO	Address Registration Option
ARP	Allocation and Retention Priority
ARPU	Average Revenue Per User
ARRA	American Recovery and Reinvestment Act
ATM	Automated Teller Machine
B2B	Business-to-Business
B2C	Business-to-Consumer
BGA	Ball Grid Array
BOM	Bill of Materials
CA	Certificate Authority
CAPEX	CAPital EXpenditure
CAT	Card Application Toolkit
CBC	Cell Broadcast Center
CCF	CDMA Certification Forum
CDMA	Code Division Multiple Access
CDR	Charging Data Record
CEN	Comité Européen de Normalisation
CENELEC	Comité Européen de Normalisation Électrotechnique
CHAP	CHallenge Authentication Protocol
CID	Connection IDentifier
CIDR	Classless Interdomain Routing

CM	Configuration Management
CoAP	Constrained Application Protocol
CRL	Certificate Revocation List
CSD	Circuit-Switched Data
CSP	Communications Service Provider
DA	Device Application
DAD	Duplicate Address Detection
DAO	DODAG Advertisement Object
DAR	Discovery Address Request
DHCP	Dynamic Host Configuration Protocol
DIO	DODAG Information Object
DIS	DODAG Information Solicitation
DLMS	Device Language Message Specification
DM	Device Management
DNS	Domain Name Server
DODAG	Direction Oriented Directed Acyclic Graph
DoE	Department of Energy
DPWS	Devices Profile for Web Services
DSCL	Device Service Capabilities Layer
DSCP	Differentiated Services Control Point
DTLS	Datagram Transport Layer Security
DTMF	Dual Tone Multi-Frequency
EAB	Extended Access Barring
EAL	Evaluation Assurance Level
EAP	Extensible Authentication Protocol
ECC	Elliptic Curve Cryptography
EEPROM	Electrically Erasable Programmable Read-Only Memory
ESO	European Standard Organisation
ETSI	European Telecommunications Standards Institute
EUSD	European Universal Service Directive
EV	Electric Vehicle
EVDO	Evolution-Data Optimized
FCC	Federal Communication Commission
FDMA	Frequency Division Multiple Access
FOTA	Firmware-Over-The-Air
FP8	8th Framework Program
FQDN	Fully Qualified Domain Name
FTP	File Transfer Protocol
FTTH	Fibre To The Home
GBA	Generic Bootstrapping Architecture
GCF	Global Certification Forum
GGSN	Gateway GPRS Support Node
GMSC	GSM Mobile Services Switching Center
GP	Global Prefix
GPIO	General Purpose Input Output
GPRS	General Packet Radio Service

GPS	Global Positioning System
GPSK	Generalized Pre-Shared Key
GRE	Generic Routing Encapsulation
GSCL	Gateway Service Capabilities Layer
GSM	Global Systems *for* Mobile *Communications*
GSMA	GSM Association
GW	GateWay
H2H	Human-to-Human
HCI	Host Controller Interface
HLIM	Hop Limit
HLR	Home Location Register
HSPA	High Speed Packet Access
HSS	Home Subscriber Service
HTTP	HyperText Transfer Protocol
IBAKE	Identity-Based Authenticated Key Exchange
IBE	Identity Based Encryption
IC	Integrated Circuit
ICCID	Integrated Circuit Card Identifier
ICMP	Internet Control Message Protocol
ICS	In-Vehicle System
ICT	Information and Communication Technologies
IDE	Integrated Development Environment
IETF	Internet Engineering Task Force
IID	Interface IDentification
IKE	Internet Key Exchange
IMEI	International Mobile Equipment Identifier
IMEISV	IMEI Software Version
IMS	IP Multimedia Subsystem
IMSI	International Mobile Subscriber Identity
IoT	Internet of Things
IOT	Interoperability Testing
IP	Internet Protocol
IPHC	IP Header Compression
IPSEC	Internet Protocol Security
ISC	IMS Service Control
IS-IS	Intermediate System to Intermediate System
ITU	International Telecommunication Union
J2EE	Java 2 Enterprise Edition
KGF	Key Generation Function
LAN	Local Area Network
LBR	LoWPAN Border Router
LDAP	Lightweight Directory Access Protocol
LGA	Line Grid Array
LLN	Low Power and Lossy Network
LTE	Long Term Evolution
M2M	Machine-to-Machine

M2ME	M2M Equipment
MAC	Medium Access Control
MAS	M2M Authentication Server
M-BUS	Meter BUS
MCM	Multichip Module
MEID	Mobile Equipment Identifier
MIMO	Multiple-Input, Multiple-Output
MME	Mobility Management Entity
MMI	Man-Machine Interface
MNC	Mobile Network Code
MNO	Mobile Network Operator
MO	Management Object
MP2P	Multi-Point To Point
MPLS	Multiprotocol Label Switching
MRHOF	Minimum Rank with Hysteresis Objective Function
MSC	Mobile Switching Center
MSF	M2M Service Bootstrap Function
MSIM	M2M Service Identity Module
MSIN	Mobile Subscriber Identification Number
MSISDN	Mobile Station Integrated Services Digital Network
MTC	Machine Type Communications
MTC-GW	Machine Type Communication Gateway
MTU	Maximum Transfer Unit
MVNO	Mobile Virtual Network Operator
NA	Network Application
NAPT	Network Address and Port Translation
NAS	Non Access Stratum
NAT	Network Address Translator
NBMA	Non-Broadcast Multiple-Access
NC	Node Confirmation
NDIS	Network Driver Interface Specification
NDP	Neighbor Discovery Protocol
NFC	Near Field Communications
NGC	Network Generic Communication
NGN	Next Generation Networks
NIST	National Institute of Standards and Technology
NR	Node Registration
NRPCA	Network-Requested PDP Context Activation
NS	Neigbour Solicitation
NSCL	Network Service Capabilities Layer
NSEC	Network Security Capability
NTACS	Narrowband Total Access Communication System
NUD	Neighbor Unreachability Detection

OC	Option Count
OCS	Online Charging Services
OCSP	Online Certificate Status Protocol
OFDMA	Orthogonal Frequency Division Multiple Access
OLSR	Optimized Link State Routing
OMA	Open Mobile Alliance
OPEX	OPerational EXpenditure
OS	Operating System
OSPF	Open Shortest Path First
OTA	Over-The-Air
OTASP	Over-The-Air Service Provisioning
P2MP	Point to Multi-Point
PAK	Password-Authenticated Key Exchange
PAN	Personal Area Network
PAP	Password Authentication Protocol
PCB	Printed Circuit Board
PCM	Pulse Code Modulation
PCRF	Policy and Charging Rules Function
PDA	Personal Digital Assistant
PDN	Packet Data Network
PDP	Packet Data Protocol
PDP-C	Packet Data Protocol Context
PDSN	Packet Data Serving Node
PGW	PDN (Public Data Network) Gateway
PIN	Personal Identifiaction Number
PKI	Public Key Infrastructure
PLC	Power Line Communication
PM	Performance Management
PoS	Point of Sale
PS	Packet Switched
PSAP	Public Safety Answering Point
PTCRB	PCS Terminal Certification Review Board
PWM	Pulse Width Modulation
QoS	Quality of Service
RA	Registration Authority
RA	Router Advertisement
RAN	Radio Access Network
RAT	Radio Access Technologies
REST	REpresentation State Transfer
RF	Radio Frequency
RFC	Request For Comment
RFID	Radio Frequency IDentification
RIP	Routing Information Protocol

RNC	Radio Network Controller
RoI	Return on Investment
RPL	Routing Protocol for LLN (Low Power and Lossy Network)s
RRC	Radio Resource Control
RSSI	Received Signal Strength Indicator
SAC	Source Address Compression
SAM	Source Address Mode
SC	Service Capabilities
SCADA	Supervisory Control and Data Acquisition
SCP	Smart Card Platform
S-CSCF	Serving Call Session Control Function
SDK	Software Development Kit
SDO	Standards Development Organization
SGSN	Serving GPRS Support Node
SGW	Serving Gateway
SID	Subnet IDentifier
SIM	Subscriber Identification Module
SLAAC	StateLess Auto Address Configuration
SLLAO	Source Link-Layer Adress Option
SMS SC	SMS Service Center
SMS	Short Message Service
SMS-C	Short Message Service Center
SMT	Survace Mount Technology
SMTP	Simple Mail Transfer Protocol
SNMP	Simple Network Management Protocol
SNR	Serial NumbeR
SOA	Service Oriented Architecture
SOAP	Simple Object Access Protocol
TACS	Total Access Communications System
TCG	Trusted Computing Group
TCO	Total Cost of Ownership
TCP	Transmission Control Protocol
TDMA	Time Division Multiple Access
TD-SCDMA	Time Division Synchronous Code Division Multiple Access
TIS	Total Isotropic Sensitivity
TLS	Transport Layer Security
TLS-PSK	Transport Layer Security Pre-Shared Key
TLV	Type Length Value
TRP	Total Radiated Power
TTM	Time-to-Market
UART	Universal Asynchronous Receiver/Transmitter
UATI	Unicast Access Terminal Identifier
UDC	Universal Decimal Classification

UDDI	Universal Description Discovery and Integration
UDP	User Datagram Protocol
UE	User Equipment
UICC	Universal Integrated Circuit Card
ULA	Unique Local Address
UMB	Ultra-Mobile Broadband
UMTS	Universal Mobile Telecommunications System
USB	Universal Serial Bus
USIM	Universal Subscriber Identity Module
VLR	Visitor Location Register
VPN	Virtual Private Network
WAN	Wide Area Network
WLAN	Wireless Local-Area Networks
WPAN	Wireless Personal Area Network
WSDL	Web Service Description Language
WSN	Wireless Sensor Network
WWAN	Wireless Wide-Area Networks
xDSL	x Digital Subscriber Line

1

Introduction to M2M

Emmanuel Darmois and Omar Elloumi
Alcatel-Lucent, Velizy, France

M2M (Machine-to-Machine) has come of age. It has been almost a decade since the idea of expanding the scope of entities connected to "the network" (wireless, wireline; private, public) beyond mere humans and their preferred communication gadgets has emerged around the notions of the "Internet of Things" (IoT), the "Internet of Objects" or M2M. The initial vision was that of a myriad of new devices, largely unnoticed by humans, working together to expand the footprint of end-user services. This will create new ways to care for safety or comfort, optimizing a variety of goods-delivery mechanisms, enabling efficient tracking of people or vehicles, and at the same time creating new systems and generating new value.

As with every vision, it has taken time to materialize. Early efforts concentrated on refining the initial vision by testing new business models, developing point solutions to test feasibility, and also forecasting the impact of insufficient interoperability. Over the past few years, the realization that there are new viable sources of demand that can be met and monetized has created the push for a joint effort by industry to turn a patchwork of standalone elements and solutions into a coherent "system of systems", gradually turning the focus from the "what" to the "how" and developing the appropriate technologies and standards.

This chapter introduces the M2M concept and proposes a definition from the multitude of definitions available today. It outlines the main characteristics of the emerging M2M business and presents a high-level view of the M2M framework that is further analyzed and dissected in subsequent chapters. Moreover, this chapter analyzes some of the main changes that have occurred recently and that have largely enabled the development of M2M, namely the emergence of regulation and standards as market shapers. The role of

M2M Communications: A Systems Approach, First Edition.
Edited by David Boswarthick, Omar Elloumi and Olivier Hersent.

standards is one of this book's central themes and a presentation of the main actors and the latest status of related work is provided as a guide through this complex ecosystem.

The reader will finally be introduced to the structure and content of this book, which is actually the first of a set of two. In the hands of the reader in paper format or on an eBook reader after being loaded by an M2M application, the first book *M2M Communications: A Systems Approach* essentially introduces the M2M framework – requirements, high-level architecture – and some of its main systems aspects, such as network optimization for M2M, security, or the role of IP.

The second book *Internet of Things: Key Applications and Protocols* will address more specifically the domain in which the "Internet of Objects" will be acting, namely the M2M area networks, in particular the associated protocols and the interconnection of such networks. It will also analyze, from this perspective, some of the future M2M applications, such as Smart Grids and Home Automation.

1.1 What is M2M?

Many attempts have been made to propose a single definition of the M(s) of the M2M acronym: Machine-to-Machine, Machine-to-Mobile (or vice versa), Machine-to-Man, etc. Throughout this book, M2M is considered to be "Machine-to-Machine". This being decided, defining the complete "Machine-to-Machine" concept is not a simple task either: the scope of M2M is, by nature, elastic, and the boundaries are not always clearly defined.

Perhaps the most basic way to describe M2M is shown in Figure 1.1 (the "essence" of M2M). The role of M2M is to establish the conditions that allow a device to (bidirectionally) exchange information with a business application via a communication network, so that the device and/or application can act as the basis for this information exchange. In this definition, the communication network has a key role: a collocated application and device can hardly be considered as having an M2M relationship. This is why M2M will often be a shortened synonym for M2M communications, which is itself a shortened acronym for M2(CN2)M: Machine-to-(Communication-Network-to-)Machine).

In itself, this description still does not fully characterize M2M. For instance, a mobile phone interacting with a call center application is not seen as an M2M application because a human is in command. Some of the more complex characteristics of the M2M relationship are discussed below in order to clarify this.

In many cases, M2M involves a group of similar devices interacting with a single application, as depicted in Figure 1.2. Fleet management is an example of such an application, where devices are, for example, trucks, and the communication network is a mobile network. In some cases, as shown in Figure 1.3, the devices in the group may not directly interact with the application owing to having only limited capacities. In this scenario, the relationship is mediated by another device (e.g., a gateway) that enables some form of consolidation of the communication. "Smart metering" is an example of such an application where the devices are smart meters and the communication network can be a mobile network or the public Internet.

To take this into account, the term "M2M area network" has been introduced by the European Telecommunication Standards Institute (ETSI). An M2M area network provides physical and MAC layer connectivity between different M2M devices connected

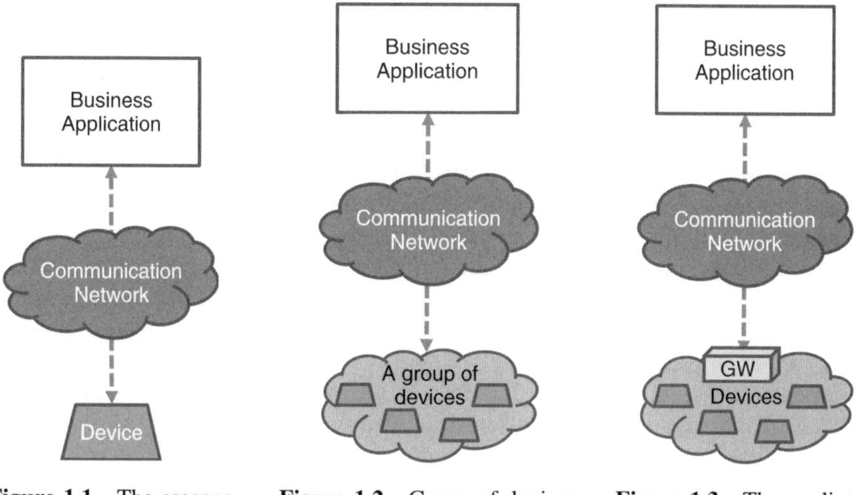

Figure 1.1 The essence of M2M.

Figure 1.2 Group of devices in an M2M relationship.

Figure 1.3 The mediated M2M relationship.

to the same M2M area network, thus allowing M2M devices to gain access to a public network via a router or a gateway.

M2M's unique characteristic is largely due to the key role of the end-device. Devices are not new in the world of information and communication technologies (ICT), but with M2M that market is seeing a new family of devices with very specific characteristics. These characteristics are further discussed below, particularly their impact on the requirements for applications and networks that have not until now been fully taken into account.

- **Multitude** – This is the most advocated change brought about by M2M. It is generally agreed that the number of "devices" connected in M2M relationships will soon largely exceed the sum of all those that directly interact with humans (e.g., mobile phones, PCs, tablets, etc.). An increased order of magnitude in the number of devices results in significantly more pressure on applications architectures, as well as on network load, creating in particular scalability problems on systems that have been designed to accommodate fewer "actors" and far greater levels and types of traffic. One of the early instances of such problems is the impact of M2M devices on mobile networks that have not been designed with this set of devices in mind and are in the process of being adapted to allow large numbers of devices with non-standard usage patterns (this will be discussed later in this chapter).
- **Variety** – There are already a particularly large number of documented possible use cases for M2M that apply to a variety of contexts and business domains. The initial implementations of M2M applications have already led to the emergence of a large variety of devices with extremely diverse requirements in terms of data exchange rate, form factor, computing, or communication capabilities. One result of the wide variety is heterogeneity, which is in itself a major challenge to interoperability. This can be a major obstacle to the generalization of M2M. It is also a challenge for the frameworks

on which M2M applications have to be built, in order to define and develop common-enabling capabilities.
- **Invisibility** – This is a strong requirement in many M2M applications: the devices have to routinely deliver their service with very little or no human control. In particular, this is preventing humans from correcting mistakes (and also from creating new ones). As a result, device management more than ever becomes a key part of service and network management and needs to be integrated seamlessly.
- **Criticality** – Some devices are life-savers, such as in the field of eHealth (blood captors, fall detectors, etc.). Some are key elements of life-critical infrastructures, such as voltage or phase detectors, breakers, etc, in the Smart Grid. Their usage places stringent requirements upon latency or reliability, which may challenge or exceed the capabilities of today's networks.
- **Intrusiveness** – Many new M2M devices are designed with the explicit intention to "better manage" some of the systems that deal with the end-users' well-being, health, etc. Examples are the eHealth devices already mentioned, smart meters for measuring and/or controlling electrical consumption in the home, etc. This in turn leads to issues of privacy. In essence, this is not a new issue for ICT systems but it is likely that privacy may present a major obstacle in the deployment of M2M systems. This may occur when the large deployment of smart meters demands prior arbitration between the rights of end-users to privacy and the needs of energy distributors to better shape household energy consumption.

In addition to the above-listed characteristics and their impact on the architecture of M2M systems, it is important to consider the other specificities of M2M devices that put additional constraints on the way they communicate through the network. This may require new ways to group the devices together (the "mediated" approach mentioned in Figure 1.3). Among other things, devices can be:

- **limited in functionality** – Most M2M devices have computational capabilities several orders of magnitude below what is currently present in a modern portable computer or a smart phone. In particular, devices may be lack remote software update capabilities. One of the main reasons for this design choice is cost, often because the business model requires very competitively priced devices (e.g., smart meters in many cases). Limited functionality also results from rational decisions based on the nature of the exchanged information and performable actions: most sensors are not meant to be talkative and operationally complex.
- **low-powered** – Although many M2M devices are connected to a power network, many of them have to be powered differently (often on batteries) for a variety of reasons. For instance, a large number of them are, or will be, located outdoors and cannot be easily connected to a power supply (e.g., industrial process sensors, water meters, roadside captors). This will reduce the amount of interaction between such devices and the M2M applications (e.g., in the frequency and quantity of information exchanged).
- **embedded** – Many devices are, and will be, deployed in systems with specific (hostile, secure) operating conditions that will make them difficult to change without a significant impact on the system itself. Examples are systems embedded in buildings or in cars that are hard to replace (e.g., when they are soldered to the car engine, as is the case with some M2M devices).

- **here to stay** – Last but not least, many of the new M2M devices are and will be deployed in non-ICT applications with very different lifetime expectancy. The rate of equipment change in many potential M2M business domains may be lower than in the ICT industry. This may be linked to cost issues due to different business models (e.g., no subsidization of devices by the operators), to the fact that they are embedded, but also to the complexity of evolution of the industrial process in which the device is operating (e.g., criticality of the service makes changing equipment in a electricity network very difficult, which leads to long life cycle of equipment in the field).

Two final remarks regarding the scope of M2M and the difficulty of defining clear-cut boundaries.

Firstly, a separation between "regular" ICT applications versus M2M applications is to a large extent purely artificial since, in some cases, devices are able to operate both in "regular" and M2M modes. A classical example of this is Amazon's Kindle™. Although it is a "regular" ICT device centered on both human-to-machine function (enabling eBooks) and interface (the eBook reader), it is also an M2M device in its role of providing an eBook to an end-user. When the end-user has decided to buy an eBook and clicks to get it, the Kindle™ device enters M2M mode with a server (providing the appropriate file with the appropriate format) and a network (a "regular" mobile network). This is perfectly transparent to the end-user, thanks to a set of enablers, including the SIM card in the device, the secure identification of the device by the network, and the pre-provisioning of the device in the operator network.

Secondly, it is important to outline some differences between M2M devices and what is referred to as "Things" or "Objects" in the so-called "Internet of Things" (IoT). Actually, M2M and IoT largely overlap but neither is a subset of the other and there are areas that are particularly specific to each:

- IoT is dealing with Things or Objects that may not be in an M2M relationship with an ICT system. An example of this is in the supermarket where radio-frequency identification (RFID) "tagged" objects are offered to the customer. These objects are "passive" and have no direct means with which to communicate "upstream" with the M2M application but they can be "read" by an M2M scanner which will be able to consolidate the bill, as well as making additional purchase recommendations to the customer. From this perspective, the M2M scanner is the "end point" of the M2M relationship.
- There are M2M relationships initiated by devices that are to be seen as direct human–machine interface extensions of a person (e.g., the above-mentioned end-user Kindle™) rather than as Things (e.g., the end-user refrigerator).

In the longer term, it is quite likely that the rather artificial distinctions, on the one hand, between traditional and M2M communication types and, on the other, between IoT and M2M domains will become further blurred with the advance of M2M and its ability to integrate more objects within existing systems.

1.2 The Business of M2M

After a decade of gradual development, there is a vast quantity of documented use cases for M2M, some of which have never progressed past the drawing board, although some have

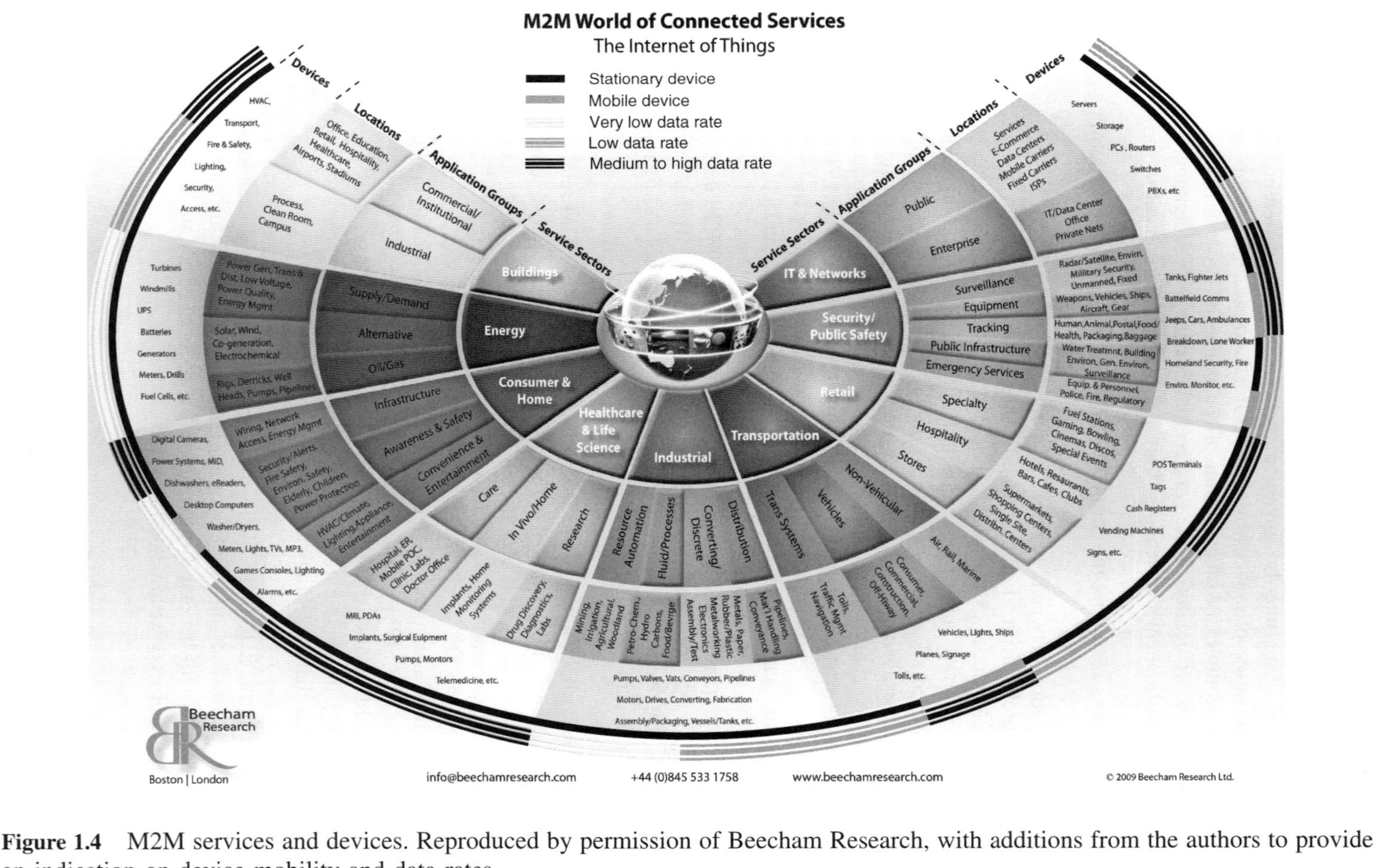

Figure 1.4 M2M services and devices. Reproduced by permission of Beecham Research, with additions from the authors to provide an indication on device mobility and data rates.

been subject to prototypes, early implementations, and commercial deployments. Only a few have led to the creation of significant business models in terms both of the revenue generated and the ecosystem of solid actors. However, the situation is evolving rapidly.

Figure 1.4 is an illustration by Beecham Research of the potential of M2M business that describes the major sectors that are applicable to M2M. From this perspective, the potential impact of M2M can be considered as important, and it is essential to fully understand the current status of M2M, as well as what has prevented it from emerging faster and what can be done to accelerate its emergence. This will be addressed at length in the first part of this book.

Another interesting aspect of Figure 1.4 is that it links the business domains with the associated devices from two major points of view: data rate and mobility. The former has been addressed above as one of the facets of the variety of M2M devices. The latter is also important since it is critical in some of the M2M applications that have recently emerged such as smart metering. Figure 1.4, in particular with the indication on the mobility of the device, suggests that there are potentially many more stationary devices involved than mobile devices. However, at this stage of M2M developments, it is mainly solutions based on wireless and cellular networks that have been investigated and deployed, as opposed to those based on wireline networks. One reason for this is that they benefit from some of the enabling aspects of cellular networks, for example the possibility of deploying M2M devices as mobile devices (e.g., by embedding a SIM card within the device), built-in authentication and security and easier deployment in industrial settings. Massive deployment of M2M applications over cellular networks drives clearly the need to optimize those networks for stationary or low mobility M2M devices as a means to reduce the overall connectivity cost.

Rather than trying to establish the list of all M2M domains and to provide application examples, this section investigates the maturity of each major M2M business in order to identify both obstacles to its emergence and possible enablers.

Figure 1.5 provides a staged view for M2M industry maturity. Three stages for M2M deployments are depicted with a far-reaching view (20 years).

The current "emergent" phase of M2M is cellular network-centric, where most applications are in the area of telemetry and fleet management. They mostly use existing cellular network infrastructure, while addressing predominantly business-to-business (B2B) applications. Considering that we are about to reach the end of this phase, it is important to see if the conditions are being met in order to enter the next phase ("transition") where a larger proportion of the M2M market will be developed, in particular business-to-consumer (B2C) applications that will be more demanding than the earlier B2B applications.

Despite progress in M2M technologies, there remain many challenges, the most pressing ones being:

- **fragmentation of solutions** – In the vast majority of cases, the solutions developed and implemented to date have been addressing specific vertical applications requirements in isolation from all others. This has created "silo" solutions based on very heterogeneous forms of technology, platforms, and data models. Interoperability is in general very limited or non-existent. Overcoming this challenge requires effort on at least two fronts. First, it is essential to define more comprehensive standards, in particular regarding data models. In addition, it is important to have service platforms that can be reused for multiple applications, avoiding the necessity to completely redesign solutions per application due to the lack of common capabilities.

- **network misalignment** – As already stated, communication networks have been designed with many requirements that differ substantially from those of M2M. One example of this is mobile networks that have not been designed to take into account the large numbers of devices generating very small amounts of data transport and potentially a very significant overload of the control and connectivity planes (not to mention that stationary devices do not require roaming capabilities). Some optimization of M2M traffic needs to be done which will essentially take place within standardization. Another example is latency. In some applications (e.g., in Smart Grids), latency requirements are largely below 10 ms, more than one order of teleprotection magnitude below those of voice over internet protocol (VoIP).
- **security** – As already outlined, some of the most promising M2M applications (eHealth, Smart Grids) are safety-critical and must be made robust against a large variety of security threats. This demands that security requirements be precisely understood and developed through standards and certification.
- **privacy** – To develop and resolve the sensitive issue of privacy, both regulation (an essential precondition) and standardization are required.
- **service capabilities** – In order to deal with the fragmented market, it is necessary to outline capabilities that can be reused across several applications. The history of ICT networks shows that this always requires that some separation be made between different architectural layers. In particular, separating applications from service capabilities (e.g., device management) and network capabilities (e.g., policy) will be key.
- **testing, certification** – A large number of M2M solutions will have to be developed outside of the traditional service silos, integrated with other M2M or traditional applications. This will require a larger degree of interoperability and vendor compliance, which will in turn necessitate the organization of (interoperability) testing and certification of devices and equipment. This will be the role of industry and/or standards organizations or forums.

Taking into account the above challenges, Figure 1.5 also outlines three major maturity accelerators for M2M, namely:

- **high-level frameworks** – This refers to an emerging set of standards-based architectures, platforms, and technologies integrated in a way that allows for the development of "non-silo," future-proof applications. These frameworks allow, in particular, for economies of scale that will change the dynamics of M2M business models.
- **policy and government incentives** – Based on the realization that some M2M challenges may not be addressed by the industry alone, public authorities and governments have started to play an active role both in stimulating the investment by setting up ambitious incentive programs and in policy-making. This is, in turn, drives more investment in the wider M2M ecosystem, as well as creating more trust in the viability of the M2M industry.
- **standards** – A large number of credible industrial partners, large and small, from various industries (including but not only ICT) have started to work together in order to create the new standards required to address M2M at the global system level.

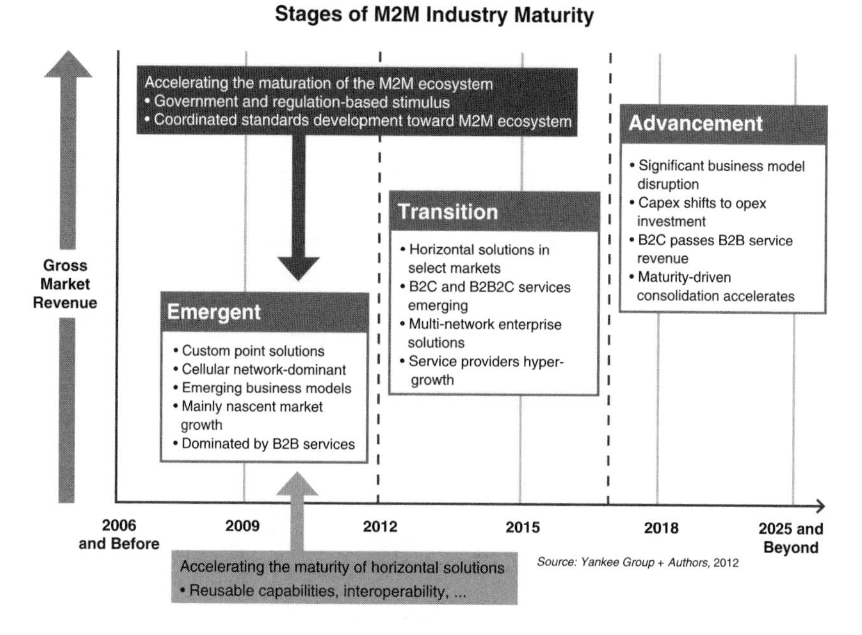

Figure 1.5 Stages of M2M industry maturity. Reproduced by permission of the Yankee Group.

This chapter and more broadly this book, will focus on these aspects, in particular standards, with the a priori view that these accelerators are already engaged and have already begun to change the landscape of M2M.

1.3 Accelerating M2M Maturity

1.3.1 High-Level M2M Frameworks

The largest part of the challenge facing the M2M actors is to transform vertical silos into a set of easily developable and incrementally deployable applications. Figure 1.5 shows that the transition to the phase two of M2M maturity will be marked by the advent and deployment of horizontal platforms.

What is meant by "horizontal" is a coherent framework valid across a large variety of business domains, networks, and devices, that is, a set of technologies, architectures, and processes that will enable functional separations, in particular application and network layers, as depicted in Figure 1.6.

Such a platform will be based on a set of capabilities in the form of software modules that are offered to the M2M applications in order to accelerate their development, test, and deployment life cycles. The confirmed need for, and possibility of, defining common service capabilities is the result of a thorough analysis within the industry of several M2M application Use Cases along with their related requirements.

The development and deployment of M2M applications could benefit from a set of building blocks that are carefully designed, tested, and optimized, irrespective of the type

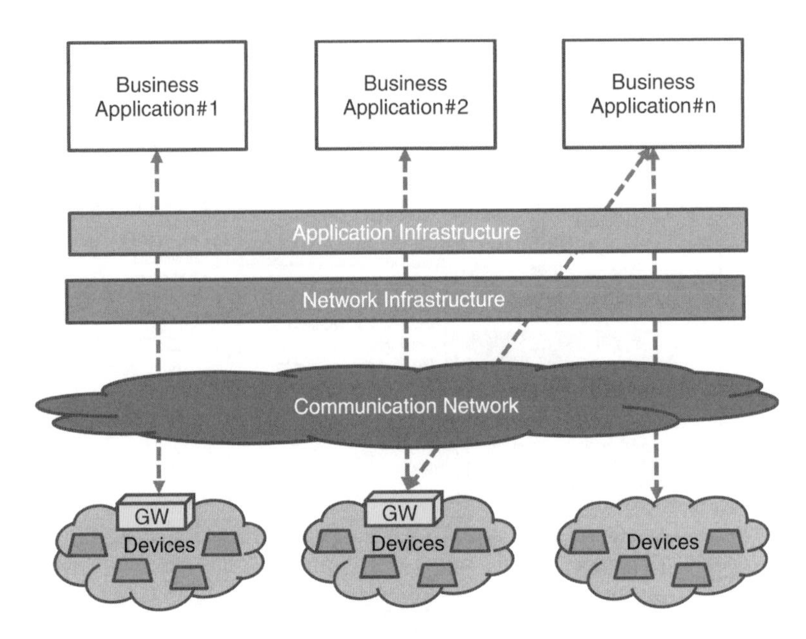

Figure 1.6 An M2M challenge: emergence of the M2M service layer.

of M2M application being deployed. Throughout the remainder of the book, these building blocks will be referred to as M2M service capabilities.

The further development of that concept allows the M2M application to primarily focus on the business logic, leaving the M2M service capabilities with all other aspects of the applications, such as device activation, device monitoring, device localization, data storage, and mediation to the horizontal platform (running the M2M service capabilities), to name but a few.

Once a set of solid M2M service capabilities has been specified, the logical next step is to expose them to the M2M applications through the use of application programming interfaces (APIs). This will be addressed in the following sections on standards.

1.3.2 Policy and Government Incentives

After initial slow progress, public authorities and governments have now realized that they have a key role to play in the take-off of M2M communications, especially because M2M is an integral element in many of the new systems that are deemed to be essential for the future of their countries or regions. Several lessons have been learnt and measures applied in the definition of incentives regarding, in particular, the role of standards to allow for infrastructure optimization, or the positive effect of economies of scale and reusability of service capabilities as a major enabler for mainstream deployments. A number of policy and government incentives have recently been put in place that play an important – sometimes pivotal – role for the following reasons:

- **Economic incentives** provide an attractive and stable framework that creates additional opportunities for investment in new projects and operational deployments. The most

notable example is the American Recovery and Reinvestment Act (ARRA) signed into law in 2009 by President Obama that allocates more than $27 billion to energy efficiency and renewable energy research and investment in the form of loan guarantees, R&D grants, workers' training, etc.
- **Regulation** provides precise directions for the development of the set of standards applicable or to be enforced within a country or a region. Examples include the European Commission mandates for smart metering [M/441] or for ICT applied to RFID and systems [M/436]. The other notable example is the US Energy Independence and Security Act of 2007 (EISA) where the National Institute of Standards and Technology (NIST) is assigned the "primary responsibility to coordinate development of a framework that includes protocols and model standards for information management to achieve interoperability of Smart Grid devices and systems..." [EISA].
- **Funding** of cooperative research and development projects can be either a step leading to standards development or a means of developing proof of concepts, validating existing standards. An important example is the 8th Framework Program (FP8) of the European Union.

1.3.2.1 Impact of Regulation on M2M Markets and Standards

Regulation plays an important role for M2M market growth, once it mandates new services that in turn mandate the use of M2M architectures (e.g., to achieve a certain compliance level in global systems) or the use of M2M technology and equipment. This is true in various parts of the world, and Europe can be taken as an example of how regulation mandating M2M technology is making great progress.

A prime example is climate change. The European Union has passed a new regulation for the energy sector by putting forward the "20-20-20" objective as a global target for EU Member States, that is, 20% reduction in emissions, 20% renewable energies, and 20% improvement in energy efficiency by the year 2020. As part of this objective, smart metering is becoming mandated across various EU Member States. France, for instance, has adopted a law mandating that all new houses must be equipped with smart meters starting in 2012.

The other notable example is eCall, the project of the European Commission intended to bring rapid assistance to drivers involved in a collision anywhere in the European Union. The aim is to deploy vehicles with "black box" cellular equipment that will send accident information (based on airbag deployment and impact sensor information) as well as GPS coordinates to local emergency agencies. Aside from early implementation by different car manufacturers such as BMW, PSA, and Volvo, the European Commission expects implementation by 2014. Once eCall has been massively deployed, other telematics services, such as traffic information, can also be installed and leverage existing eCall equipment.

The current wave of standards activities should help the migration to the next M2M phase ("Transition") where horizontal architectures become massively deployed, and total cost of ownership provides an attractive pricing model for the emergence of B2C M2M applications. This transition phase clearly benefits from different incentives in the USA, Europe, and Asia aimed at both funding research and prototype projects and also catalyzing the development of the interoperability standards to be used as a basis for global deployments.

1.3.2.2 Impact of Government Initiatives on M2M Standards

The eCall project is a typical example where a standard has been specifically developed to respond to a regulatory requirement: 3GPP (Third-generation Partnership Project) has developed two specifications to address eCall requirements (3GPP TS 26.267: "eCall Data Transfer; In-band modem solution; General Description" and 3GPP TS 26.268: "eCall Data Transfer; In-band modem solution; ANSI-C reference code"). These two standards specify the eCall in-band modem for the reliable transmission of accident data from an In-Vehicle System (IVS) to the Public Safety Answering Point (PSAP) via the voice channel of cellular and PST networks.

The European Commission has also issued the [M/441] Mandate on smart metering asking CEN, CENELEC, and ETSI, the three European Standard Organisations (ESOs), to provide the set of standards needed to deploy interoperable smart metering systems. The EC mandates also set target dates for the delivery for the needed standards, placing considerable time pressure for their delivery. The smart metering mandate has served as a catalyst and a major use case for the ETSI M2M Technical Committee whose members have rapidly agreed not to take the direction of creating yet another vertical architecture targeted at smart metering per se. The approach taken has been to address the [M/441] mandate requirements as only one part of broader M2M application requirements, thus enabling a strong push of the industry toward a horizontal architecture (the model where capabilities reusable in different M2M market segments are exposed to M2M applications via open APIs) as the only sustainable and cost-effective model.

In the USA, as part of the implementation plan of the American Recovery and Reinvestment Act [ARRA], the NIST was awarded $10 million in funds transferred from the Department of Energy (DOE) to help develop a comprehensive framework for a nationwide, fully interoperable Smart Grid for the US electric power system. Consequently, NIST has developed the Smart Grid framework (NIST SG-FW) which provides a conceptual reference model for Smart Grids – identifying domains, actors, and interfaces – and defined 17 priority action plans (PAPs). The aim of these PAPs is to evaluate the standards gaps for which resolution is most urgently needed to support one or more of the Smart Grid priority areas. The PAPs specify organizations that have agreed to accomplish defined tasks with specific deliverables. Ultimately the objective is to make the necessary submissions to institutional standards organizations so as to deliver a coherent set of standards for the US Smart Grid deployments. There is particular focus on interoperability (with the creation of the NIST Smart Grid Interoperability Panel) that is seen by [NIST SG-FW] as an essential means to protect the Smart Grid investments:

> Deployment of various Smart Grid elements, including smart sensors on distribution lines, smart meters in homes, and widely dispersed sources of renewable energy, is already underway and will be accelerated as a result of Department of Energy (DOE) Smart Grid Investment Grants and other incentives, such as loan guarantees for renewable energy generation projects. Without standards, there is the potential for technologies developed or implemented with sizable public and private investments to become obsolete prematurely or to be implemented without measures necessary to ensure security.
>
> (NIST Smart Grid Interoperability Framework)

1.4 M2M Standards

Unlike in several other ICT segments where it may be possible to deploy operational systems despite a lack of standards, several M2M market segments demand strong standards to ensure long-term investment protection. For several M2M applications, including smart metering or Smart Grids, there is an expectation that the installed equipment will be deployed for more than 20 years. While such a lifetime may appear unrealistic (or at least unusual) for traditional Telco deployments, the infrastructures deployed by utilities have very long deployment cycles that could dramatically influence their design and henceforth the related standards.

While there is common agreement that the market is still lacking standards for M2M, the situation has been evolving, although the level of maturity of M2M standards still varies, depending on the standards segment. It is now becoming relatively clear what needs to be done and in which technical and geographic areas.

1.4.1 Which Standards for M2M?

The different areas where standards are needed for M2M are broadly classified below.

1.4.1.1 Data Models

Data models explicitly determine the structure of data exchanged primarily between M2M applications but also with other entities within an M2M system. The logic behind the use of, eventually standardized, data models is that if the same data structures are used to store and access data, then different applications can exchange data in an interoperable fashion. Data models for M2M are application and business-logic specific. It goes without saying that a data model built from scratch for a meter designed to provide reporting on consumption data to a utility application will not be useful as a sensor designed to report data on patient health monitoring.

1.4.1.2 M2M Area Networks

The term M2M area network was used for the first time in the ETSI TS 102 690 Technical Specification [TS 102 690]. An M2M area network is a generic term referring to any network technology providing physical and MAC layer connectivity between different M2M devices connected to the same M2M area network or allowing an M2M device to gain access to a public network via a router or a gateway. Examples of M2M area networks include: Wireless Personal Area Network (WPAN) technologies such as IEEE 802.15.x, ZigBee, KNX, Bluetooth, etc. or local networks such as power-line communication (PLC), meter bus (M-BUS), Wireless M-BUS, etc.

While several M2M area networks are based on wireless RF technologies, other wireline-based technologies are also considered. The most notable example beyond PLC is the G.hn family of standards [G.hn] which has been designed with the aim of providing multiple profiles adapted for both multimedia/bandwidth-hungry applications and low-complexity/lower-bandwidth terminals. This latter option fits naturally with M2M applications such as home energy management. At the time of the writing of this book, ITU-T Study Group 15 initiated work known as G.hnem (home network energy management) to specify how G.hn can be used for Smart Grid applications such as advanced metering infrastructure (AMI) [G.hnem].

As outlined above, critical requirements for the design of M2M area networks relate directly to the nature of the M2M devices themselves, for example:

- low CPU;
- limited memory;
- low data rate;
- battery-operated, low power;
- low cost;
- small size (placing further constraints on the battery size).

The Internet Engineering Task Force (IETF) has adopted the term constrained devices for devices that qualify for one or more of the above criteria. Constrained devices place new and challenging requirements on the communication protocols that are supported by the device. As an example, it is often expected that battery-operated M2M devices will have a battery life of 10–15 years. The only way this could be achieved is through self-switching the device into a "sleep mode" when there is no need to send or receive data. As such, network applications, for instance, cannot rely on the device being constantly available and able to send or receive data. Examples of IP-based communication protocol evolutions to cope with constrained devices include the work done in the IETF 6LoWPAN (IPv6 over low power and lossy networks) Working Group. The 6LoWPAN WG group has defined encapsulation and header compression mechanisms that allow IPv6 packets to be sent and received over IEEE 802.15.4-based networks. IEEE 802.15.4 maximum transfer unit (MTU) is limited to 127 bytes. Taking into account frame overhead and optional security headers, very little is left for upper layers, that is, TCP/IP and application payload, unless protocol overhead is optimized. In essence, the 6LowPAN work allows IP to be used all the way to constrained devices, a desirable feature to allow for an end-to-end IP-based communication.

1.4.1.3 Access and Core Network Optimizations for M2M

As opposed to M2M area networks where purpose-built (and often competing) standards have been defined, there is no need to design still further access and core networks for M2M operations. Telecommunications operators consider that improvements and enhancements to access and core networks can be achieved in order to cope with the additional M2M traffic. In particular, cellular networks providing circuit-based services (namely voice or SMS) and data services have been optimized for personal communications. After an initial deployment phase of M2M services, mostly driven by B2B applications such as telemetry or fleet management, cellular operators came to the conclusion that their networks need to become "M2M-enabled."

One key element of this adaptation is the special nature of M2M traffic. As already noted in Figure 1.5 on the M2M business, around 90% of the M2M devices across all applications are stationary. In 3GPP and 3GPP2 wireless access, the network has procedures in place to track the location (cell or cell group) of the device. For naturally stationary devices such as smart meters, constantly keeping track of the device location becomes cumbersome and consumes valuable radio resources on the air interface.

The other notable characteristic of a large variety of M2M devices is that they generate low volumes of data. As an example, a utility smart meter is required to generate meter data of around 200–500 bytes every hour (maybe slightly more frequently during peak hours). In a cellular network, sending data requires the establishment of a data bearer within the access network, which means several handshake messages back and forth between the device and other entities in the access and core network (for access to radio resource, authentication/security procedures, acquiring IP address, enforcement of bearer QoS parameters, confirmation, etc.). Data bearer establishment and its teardown after use require more than 20 handshakes (but not all are originated/terminated at the terminal), not including the often-used TCP transport protocol's three-way handshake, acknowledgments, and connection release.

Clearly, cellular access and core networks have not been designed to cope with data-traffic models where the control plane traffic becomes dominant (larger than 80%) compared to the actual application payload traffic.

These two examples make it very clear why it is important to optimize access and core networks for M2M traffic. This need is further amplified by new, emerging business models for M2M where the average revenue per user (ARPU) is often 10–15 times lower than in the case of personal communications. These new business models are also completely changing the paradigms, for example, how to perform charging and billing. As opposed to personal communications where charging and billing are performed for each device subscription, M2M often requires charging to be performed on a network application basis (the utility back-end application, the central fleet management application, etc.). A concrete example would be to send a single network usage bill for all utility smart meters connected to a network operator, as opposed to a bill per connected smart meter.

Network operators and equipment vendors alike have initiated work on access and core network optimization for 3GPP and 3GPP2 cellular systems. However, since the resulting standards will take time before being deployed in operational networks, operators have adopted *a two-step approach* to cope with the growth of M2M traffic:

- **Step 1** – Re-architect access and core networks so as to adapt better to the fundamental characteristics of M2M traffic while avoiding impacting the high-revenue services related to personal communications. Some of the considered scenarios include the deployment of dedicated equipment (home location register (HLR), gateway GPRS support node (GGSN)) and traffic isolation. In summary, Step 1 is the set of de facto "best current practices" for network architecture, allowing operators to make best use of the current toolbox of standards and products. All of this is to be done by taking into account the fundamental characteristics of M2M traffic, such as low data, non-predictive, low priority, and burstiness.
- **Step 2** – Progressively deploy new equipment, software upgrade, and network solutions that are optimized for M2M traffic types, based on the developing work on M2M standards in 3GPP and 3GPP2. While Step 1 is an intermediary step, Step 2 is believed to provide the longer-term fix that is essential for the massive growth M2M business.

To a large extent, Step 1 is the result of an initial ad hoc phase (or Step 0) where M2M modules have been deployed in cellular networks and handled as if they were mobile handsets. Several lessons have been learned in terms of the impact, sometimes harmful, of M2M traffic on the operator's network. Over time, operators have come to the conclusion that M2M would require a different approach from that of mobile personal communication services.

1.4.1.4 Horizontal Service Platforms and Related APIs

As outlined above, the emergence of the next phase of M2M business will rely on the deployment of horizontal platforms implementing a set of service capabilities, that is, software modules that are exposed to the M2M applications in order to expedite their development, test, and deployment life cycles. Examples of service capabilities include device activation, device monitoring, device localization, data storage, and Mediation to the horizontal platform (running the M2M service capabilities). Standardization will be used to specify a set of typical M2M service capabilities that can be exposed to the M2M applications through the use of APIs.

Telecom operators have designed several API sets in the past, but their level of adoption and deployment has often suffered from lack of visibility and expertise within the IT and application developer community. Today, it is becoming increasingly clear that a successful Telecoms application enablement strategy mandates the use of IT-friendly APIs inspired by the Web 2.0 framework. Representation state transfer (REST) APIs have in the past enjoyed wide-scale market adoption among the IT industry, and constitute, in the view of the authors of this book, a recipe for the success of new API work. As an example, in the context of M2M APIs, REST-based APIs using HTTP protocol are being specified by ETSI M2M.

1.4.1.5 Certification for M2M Modules and Terminals

Certification refers to the confirmation of certain characteristics (usually based on a standard) of items of equipment. Generally, certification is provided by some form of external review, education, or assessment performed by an independent entity. It has become an important requirement for deploying an item of equipment (in particular, terminals) in operational environments. Certification can be divided into:

- **mandatory regulatory certification** – European Member States mandate the following certifications for M2M modules: RoHS, WEE/RAEE, and R&TTE directives which pertain respectively to reducing risks of hazardous substances, GSM radio spectrum, electromagnetic compatibility, and low-voltage equipment. In the USA, all devices must comply with the Federal Communication Commission (FCC). Additionally, communication devices must be PTCRB-certified.
- **voluntary certification** – This is usually mandated by operators for devices deployed in their networks. The Global Certification Forum (GCF) runs an independent certification program aimed at ensuring compliance with 2G and 3G wireless standards. The program also mandates device testing on five GCF-qualified networks.

Additionally, certain operators impose additional certification for devices deployed in their networks.

Ultimately, the certification process aims at both ensuring that equipment adheres to certain environmental and electromagnetic compatibility characteristics, but also does not cause any harm when deployed in operational networks.

Besides the certification programs pertaining to devices directly connected to an operator access networks, several other certification programs exist for devices using short-range radio technologies. The most notable ones are ZigBee, Bluetooth, and KNX. In the case of KNX (provided as an example), certified products must show compliance to the following set of standards:

- quality system in accordance with ISO 9001;
- European standard EN 50090-2-2 (covering such aspects as EMC, electrical safety, and environmental conditions of bus products) and an appropriate product standard;
- Volumes 3 and 6 of the KNX specifications, the former being a toolbox of the KNX protocol features, the latter listing the permitted profiles of the KNX stack based on the toolbox as mentioned above;
- KNX interworking requirements as regards standardized data types and (optionally) agreed functional blocks.

As standards for M2M evolve, certification programs are also expected to evolve, in order to cope with these evolutions.

1.4.1.6 The Standards Organizations Ecosystem for M2M

Several standards organizations focus on one or multiple aspects of M2M, making duplication sometimes inevitable. Detailing the different initiatives surrounding M2M would almost necessitate a book on its own. This is particularly true when it comes to M2M area networks and application data models, where multiple standards exist (Part II of this book is dedicated to describing these two aspects).

Figure 1.7 provides some examples of M2M area networks, namely:

- **ZigBee alliance** – A set of specifications of communication protocols (but also data models) using small low-power radio devices based on the IEEE 802.15.4. Target applications include switches with lamps, electricity meters with in-home displays, consumer electronics equipment, etc.
- **KNX** – A purpose-built radio frequency standard developed for home and building control. It is approved as an International Standard (ISO/IEC 14543-3) as well as a European Standard (CENELEC EN 50090 and CEN EN 13321-1) and Chinese Standard (GB/Z 20965).
- **Home grid** – Based on the ITU-T specification suite known as G.hn that is designed to provide communications within the home environment, making use of existing wires such as powerline, coax cable or copper pairs.
- **IETF protocol suite** – Based on 802.15.4 providing a specification for L1 and L2 WPAN, the IETF has developed a set of protocols aimed at bringing native IP support to constrained devices.

Figure 1.7 provides an example of data models used in the context of smart metering applications, the most notable of which is the device language message specification

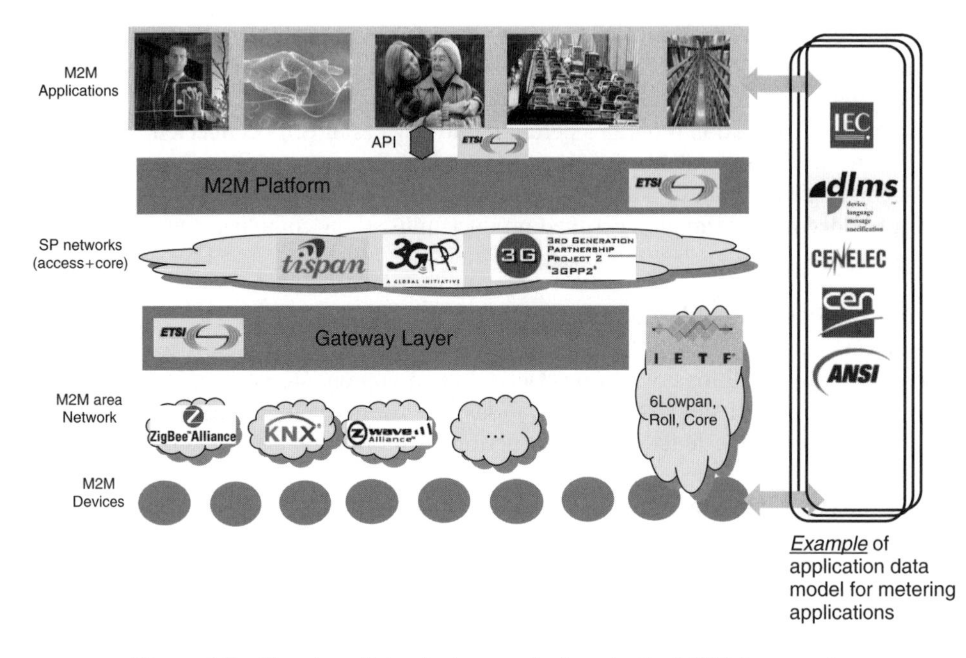

Figure 1.7 Mapping of standards organizations to the M2M framework.

(DLMS) that has been adopted at the European level by CEN and CENELEC and by IEC at the international level. ANSI C12.18 suite is the US counterpart for the DLMS data model. Standards organizations working on M2M area networks also frequently produce vertical application specific data models. This is the case for ZigBee and KNX.

When it comes to service provider access and core network optimizations, 3GPP and 3GPP2 are the natural organizations for cellular systems, while ETSI TISPAN provides the equivalent set of standards for wireline NGN (next generation network). However, it is commonly agreed that wireline standards will see less activity in this area.

3GPP TS 22.368 [TS 22.368] lists both common service requirements and specific service requirements for machine type communications (MTC). This specification has been elaborated with marked involvement of operators, cellular system vendors, and M2M module vendors and will be the basis for the ongoing work on network improvement for M2M. The level of maturity of this work is low compared, for instance, to the work already done on the data models.

Above and beyond the access and core networks, the ETSI M2M Technical Committee and the TIA TR50 Committee on Smart Devices are developing a service aimed at providing service capabilities to be exposed to M2M applications via open APIs. Neither group is starting the work from scratch and both are looking at means by which to endorse building blocks from existing standards, the most notable being the Open Mobile Alliance (OMA) device management and the Broadband Forum TR069 protocols. These two protocols have been designed to provide remote device configuration-related functions such as: configuration management, performance management, fault management and firmware, and software upgrade.

ETSI M2M has fully endorsed the HTTP/REST approach for its API work. Reuse of existing guidelines and API work from OMA has been identified as a target by ETSI M2M experts.

1.5 Roadmap of the Book

The authors hope that this introduction has captured the interest of the reader and paved the way for the following two parts.

Part I (Chapters 2 and 3) addresses the current state of the M2M landscape.

Chapter 2 describes in greater depth the business of M2M, by analyzing the market landscape with a realistic outlook as to whether M2M penetration can be achieved, given the recent evolution outlined in this introduction. Market drivers and market barriers will be addressed with possible roll-out scenarios.

Chapter 3: Feedback and lessons learned from early market deployments. This chapter provides a set of scenarios and related network configurations for an operational deployment of M2M over a 3GPP commercial network. It provides as a conclusion a list of lessons learned, as well as recommendations for future standards work.

Part II (Chapters 4–8) will address M2M architectures and protocols.

Chapter 4 analyzes M2M requirements using a Use-Case-driven approach to service requirements. It then introduces the high level architecture, focusing in particular on network, services, M2M traffic models, end-to-end architectures for M2M, vertical versus horizontal architectures and the M2M service deployment platform (SDP).

Chapter 5 introduces the M2M services architecture capabilities defined by ETSI. More specifically, it presents the M2M capabilities in the services domain, in the gateway and in the device. Specific points are made on the protocols and APIs. This is followed by considerations on interoperability and the REST-based architecture and the impact on device management protocols.

Chapter 6 addresses Access and Core Network optimizations for M2M. After introducing the problem, it outlines how it is handled in various standardization actions within 3GPP.

Chapter 7 investigates the role of IP in M2M, and in particular the IETF approach. It analyzes the IP protocol stack (6LoWPAN), routing for M2M area networks (roll) and REST for constrained devices (core).

Chapter 8 deals with horizontal security architecture for M2M.

Chapter 9 addresses M2M terminals and modules.

Chapter 10 presents standards evolutions in SIM cards and their impact on M2M.

Chapter 11 provides a set of concluding remarks as well as views on future developments.

References

[ARRA] American Recovery and Reinvestment Act of 2009, Public Law 111-5, http://www.gpo.gov/fdsys/pkg/PLAW-111publ5/content-detail.html.

[EISA] EISA (2007) Energy Independence and Security Act of 2007.

[G.hn] ITU-T Recommendation G.9960. Next Generation Home Networking Transceivers Foundation (Copperpair, Powerline and Coax PHY Layer).

[G.hnem] ITU-T DRAFT (2009) ITU-T DRAFT Recommendation Home Network Energy Management (Smart Grid PHY for Powerline).

[M/436] European Commission (2008) Standardisation Mandate to the European Standardisation Organisation CEN, CENELEC and ETSI in the Field of Information and Communication Technologies Applied to Radio Frequency Identification (RFID) and Systems, M/436.

[M/441] European Commission (2009) Standardisation Mandate to CEN, CENELEC and ETSI in the Field of Measuring Instruments for the Development of an Open Architecture for Utility Meters Involving Communication Protocols Enabling Interoperability, M/441.

[NIST SG-FW] NIST (2010) NIST Framework and Roadmap for Smart Grid Interoperability Standards, Release 1.0, NIST Special Publication 1108, http://www.nist.gov/public_affairs/releases/upload/smartgrid_interoperability_final.pdf.

[TS 22.368] (2009) Technical Specification Group Services and System Aspects; Service requirements for Machine-Type Communications (MTC); Stage 1; Release 10.

[TS 102 690] (2011) ETSI Technical Specification, M2M Functional Architecture, Stage 2 Specification.

Part One

M2M Current Landscape

2

The Business of M2M

Harish Viswanathan
Alcatel-Lucent, New Jersey, USA

The term M2M is quite broad and refers to many different vertical markets, diverse communication technologies and potentially a large geographical scope. M2M has the potential to enhance existing processes in many industries, such as healthcare, automotive, manufacturing, energy, retail, and public safety, to name but a few. Similarly, in terms of technology, M2M applies to numerous short-range wireless communication technologies, such as ZigBee [1] or Zwave [2], that connect sensors via a more intelligent gateway or router node, to wireline technologies, such as X10 [3], to wide-area supervisory control and data acquisition (SCADA) networks designed for monitoring and control of the power grids, to radio frequency identification (RFID) tags employing near-field communication technologies, and to aspects of wide-area mobile wireless networks, such as CDMA, GPRS, UMTS, and LTE networks and satellite communication networks. As a result, estimating the number of M2M devices currently deployed, and their potential for growth, is no simple task. The geographical scope of M2M solutions can be both local and global, as in the case of multinational trucking/shipping companies tracking their fleet across a continent, or a small apartment with remotely controllable appliances networked to an energy-management control application running on a local home server. Despite such a broad scope, there is indeed a common denominator linked to the notion of communication without human intervention between devices and applications that govern their behavior.

The variety of applications, technologies, and other dimensions of M2M makes the business models of M2M highly complex. The M2M value chain has numerous players with little consolidation, resulting in a fragmented marketplace for M2M. The plethora of standards that address partial aspects and solutions, with insufficient or incomplete end-to-end standards solutions contributes further to market fragmentation. As a result,

M2M Communications: A Systems Approach, First Edition.
Edited by David Boswarthick, Omar Elloumi and Olivier Hersent.

multiple business models for M2M deployment still prevail and will continue to do so for several years.

This chapter is structured as follows. It begins with a description of the major vertical markets and provides specific applications of M2M within each of these markets. It then discusses the key value-chain components with examples of industry players for each one, followed by market-size projections for each of the main markets. Thereafter, the three different business models that are commonly seen in wide-area wireless M2M deployments are presented, concluding with the provision of several guidelines on aspects of market evolution.

2.1 The M2M Market

To a large extent, M2M is not a market in itself; it is rather an extension of multiple vertical markets that benefit from M2M communications. The major markets where M2M plays a significant role are healthcare, transportation, energy, security and surveillance, public services management (safety and traffic), retail, point of sale, vending, building control/management, industrial automation and control, home automation and control, and agriculture. We further detail some typical applications and Use Cases in the domains of healthcare, transportation, and energy, and the nature of the technology employed in each of these markets. The other vertical markets also have similar underlying applications, although they are not detailed in this chapter.

2.1.1 Healthcare

- Remote patient monitoring
  - Obtain heart rate, blood glucose levels, and other bodily functional parameters through a wide area network (WAN).
- Homecare/assisted living
  - Remotely monitor the safety and well-being of people needing care and assistance.
  - Monitor and send reminders to patients for daily activities, such as taking medication.
  - Treat or assist patients through video consultation.
- Asset tracking
  - Track high-value assets such as intravenous pumps, wheel chairs, and stretchers within hospitals.
  - Tag medications with RFIDs to eliminate errors in hospitals when administering medication.

Remote patient monitoring typically involves the use of Bluetooth connectivity between the monitoring device and the cell phone acting as a gateway, which is in turn used for connectivity to the WAN.

Homecare applications use a variety of sensors within the home that communicate using a LAN (also referred to as M2M area network in ETSI M2M specifications) and connect to the Internet through a home gateway and the broadband connection.

Asset tracking of medical devices is typically employed in hospitals and is based on WiFi localization, RFIDs, and occasionally infrared communications.

2.1.2 Transportation

- Fleet management
  - Ascertain location of vehicles, send dispatch notification to individual vehicles, send group dispatch notifications to a group of vehicles regarding certain incidents in the neighborhood, and collect statistics from vehicles regarding usage and maintenance.
- Vehicle maintenance
  - Obtain various operating parameters from the vehicle for the purpose of proactively diagnosing mechanical issues.
  - Push software upgrades to vehicles, thus avoiding expensive recalls to the car dealer.
  - Car dealers can push reminders and special deals to vehicle owners for routine maintenance or promotional offers.
  - Electric vehicle (EV) charging-related applications, such as providing location of charging stations, managing the charging process and parameters in the car, if any, and generating relevant records for billing purposes.
- Insurance
  - Remotely monitor location and usage data and driver behavior data to provide different insurance premiums to reflect different profiles of driver.
- Infotainment
  - Provide media (audio, video map information, etc.) to the vehicle.
- Theft prevention
  - Remotely disable the car if reported as stolen and locate the car if lost.
- Emergency-call support
  - Automatically upload parameters (speed, photos, location, etc.) in the event of an accident and establish a voice connection.
- Navigation
  - Optimize routing options based on specific criteria and provide directions and information on surrounding areas, including points of interest.
- Toll
  - Automatic payment of tolls while driving on specific roads; tolls may be based on the time of day when the road is used and on other factors, such as pollution levels.
- Asset tracking
  - Track objects within trucks or other forms of transportation.

Almost all transportation applications involve the use of commercial cellular networks due to the inherent mobility requirement of vehicular communications. Typically, a modem is embedded in the vehicle that uses mobile communications in order to connect to the application server in the network.

2.1.3 Energy

- Smart metering
  - Automatic collection of consumption, diagnostic, and state information data from water- or energy-metering devices. Data communicated to a central database for billing, troubleshooting, analysis, and load management.

- Demand response
  - Adjust and distribute dynamic price information for peak-load minimization.
- Consumer-managed applications (independent of the utility provider)
  - Monitor usage of electricity for personal appliances and report to a central entity that provides management services.
- Alternative energy source (solar panels, wind, etc.)
  - Monitoring, maintenance.
- EVs
  - Charging control, billing information, and usage of the EV as a grid storage device.

Energy applications use a mixture of home area networks, local communication technologies, such as powerline communications or meshed Radio frequency (RF) networks to the nearest substation, and wireline networks to the central location where the data is processed. Some smart meters are also based on cellular wireless networks. Large utilities tend to deploy their own networks for their communications needs.

2.2 The M2M Market Adoption: Drivers and Barriers

There is expected to be phenomenal growth in M2M over the next 10 years.

- What are some of the factors driving this growth?
- Will this really pan out, or is this simply hype, as with certain other technologies?

It should be noted that the M2M concept is neither revolutionary nor new. It has been in use in niche applications for decades. However, what is different now is the potential for mass adoption across different industries, propelled by a number of converging factors that seem to be causing a "perfect storm." Nevertheless, there are still significant barriers to growth that need to be overcome before the M2M market can reach its full potential.

The primary drivers that we currently see are as follows.

- **Diminishing prices for devices and communication costs** – Advances in semiconductor and radio technologies coupled with more mature wide-area communications protocols have resulted in falling communication module prices. With voice revenues reaching saturation point in most countries, network operators are looking to new areas, including data revenues, for future growth. One inherent quality of M2M traffic is that the data exchanged is primarily composed of a high number of small payload transactions, making it highly profitable per byte of information carried. For example, it costs approximately 20 cents to send a 140-byte SMS, while a 5 GB-per-month data plan costs about $50, making SMS 100 000 times more profitable per byte of data, even if such pricing models are subject to evolution when it comes to wide-scale deployments. Network operators are also eager to add as many machine device subscriptions to their networks as possible. Flexible and attractive pricing schemes are offered to generate positive return on investment (RoI) for M2M deployments, thus fostering their growth.
- **Widespread deployment of wireline and wireless IP networks** – IP has become the de facto standard for network communications across different types of network. Most communication service providers worldwide have deployed IP networks at a national

and international level. The use of same-core networking technology significantly simplifies deployment and maintenance of devices and applications. For most large M2M deployments involving both wireline and wireless devices, a single application that deals with all devices homogeneously will imply lower total cost of ownership resulting in faster RoI.

- **Ubiquitous coverage provided by commercial networks** – In the past, companies had to rely on their own networks to meet the needs of M2M applications. Early commercial networks did not provide sufficient coverage in all geographical areas nor did they meet all of the demanding performance requirements of some M2M applications. With the advent of wireline and wireless broadband, network reach is now more available than ever, with a choice of multiple operators. Advanced technologies that provide good latency and quality of service (QoS) can be used by companies without substantial upfront investment in network costs.
- **Clear regulatory requirements and green technology investments** – Global warming has created a new awareness of the need for greener technologies. Governments around the world are pushing for efficiencies in the distribution and consumption of energy. It is well understood that information and communication technologies (ICT) can play a significant role in reducing carbon emissions by carefully monitoring and controlling energy consumption. Remote monitoring and control is a key application of M2M technologies and this is driving M2M usage in green technologies.

The following are some of the primary barriers to the large-scale roll-out of M2M solutions.

- **Numerous incomplete standards leading to market fragmentation** – There are a large number of standards addressing the same problems, none of which are complete enough to address the end-to-end solution needs. For example, ZigBee, Zwave, Wireless HART, IETF 6LowPAN/ROLL all address short-range communication between sensors and a gateway or a router. Application-level standards exist in some verticals such as healthcare and smart metering, but again there are multiple implementation options. It is simply not possible for a consumer to add an off-the-shelf sensor from a retail store to their home network and expect it to work with an existing home control application from some vendors. Most applications connect to their devices using proprietary technologies. A substantial amount of system integration work is thus required, resulting in higher costs. Innovation is also stifled by a lack of an overall framework within which to invent new concepts.
- **Global regulatory hurdles** – Regulation can also delay global deployments because of different rules in different countries. Regulation may be at the application level, such as in how healthcare data may be acquired, stored, and disseminated, or even at the communication level. In addition, certification may have to be acquired in each country or region separately, resulting in higher costs.
- **Security and privacy** – M2M generally involves the collection of data from items that people own, and often in their own homes. There is a natural tendency to fight data collection from the perspective of privacy. For example, if all the things that people buy are tagged with RFIDs, including their wallets, it is possible to extract information about people's movements and spending behavior. The general backlash against technology owing to privacy concerns may delay or even prevent numerous M2M deployments.

- **Carrier portability** – One concern facing many companies in the case of wide-area wireless M2M is the inability to easily switch operators. Several types of device are deployed with a soldered SIM that cannot be removed, therefore preventing any changing of network operator. Even if the SIM is removable, it will be very expensive to send personnel to change the SIM card in many thousands of M2M devices. This makes it difficult for the company to switch operators before the end of the life cycle of the device, which in some cases can be up to several decades. Potential solutions to this problem are to involve multivendor network operators who have bandwidth relationships to multiple network operators. However, with network operators directly approaching corporate customers, lack of portability could be a stumbling block.
- **Network operator and company mismatch** – The life cycle of many M2M services can be up to 15 years, for example, as in the case of smart meter deployments. This means that companies such as utilities want network operators to guarantee availability of a specific technology for a very long period of time. This is, however, difficult for network operators that need to upgrade the technology frequently in order to make better use of the limited available spectrum. Thus, there is a potential mismatch in technology expectations among the parties involved in a service deployment.
- **Technology challenges** – Last but not least, as M2M takes off, there are likely to be a number of technical challenges that will have to be overcome, for example:
  - device management;
  - network scalability;
  - device authentication on the provider's network (initial and repeat request);
  - subscriber network and application policy;
  - charging rules.

2.3 The M2M Value Chain

Figure 2.1 shows the value chain for typical deployment involving a WAN, such as in the transportation domain. The list of companies in Figure 2.1 is provided as an example and is not meant to be comprehensive. The device end of the value chain consists of the manufacturers of the machine in which the module is embedded. For example, the car is the "machine" in the case of a connected car application, or the meter is the "machine" in which a communication module is embedded. The machine includes the sensory and actuation capability and may be part of the same integrated circuit as the communication module. The value-chain players for this segment are varied and depend on the particular vertical industry.

The next important sector is the communication module or the chipset manufacturers. Modules include cellular modules as well as short-range communication technologies, such as ZigBee or Zwave. Also included in this category are the vendors or manufacturers of standalone modem or gateway devices to which sensor devices may be attached directly or through another local communication interface. Examples of companies that provide M2M communication modules are provided in Figure 2.1. Most of these companies provide modules for mobile cellular connectivity.

Commercial networks are becoming increasingly popular for use in M2M. When commercial networks are used, network operators or communication service providers form the core of the value chain from the point of view of providing the connectivity and

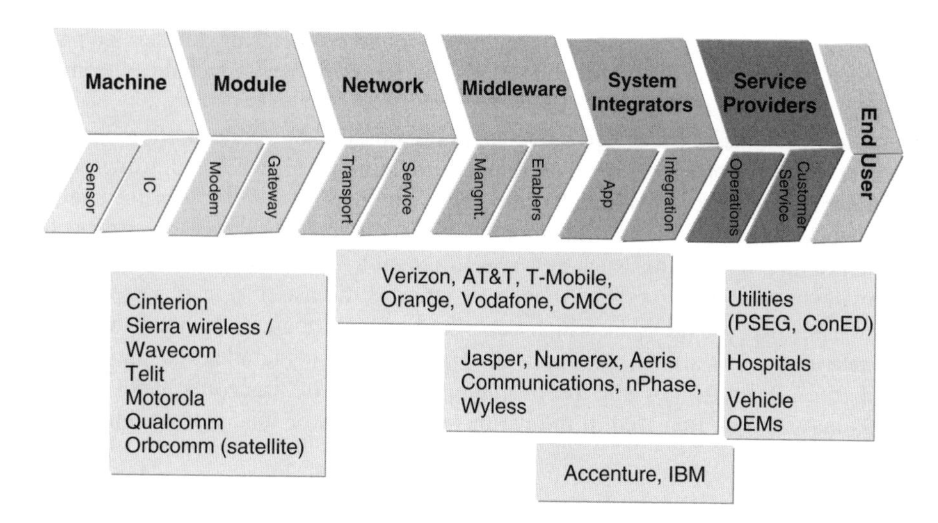

Figure 2.1 A typical M2M value chain.

transport of the data between the device and the Internet-based application. Large companies such as electricity utilities may deploy their own infrastructure to transport their own data. In such cases, some of these components in the value chain are absorbed within the companies or the service provider component. Similarly, when the geographical scope of the solution is limited to inside a building, then there may be no network operator involved since communication will involve the LAN within the building.

Middleware includes the bulk of the M2M service capabilities that are horizontal and applicable to many different applications. Some of the capabilities are device management, diagnostics, messaging capabilities such as SMS, and automated device activation and service provisioning. The middleware is typically part of an M2M service platform and is provided either by the large network operators or by special M2M providers, such as Jasper Wireless, Wyless, or Aeris Communications. Some network operators partner with the platform providers who provide their platform to the network operators in hosted software as a service model.

System integrators are companies that bring together all the capabilities needed for an end-to-end solution, and perform hardware and software integration of the different pieces of the solution. They employ subcontractors as necessary to build the dedicated application software. Typically, a significant amount of integration work is needed as each M2M solution tends to be custom-designed for the specific application at hand. System integrators often choose the machine vendor, network operator, and the M2M platform provider on behalf of the end-service provider. Examples of system integrators include companies such as Accenture and IBM.

Application service providers are entities that provide the M2M service to their customers. Customers are typically the consumer or the employees of the company to whom they are providing the service. Service providers are responsible for the day-to-day operations of the M2M service and are responsible for customer service and billing, if applicable. In the case of the remote patient monitoring solution, the service provider is typically a hospital or clinic. The hospital's IT department is providing the service to the hospital's doctors, nurses, and patients. In the case of smart metering, the service provider would be the utility company. In some cases, the service provider could outsource the service to a managed service provider.

Multiple variations in the above value chain are possible and indeed often observed in the market. Network operators are increasingly providing their own home-grown solutions or a partner middleware or M2M platform. They are also involved in system integration and sometimes also provide pre-integrated solutions together with their partners. In the case of consumer applications such as home automation and control, the network operator may provide the end-user service. Another variation is the use of private networks by the application service provider, in which case the network operator is not in the picture. A further variation is where the application provider also includes the middleware or M2M platform itself. This is true, for example, in the case of Amazon's Kindle™ service. Resellers may also be involved in the various components of the value chain.

2.4 Market Size Projections

M2M market size estimations typically begin by estimating the number of M2M devices. As discussed earlier, estimating the number of devices is tricky because of the enormous scope of M2M and the uncertain growth potential of each market. It essentially includes potential applications from all walks of life, each with their own timeline for adoption. Below are some popular market projections on the number of devices:

- 1 trillion connected devices by 2013;
- 50 billion Internet-connected devices by 2020;
- 415 million mobile Internet devices by 2014.

Why are these projections so different? The answer lies in knowing exactly what is included in the estimate. In the first estimate, it is most likely that very-short-range devices such as RFIDs/NFCs are included in the total projections. Since there can be a very large number of RFIDs that can potentially be read from anywhere on the Internet, the total number of devices according to this estimate is very large. The second estimate likely includes only the short-range devices, such as devices based on IEEE 802.15.4 technology, Bluetooth, IEEE 802.11-based local area and mesh networks, and cellular devices. The last estimate includes only devices and gateways that connect directly to the cellular network. The key point that is common to all projections is the extremely large potential of the M2M market.

For the market size in terms of dollar value, we refer to the Beecham report from 2008. Figure 2.2 below shows the market size divided into four main categories:

- Internet enablement refers to the modules and devices component of the value chain. It includes the modems and the gateways that allow connectivity to the Internet.
- The network part of the market refers to the value in the transport of bits from the location of the M2M device to the location of the application.
- System applications in the legend refer to the middleware that includes common functions across all M2M applications, such as control and diagnostics, device management, location, status, and tracking information. This includes the middleware component of the value-chain diagram in Figure 2.1.

- Finally, value-added services typically refer to all the other components in the value chain to the right of middleware, and include services such as system integration, software application development, and the day-to-day operations of maintaining and providing the application-related services to the end-user.

The total value in the applications or services segment is broken down into different verticals. The market size for each of the major verticals in USD billions, estimated in 2008 for 2012, is shown in Figure 2.3 and also obtained from the same market report from Beecham [4]. This shows that retail, security, transportation, and energy are potentially the largest markets. Healthcare, while rather small in 2012, is among the fastest-growing market segments. It should be noted that these figures encompass all the underlying network technologies, that is, local-area, wide-area, wireline, and wireless.

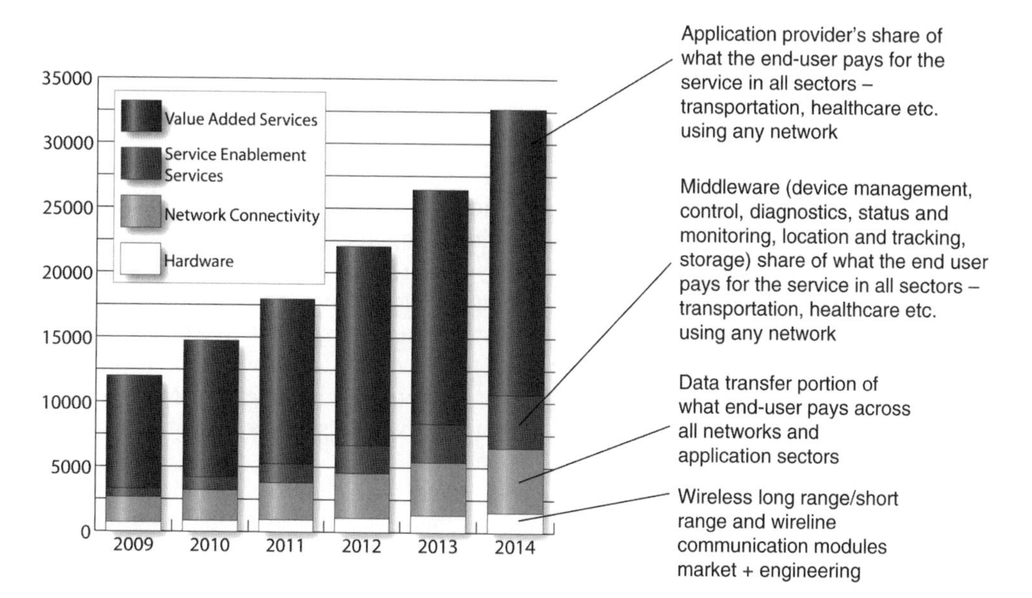

Figure 2.2 Estimated M2M revenues mapped to the different actors of the value chain. Reproduced by permission of Beecham Research.

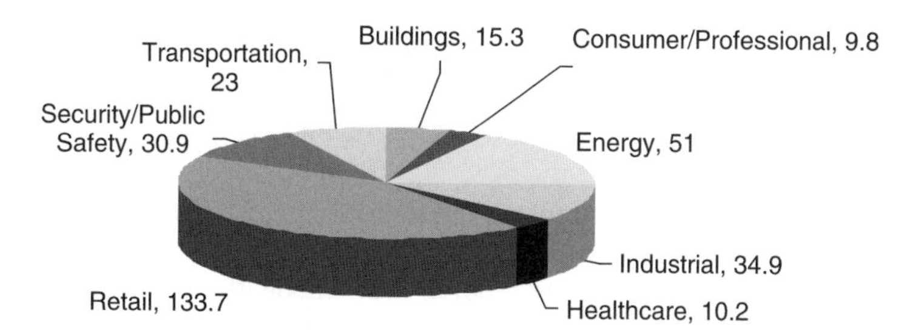

Figure 2.3 M2M market size (USD billions) by vertical segment. Reproduced by permission of Beecham Research.

2.5 Business Models

Despite the predictions of a considerable market size, M2M is still in its infancy. A number of different business models and money flows are prevalent in the market today. A typical M2M deployment may or may not involve a commercial network operator or a communication service provider. In the case of a connected car, for example, typically an embedded modem with a SIM card is installed in the car with connectivity service provided by a wireless network operator. There are other models that do not involve the commercial network operator at all. For example, large utilities may deploy their own network that is used for M2M, in addition to meeting the other needs of the utilities. Another possibility is that an M2M network may be completely local to a large corporate building, such as a hospital or a large resort hotel. Focusing attention on wide-area M2M, where there is data flow through a commercial network operator or communication service provider, we illustrate three different models in this section. These models are to be treated as examples of the nature of the M2M business to illustrate the complexity involved and are by no means exhaustive.

2.5.1 Network Operator- or CSP-Led Model

In this model, the communications service provider (CSP) plays a central role in the M2M solution. The corporate customer of the CSP directly approaches the CSP with a request for deploying an M2M service. The CSP plays the role of the system integrator and, through its partners, provides an end-to-end solution for the corporate customer. The CSP may also optionally employ the services of system integrators. The CSP selects from its partners a device vendor and an application software developer to write the application for the particular devices chosen. CSPs may have their own home-grown M2M service platform or may partner with one of the platform providers, such as Jasper Wireless, to provide the platform services on a hosted-model basis.

The value and money flow are illustrated in Figure 2.4. The corporate customer pays the CSP for initial solution development and deployment and also for ongoing network and service usage. The CSP in turn pays the device vendor for devices, an application software provider for the application software, and shares ongoing revenue with an M2M service platform provider, if one is involved. Additionally, there might be a money flow to a system integrator if one is involved.

As an example of this model, a small utility company may have a requirement to read its meters and produce meter data. AT&T, for example, has a partnership with SmartSync to provide such a solution for the utility companies.

2.5.2 MVNO-Led Model

In the early days of M2M, the number of M2M devices deployed was not substantial enough for mobile network operators to be fully involved. They were content with bandwidth agreements with specialist M2M mobile virtual network operators (MVNOs). The MVNOs in turn played the central role in the M2M ecosystem directly interacting with the end corporate customer. The MVNOs also facilitated the deployment through their

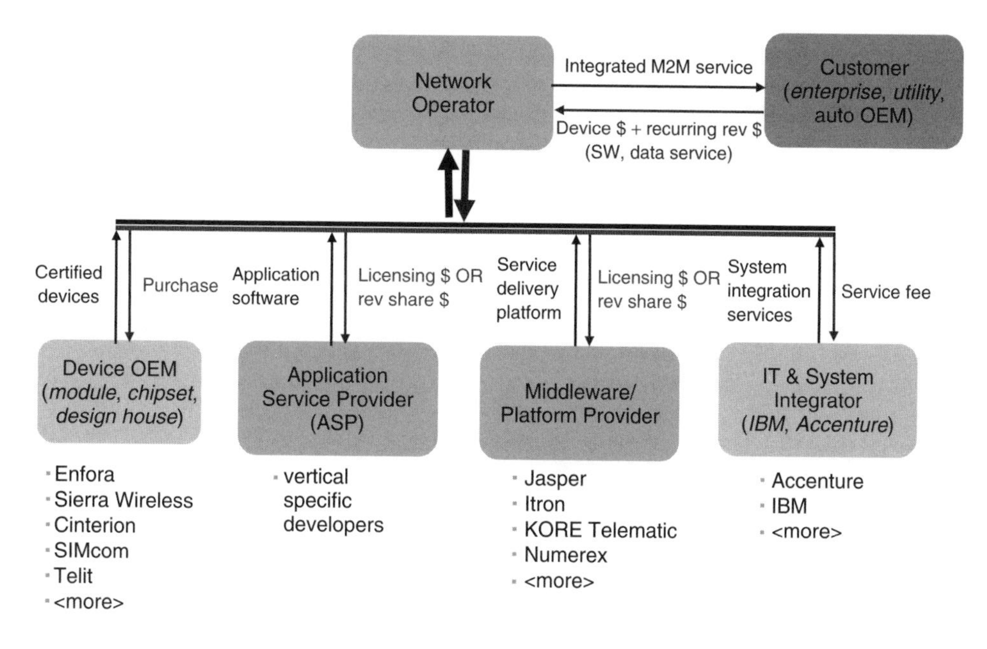

Figure 2.4 Network operator or CSP-led model.

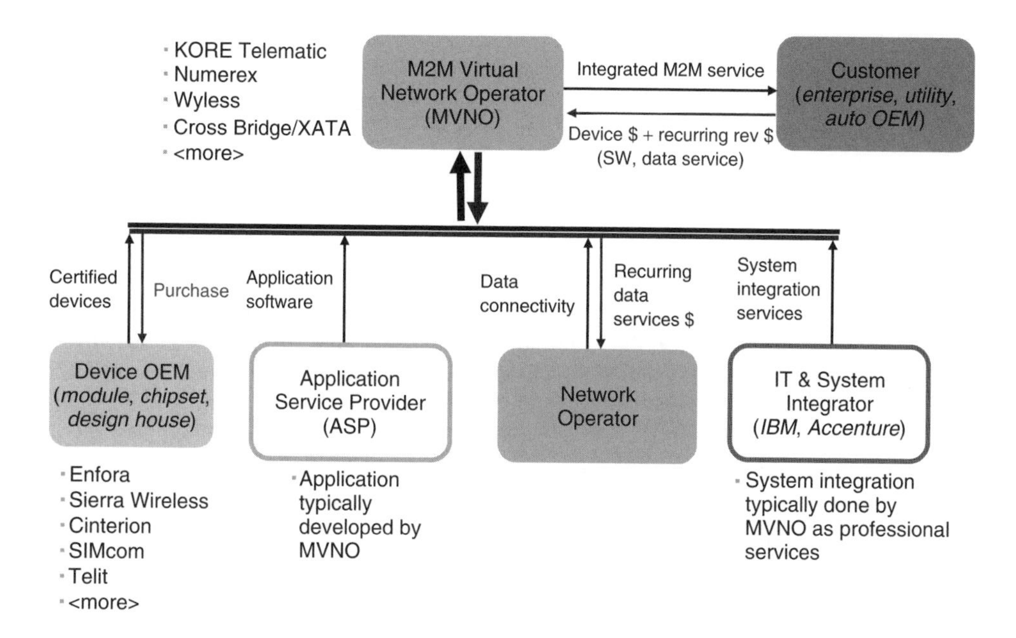

Figure 2.5 MVNO-led model.

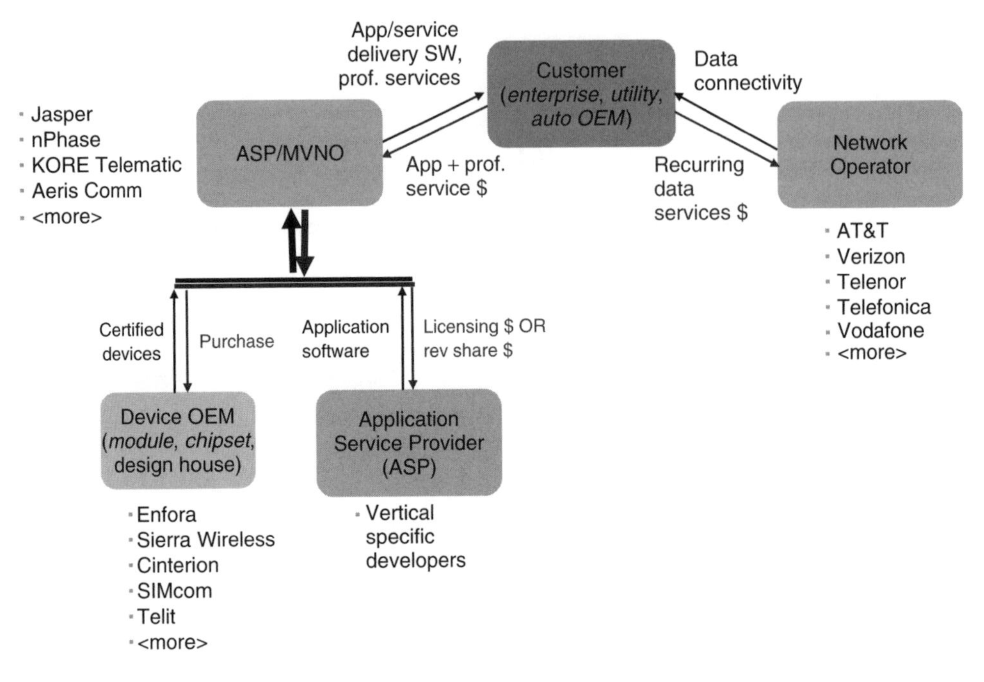

Figure 2.6 Value and money flow example 1.

platforms and device partners. They also sometimes developed the application on behalf of their customers. Figure 2.5 also shows the possibility of system integrator involvement.

2.5.3 Corporate Customer-Led Model

In this final model under discussion, illustrated in the Figure 2.6, the lead role is played by the M2M customer or service provider themselves. This is typically the case for large companies deploying a large number of devices. The company involved negotiates with selected network operators for the communication needs, and also employs a platform provider or MVNO for M2M services. The MVNO in turn may provide relevant M2M modules or devices. An example of this model is Amazon's Kindle™ e-book. Amazon has integrated the 3G module into its e-books and has bandwidth agreements with AT&T for delivering books and other information over their wireless network. The end-consumer does not need an AT&T subscription for the e-book. Amazon has deployed an M2M platform itself for this purpose.

2.6 M2M Business Metrics

How can the success of an M2M business be measured? This is, of course, quite different for an M2M application provider such as a company and for network operators. Network operators have traditionally used average revenue per user (ARPU), or essentially the monthly revenue per subscription and churn rate, as metrics for consumer voice and data connections. With M2M, the same metrics may not be entirely appropriate. While the average monthly ARPU for traditional services is around $50, the ARPU for M2M connections is much smaller at around $5, although in the best-case scenario, ARPU can be as high as $150 for some data-intensive applications, such as digital signage. However, simply looking at ARPU may be misleading. It fails to capture the fact that customer acquisition and retention costs, and even the transport costs, are much smaller for M2M connections. Even though the ARPU is small, the margins can still be fairly high. Furthermore, M2M business presence may really be a way for network operators to approach companies to win additional corporate business such as WAN or VPN and cloud services. Strategically, it is important for as many companies as possible to be present in the M2M space.

From a corporate perspective, the main M2M metric is return on investment (RoI) and some other intangibles, such as the improved customer offering, that they can provide to their customers. Another key metric related to RoI is the time it takes to see the return. Many M2M projects are abandoned at the evaluation stage because of the amount of time it takes to complete the deployment and see real benefits.

2.7 Market Evolution

One of the fundamental changes currently taking place in the market is the key role of the wireless network operators. In the past, with voice revenues being the primary focus of the wireless operators, they paid little attention to the M2M market. The wireless M2M market for network operators simply meant wholesale bandwidth agreements with M2M specialty MVNOs, who in turn served corporate customers. With voice revenues saturating, wireless operators turned their attention to data revenues to make up for the deficit and in particular M2M data revenues, which provide high margins and low-churn customers. Most operators have now created dedicated M2M sales divisions to approach the M2M corporate customers themselves. Coupled with this, they typically partner with a hosted M2M service platform provider that can automate many of the device management and billing processes. Some wireless operators have also built home-grown platforms to meet the needs of their M2M customers. The next stage of the evolution is likely to be the development of more standardized platforms provided by vendors and owned and operated by network operators as part of the network infrastructure. This will allow operators to differentiate their offering from that of their competitors and solidify their added value as the market becomes substantial.

Another evolution in the M2M market will likely be driven by emerging standards such as the ETSI Technical Committee M2M standards. With significant participation from network operators and vendors, the standards offering will establish a framework upon which an ecosystem can converge and subsequently innovate to develop new solutions. It moves M2M from being a vertical market with many different implementations to a more horizontal platform with multiple applications running on top of it. This should lower costs and thus make a more widespread adoption a real possibility.

References

1. ZigBee document 053474r17, ZigBee specification release 17, ZigBee Technical Steering Committee.
2. Zensys, SDS10242, Software Design Specification, Z-Wave Device Class Specification.
3. The X.10 specifications can be found under this link: ftp://ftp.x10.com/pub/manuals.
4. Beecham research, Worldwide Cellular M2M Services Forecast Market Brief, 02.08.2009.

3

Lessons Learned from Early M2M Deployments

Samia Benrachi-Maassam
Bouygues Telecom, Paris, France

3.1 Introduction

The current rapid growth of M2M deployments has given rise to several operational challenges that currently face network service providers whose installed infrastructure has been primarily designed and optimized for personal communications. The aim of this chapter is to share some of the most important findings and lessons learned, as well as describing the set of best current practices in terms of network architectures in order to cope with current M2M growth.

While M2M devices may use several communication technologies, including short-range RF, wireline, and cellular 2G/3G/4G, this chapter focuses on devices that use 2G and 3G communication modules to connect to a mobile network operator (MNO). Cellular technologies provide several characteristics that match the requirements of several M2M market segments. These include availability and geographical coverage, low latency and high levels of security. Additionally, network operators, and MNOs in particular, are becoming a trusted partner for M2M deployments and are increasingly providing new value-added services above and beyond basic connectivity and activation, embracing a model where they become M2M service providers.

One of the first challenges facing an MNO deploying M2M is the ability to operate a large number of M2M devices exhibiting different traffic characteristics without impacting personal communication services. It goes without saying that such a deployment

M2M Communications: A Systems Approach, First Edition.
Edited by David Boswarthick, Omar Elloumi and Olivier Hersent.

must use the existing infrastructure and technologies and should then utilise the capital expenditure (CAPEX) and operational expenditure (OPEX) MNO investments. Such a challenge mandates a clear understanding of the service requirements, as well as the traffic characteristics of each of the targeted market segments. The rest of this chapter is structured as follows. First, an overview of operational M2M deployments is provided, along with the technical architectural choices made to address the different M2M application requirements. Second, a summary of some of the challenges relating to M2M, as well as some initial architectural optimization mechanisms is introduced. Finally, the chapter provides a summary of the main lessons learned from early M2M service deployments.

3.2 Early M2M Operational Deployments

3.2.1 Introduction

This section describes several operational M2M deployment examples that use existing MNO networks. These examples have been selected in order to demonstrate the different current possible technology choices for data collection (or exchange), as well as for device triggering. Device triggering refers to the mechanisms used by an M2M Server[1] to trigger the establishment of a data bearer[2] by an M2M device. While it might be desirable for each device to be always-on and to have a permanently assigned IP address, such a setting is costly in terms of network resource usage and energy consumption. As such, for devices that transmit very small amounts of data, the use of circuit-switched (CS) domain services, such as SMS, may be overall more efficient. Roughly speaking, the following deployment examples highlight the different technology choices that are illustrated in this section:

- **Data collection and exchange** can be performed through:
  - **CS domain services such** as SMS and circuit-switched data (CSD).[3] The use of the CS domain is more targeted at M2M applications deployment examples occasionally transmitting very small amounts of data.
  - **packet-switched (PS)** bearers. The use of the PS domain is more targeted at applications needing to transmit relatively large amounts of data or requiring low network delay. A generic packet radio service (GPRS) bearer can be established on a permanent basis (always-on) or at the initiative of the M2M device, either periodically or based on a trigger, such as data availability, or an alarm.
- **Device triggering by the M2M server** can be performed through:
  - **sending a specific SMS to the device** – the device then reacts through the establishment of a GPRS bearer, also referred to as packet data protocol (PDP) context activation.
  - **an unanswered voice call** – typically, the M2M server triggers the establishment of a voice call toward the M2M device. The M2M device recognizes the calling party's number and triggers the establishment of a PDP context without answering the voice call.

---

[1] In the rest of this chapter, the terms M2M server and M2M application server are used interchangeably.

[2] A data bearer is defined by the network connectivity that allows for data exchange. Often, the following terms are used for an MNO: packet data protocol (PDP) context, generic packet radio service (GPRS) bearer and packet data network (PDN) connection.

[3] See next page for an explanation of the CSD service.

– **Network-requested PDP context activation (NRPCA)**: This allows the network, on behalf of the M2M server, to establish a PDP context. Such a feature removes the need for sending an SMS or an unanswered voice call solution. However, it is not widely deployed in operational networks because of its inherent technical constraints. Ongoing 3GPP standards work aims to optimize NRPCA mechanisms (see Chapter 6).

The Figure 3.1 provides an overview of device triggering using SMS or the unanswered call technique.

- **Steps 1 and 2**: The M2M server issues a voice call (through the MSC-S)[4] or sends an SMS to the M2M device (through the short message service-center (SMS-C). The request (voice call or SMS) is then relayed to the M2M device.
- **Step 3**: The M2M device recognizes the need to establish a PDP context.
- **Step 4**: The PDP context is established. As a result the M2M device acquires an IP address and is able to communicate with TCP/IP protocols to the M2M server.
- **Step 5**: The M2M device exchanges data with the M2M server.

Note that device triggering can be used in conjunction with a periodic PDP context bearer establishment. The device is configured to establish periodic connectivity (the periodicity is set in the M2M device by means of configuration). As such, device triggering is used when the M2M server needs to solicit the M2M device, while, for instance, periodic reporting of measured data takes place at the initiative of the M2M device.

Prior to the GPRS network deployments, the CSD has been commonly deployed as a data exchange technology for M2M applications. CSD is the original form of data transmission developed for GSM systems. The CSD technology allows for an uplink and downlink bandwidth of 9.6 kbps, which is sufficient for early/several M2M applications (sending meter data, information on the location of gas tanks, etc.).

The rest of this section provides examples on how the above-mentioned toolbox of mechanisms can be used to address the needs of different M2M applications.

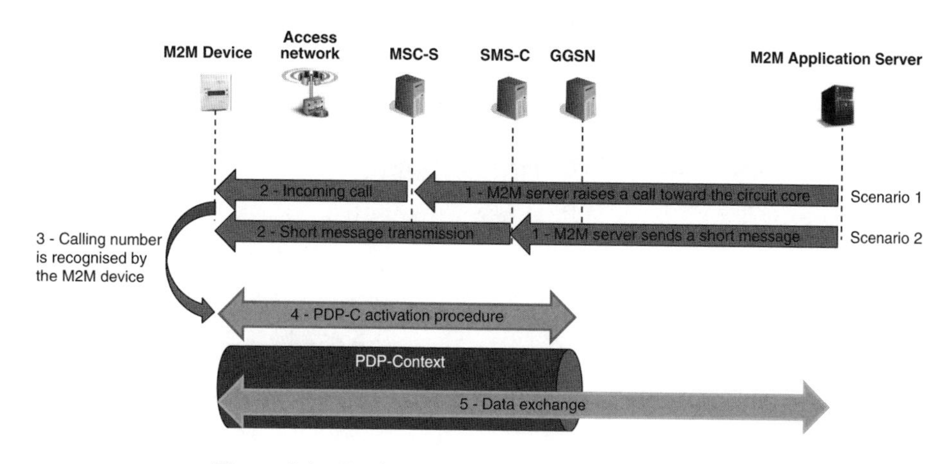

Figure 3.1 Device-triggering communication model.

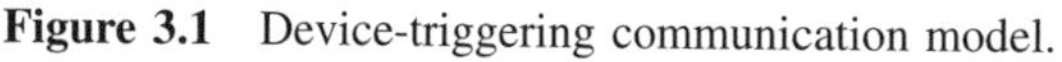

[4] MSC-S refers to the mobile switching center server. SMS-C refers to the short message service center.

3.2.2 *Early M2M Operational Deployment Examples*

3.2.2.1 Vehicle-Tracking

Vehicle-tracking-based M2M applications, such as fleet management and pay-as-you-drive car insurance, rely on the capability of the MNO to calculate and make the location of the tracked vehicle available to M2M application servers. In other scenarios, the location information is reported by the M2M terminal to the M2M application server using a communication network bearer. In this particular case, there are two types of location that can be exchanged:

- **GPS location** – provides a more accurate location but mandates the deployment of a GPS antenna and may not work under all conditions.
- **network location** – is either provided based on the base station cell that is serving a particular terminal (but provides a less accurate location) or estimated based on the triangulation technique considering signal strengths measured from base station cells.

As illustrated in Figure 3.2, the M2M server requests the geo-location of the M2M device from the location platform (Step 1). The geo-location platform polls the home location registrar (HLR) in order to contact the MSC-S that controls the M2M device (Step 2). In Steps 3–6 the geo-location server requests the MSC-S to page all the cells where the M2M device is potentially camped in order to provide the 2G/3G cell-id information. The geo-location platform then translates the 2G/3G cell-id into a geo-location position according to preconfigured mapping tables. The geo-location information is then transmitted to the M2M server over a pre-established secure VPN tunnel (Step 8) in order to ensure

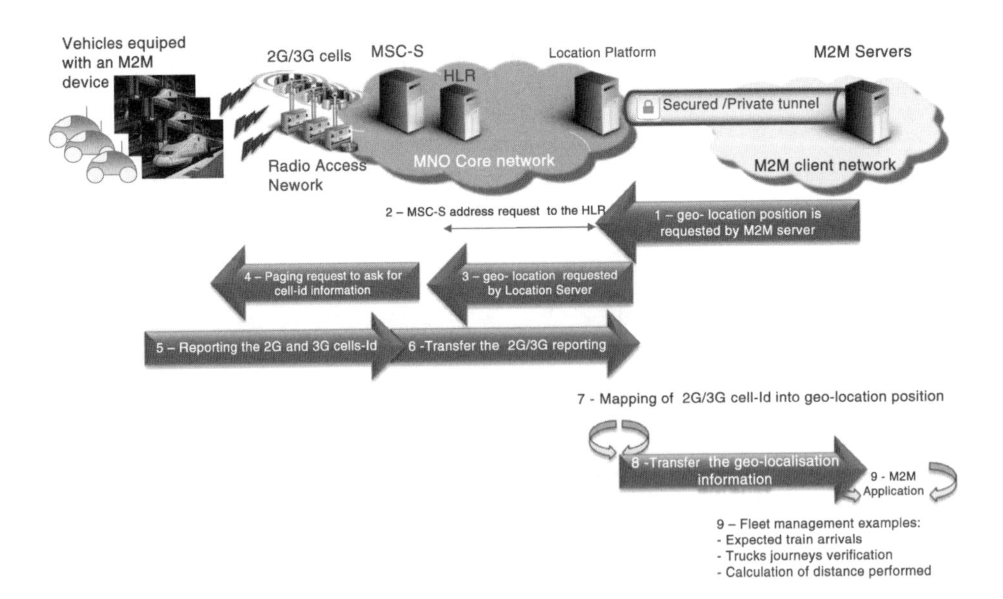

Figure 3.2 Vehicle tracking M2M applications, basic principle.

privacy. Note that the location platform can make the location information available to multiple M2M servers that have the appropriate rights for the M2M device (Step 9).

GPS and network location information can be used in a complementary fashion to provide accurate location information when needed by the M2M application server. The GPS location information can, for example, be used in the event of a car accident or other emergency situations to provide accurate information to the rescue services.

3.2.2.2 Smart Telemetry

Smart telemetry allows for the real-time collection of various data from meters such as those that provide temperature, energy consumption, or pollution levels. In this section, the following deployment scenarios are described: electrical smart metering and gas tank level monitoring.

In the smart metering example, the M2M device is configured to schedule periodic reporting of meter data, for example, every 3 hours the smart meter will report the metering information to the M2M server. Two possible solutions for the periodic reporting have been adopted in current deployments:

- **SMS solution** – the meter data is reported using SMS.
- **GPRS solution** – a GPRS bearer is established and then used to report the meter data over TCP/IP.

The choice of the SMS- or GPRS-based solution is often mandated by the needs of the M2M application provider, including the constraints relating to the deployed communication modules.

In addition to the periodic reporting of meter data, electrical smart metering (Figure 3.3) often requires the M2M server to send urgent commands to smart meters. This is particularly the case when there is a need to trigger demand/response operations, for example, temporarily switching off a piece of equipment. In this scenario, the M2M server typically triggers the sending of a specific SMS to the smart meter that is serving the targeted piece of equipment. The smart meter recognizes this signal and triggers the appropriate demand/response operations.

Another example of smart telemetry relates to gas tank level monitoring (Figure 3.4) as used by some energy companies. The use of an MNO network allows for increased

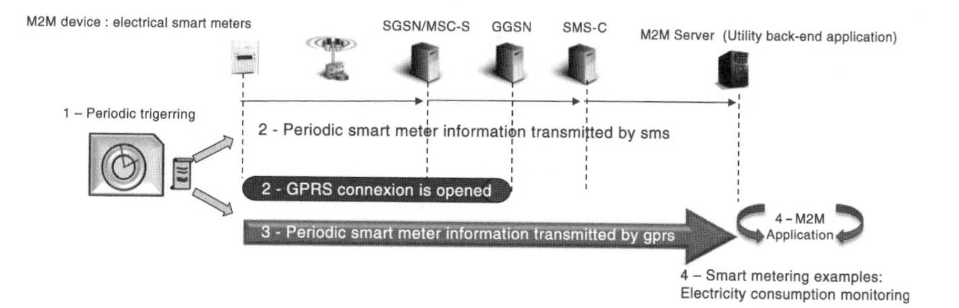

Figure 3.3 Automatic triggering in electrical smart metering.

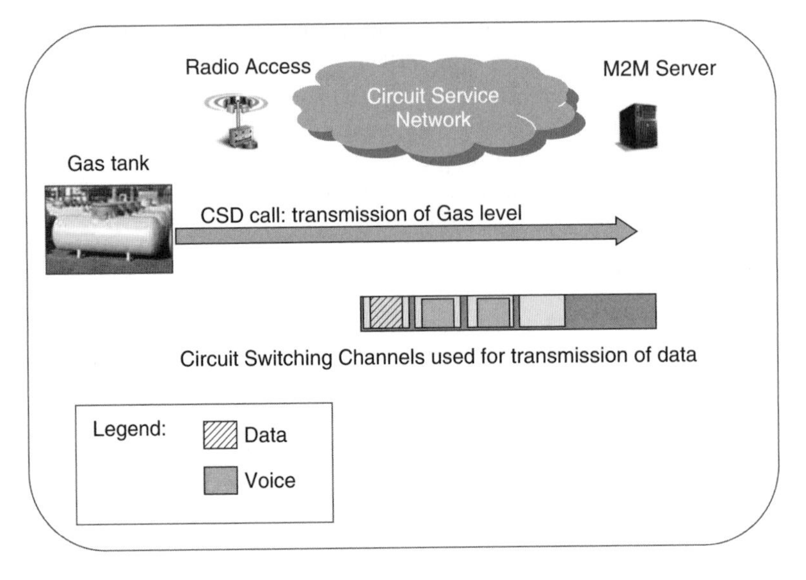

Figure 3.4 CSD for monitoring gas tank levels.

worker security, as well as a high degree of operational efficiency. Real-time monitoring of gas tank levels is performed remotely using a SIM-based communication module that is installed in the gas tank level meter. in this deployment scenario, the levels are sent, using CSD, to a central application that triggers appropriate operational actions when needed.

While CSD might be considered by the reader as being outdated, its selection is usually mandated by the end-users who often prefer to avoid major redevelopment of their existing software or costly integration of GPRS-capable communication modules.

3.2.2.3 Healthcare Monitoring

M2M eHealth applications such as remote patient monitoring, ageing independently, personal fitness, or disease management constitute a major M2M market segment. This section considers a healthy ageing application targeted at ageing populations in western countries. The service often makes use of M2M devices that are integrated into a wearable bracelet or a necklace. It allows the patient to contact an emergency center by pushing a single button on the M2M device. The M2M server recognizes the calling number (without answering) and automatically initiates a call toward the patient's M2M device. The patient is then assisted by an emergency medical technician (EMT). This procedure has been implemented to ensure all charging (and related billing) incurred is performed for the M2M server.

Additionally, in this deployment scenario, another M2M server (a maintenance center) occasionally performs software and firmware upgrades using the CSD service. Both scenarios are depicted in Figure 3.5, where two patterns of behavior are shown.

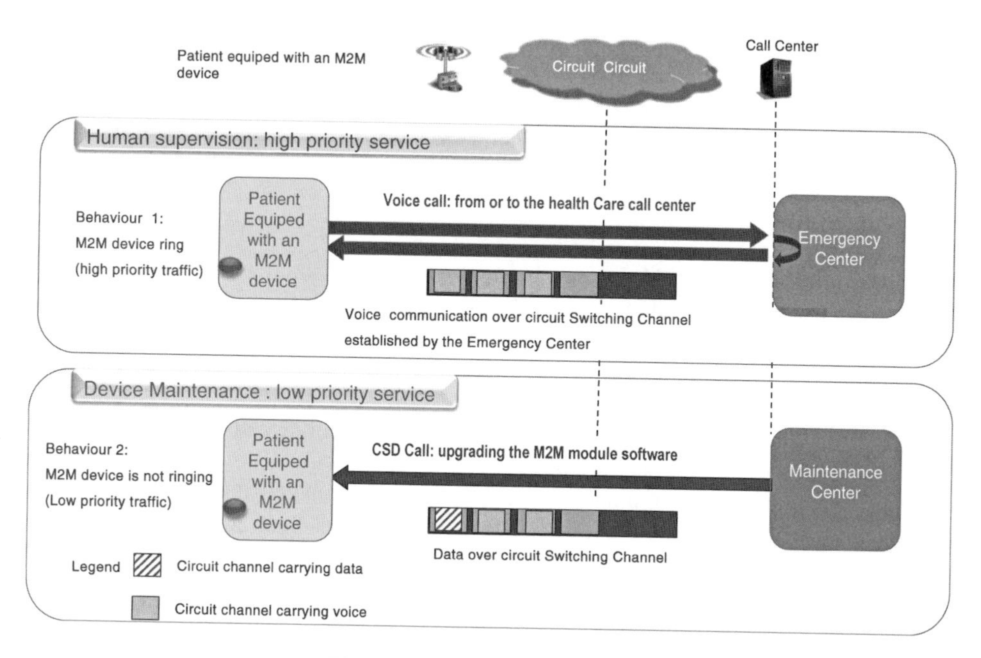

Figure 3.5 Remote healthcare.

- **M2M device behavior pattern 1**: Upon receiving a call (that remains unanswered), the emergency center issues a high-priority call toward the M2M device and establishes a communication with the EMT using the high-priority mobile station international subscriber directory number (MSISDN) of the M2M device that has been assigned two MSISDNs, that is, one for high-priority services and the other for all other services. In this behavior pattern, human supervision constitutes a critical and high-priority service for the MNO. The use of a specific MSISDN allows for the related calls to be treated as a high priority.
- **M2M device behavior pattern 2**: Since the M2M device needs regular upgrades of its software/firmware, CSD is used with a second MSISDN, allowing the network to treat this maintenance as low priority.
  In this early M2M deployment, the M2M device is assigned two MSISDNs to allow for multiple forms of network behavior, depending on the urgency of the situation:
  - The first MSISDN is used in order to issue a high-priority call or to get a high-priority call back in the event of an urgent need to establish a communication bearer. The network will use such a MSISDN to ensure high-priority handling of the high-priority calls. The other use of the MSISDN is to ensure that a high-priority call uses a free MSISDN that is not being used for low-priority CSD communication.
  - The second MSISDN is used for CSD communication that is handled as a low priority.

However, the main disadvantage is the potential increase in the consumption of MSISDN resources. As there is currently a shortage of adequate resources such as MSISDN, the development of a standardized solution that avoids the need for multiple MSISDNs is crucial for future M2M deployments.

3.2.2.4 Surveillance and Security

Surveillance and security M2M devices are deployed in residential and business premises, such as schools, public buildings, shops, to provide data, photo, and video surveillance information to security alarm applications. Cellular networks are often used as primary or backup access to provide connectivity to security-monitoring applications or premises owners.

The information exchanged primarily consists of alarm information and occasionally low- to medium-resolution video signals.

Surveillance and security applications mandate very low delays and occasionally high bandwidth when video or photo files need to be transmitted. Such a requirement clearly highlights the importance of using PS bearers. In order to avoid large-scale latency relating to the establishment of PDP contexts, an always-on connection with a permanently assigned IP address is maintained. Additionally, surveillance and security applications impose other requirements such as:

- The authentication of the M2M device is provided by the M2M application server (in addition to the MNO authentication mechanism) using the RADIUS protocol. Such an authentication is a prerequisite for the M2M devices establishing a PDP context.
- The IP address of the M2M device is assigned by the M2M application server (or an entity on its behalf) from the same subnetwork. Such a mechanism avoids the need for using public IP addressing, which is convenient for always-on bearers.
- In order to secure the authentication of the M2M devices and ensure the privacy of the data that is exchanged between the MNO and the M2M application server, an encrypted IP Security (IPSec) tunnel is established between the MNO and the company network where the M2M application server is hosted.

An illustration of the network set-up used for surveillance and security applications is provided in Figure 3.6.

Steps 1–5 illustrate the establishment of the PDP context. The M2M device is assigned an IP address following successful authentication using the RADIUS protocol. In the event of vandalism or breaches (Step 6), the M2M device automatically collects evidence (Step 7) from cameras installed in the buildings (photos or videos). In Step 8, the evidence is transmitted over TCP/IP connections through the always-on PDP context, that is, without the need to establish a network bearer, and making use of the pre-established IPSec tunnel (Step 0). An operator could perform initial checks before sending a team to the premises. The operator could raise a call to the end-user, and verify the vandalism with a picture or video feed (Step 9).

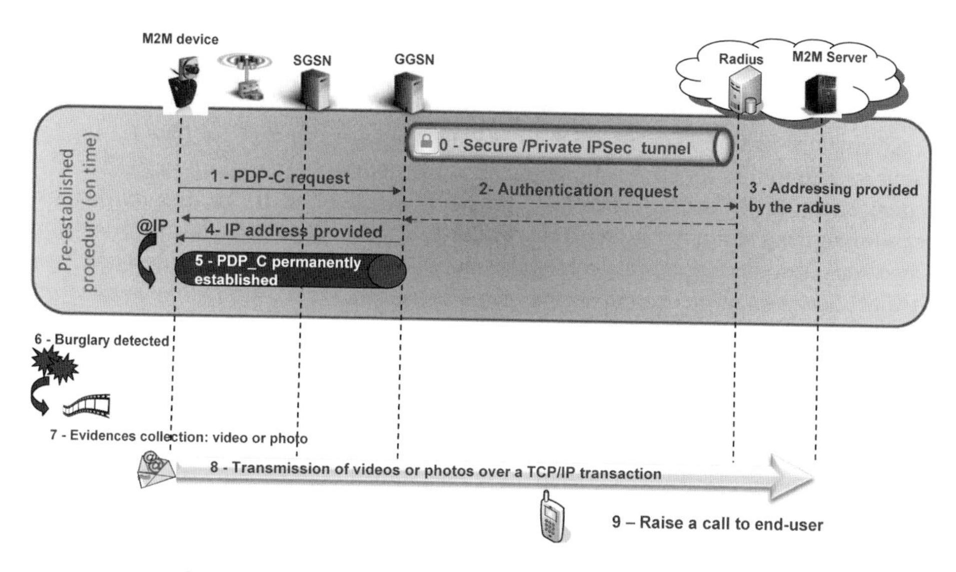

Figure 3.6 Always-on protection and remote supervision.

3.2.2.5 Point of Sale and Automated Teller Machines

Point of sale (PoS) and automated teller machines (ATMs) are widely deployed for managing the financial transaction and cash-dispensing processes. Generally, PoS/ATM terminals are equipped with wide-area network connectivity (wireless or wireline) to allow for communication with a central server to manage the payment/cash-dispensing transactions. Besides strong security and integrity requirements for the data that is exchanged,

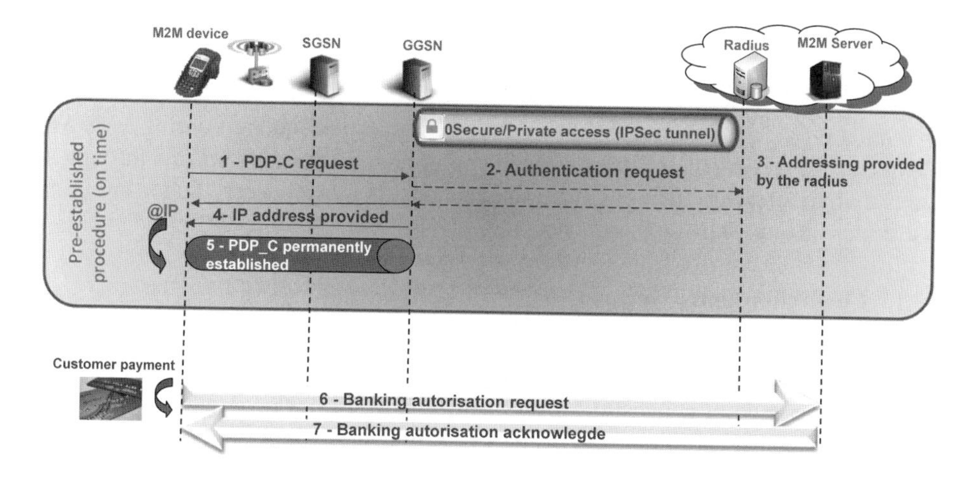

Figure 3.7 Always-on e-payment illustration.

one of the main challenges facing PoS/ATM applications is the ability to offer the near-real-time services that is mandated by the high degree of interactivity with the end-user. Such a need is addressed through the use of always-on connectivity to avoid excessive delays incurred for establishing a PDP context in case of an MNO.

Figure 3.7 describes an electronic payment deployment. Steps 0–5 illustrate the establishment of the PDP context. The M2M device is assigned an IP address following the successful authentication using the IPSec tunnel.

Once the customer proceeds to pay the bill, the terminal performs data exchanges with the M2M server in accordance with the requirements of the payment transaction (Steps 6 and 7).

3.2.2.6 General Conclusions from Early M2M Examples

One conclusion that can be drawn from early operational deployments is that the selected solutions depend essentially on the characteristics of the requirements of the targeted application. In general, the following application requirements must be considered:

- end-to-end delay and interactivity requirements;
- data volumes;
- data exchange frequency;
- server- versus client-initiated communication;
- communication module capabilities.

Table 3.1 provides a summary of the network solutions that are used, based on application requirements. While the Table 3.1 provides the general rules, final decisions often have to take into account other constraints such as end-user requirements, along with the installed legacy applications and communication modules.

3.2.3 Common Questions in Early M2M Deployments

This section provides an overview of some[5] of the issues encountered during early M2M deployments. Resolving these issues through appropriate network set-up, and through new features that are presently being defined in standards, is a key success factor in ensuring massive mainstream M2M deployments.

3.2.3.1 Congestion and Overload

As an illustration of where congestion can occur, an indication of the network devices that may be subject to congestion and overload is provided in Figure 3.8.

Typically, the issue of congestion occurs when the M2M device requests the establishment of a PDP context. In the event of the establishment of the PDP context failing, the M2M devices keep constantly trying until they establish a connection.

Due to the abnormally high number of requests, the authentication servers (RADIUS) start to overload early on in the process. When the regular human-to-human (H2H) PDP

[5] Documenting all the deployment issues encountered could fill a chapter on its own.

Table 3.1 Mapping of M2M solutions and Use Case examples

Use Case needs	CS domain/ PS domain	Connectivity mode	Description	Pros	Cons
Real-time/ interactive data	PS domain	Always-on	The M2M device has a permanently active PDP context allowing for always-on connectivity	No latency pertaining to data bearer establishment	Context maintained in the network devices for each bearer (additional CAPEX and OPEX costs)
Medium to high volume of data (but no real-time requirements)	PS domain	PDP context established for the duration of the data transfer	When the communication is server-initiated, the device triggering is performed at the initiative of the server, an SMS or an unanswered call is initiated toward the M2M device that requests PDP context establishment. Alternatively, the M2M device initiates the PDP context, e.g., when it has data to report	Avoids the need to maintain context for each PDP context on a permanent basis	CS domain is still used despite the fact that the actual data exchange is performed over the PS domain
Low volume of data and low periodicity	CS domain	Voice or SMS	Voice call/SMS generated for short information to be transmitted	No costly use of PS domain signaling; just for a small volume of data transmission	Additional overload on CS domain resources
High availability data service	CS domain	Data CSD with multiple MSISDNs	CSD call raised on a dedicated MSISDN	High priority calls are routed with a higher priority	Two MSISDNs are needed

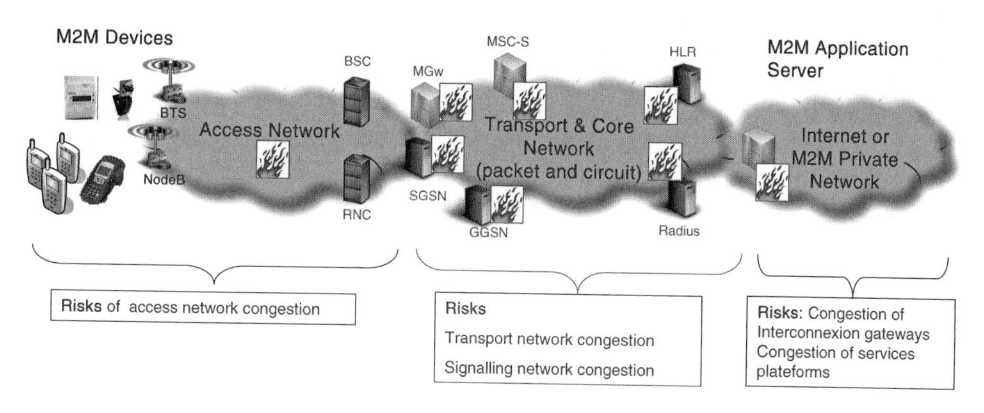

Figure 3.8 Congestion in M2M devices.

connectivity requests come to the packet core network, the authentication server is unable to reply and sends back a time-out failure response. The overload propagates back through the gateway GPRS support node (GGSN) interfaces and consequently freezes the entire packet core network in the latter stages.

There is often a need for the network administrators to momentarily stop the M2M traffic as a means to identifying and isolating faulty or malfunctioning communication modules. However, such an operation may not be easy to achieve due to the lack of means for identifying specific M2M traffic as such.

As a preliminary analysis of M2M traffic behavior versus H2H traffic that is typically transported over telecom network infrastructures, the following characteristics can be deduced:

- **Synchronized**: M2M devices are very often programmed to report data at given intervals, such as every hour, without any randomization in the establishment of the connections. Such behavior results in synchronization in terms of network resource usage and often leads to congestion.
- **Unpredictable**: In other cases, M2M modules are deployed without the operator being aware that such a deployment corresponds to M2M. In this case, it becomes difficult for the operator to predict the traffic and dimension the network accordingly.
- **Bursty**: Several devices such as surveillance and security devices usually generate a small amount of data, except when an alarm has been raised. A video camera, for instance, may generate a high volume of data when there is an alarm.
- **Uncontrollable**: A set of devices randomly connects to the network without any predictable pattern. Take, for example, a set of roaming devices – when they lose coverage with Operator A (due to radio issues), they will simultaneously attempt to roam to Operator B.

3.2.3.2 Shortage of Identification and Addressing Resources

- MSISDNs are used for identifying mobile terminals. They also provide a human-friendly means with which to reach a particular subscriber's terminal. In current MNO deployments, whether they are for M2M or other human communications, each subscription is assigned an MSISDN. The MSISDN format is specified in Recommendation E.164 of the ITU-T. While it allows for a large number of devices to be deployed, national numbering plans in different countries have restricted the number of possibilities to a few digits (based on predictions for personal communications). Let us take, for example, the case of France, where the following numbering blocks are assigned for mobile operators: 06 XX XX XX XX and 07 XX XX XX XX. Such numbering plans allow for a maximum of 200 million subscriptions, which is sufficient for a population of 65 million inhabitants but may reach its limit once M2M has been massively deployed. In early deployment phases, one of the alternatives for optimizing these resources is to use private MSISDN plans. Unfortunately, this solution has certain limitations, such as the interoperability of the service in the event of roaming. As such, this solution may not work for a connected car that is traveling abroad and must be able to roam in order to continue offering the service.

- IP addressing is used when the M2M devices establish data bearers. Depending on the deployment scenarios and requirements, both public and private IP addressing can be used. In order to optimize the MNO IP addressing resources, private IP addressing can be used but needs a specific network set-up either to build specific private networks for an M2M server and the M2M devices that it controls or to deploy a network address translator (NAT) when the M2M server is located on the Internet. Private IP addressing has the following limitations:
  - the additional cost relating to NAT deployment;
  - in the case of server-initiated communications, the M2M server may not be able to reach the M2M device unless an explicit mechanism is deployed to open pinholes in the NAT. Device triggering via SMS is therefore still needed even if the M2M device has an established PDP context.

3.2.3.3 Use of CS Domain Context for Data-Only Services

While it could be argued that an M2M device that is configured to use PS domain-only services may not need to have an MSISDN, in practice however, the 3GPP standard still mandates the need for an MSISDN since some of the procedures, such as charging, rely on the use of an MSISDN as an identifier. Additionally, the use of SMS as a device-triggering mechanism mandated the need for an MSISDN, as well as a subscription to the CS domain in addition to a subscription to the PS domain. Such a subscription to the CS domain comes at a certain cost due to the need to maintain a context in the CS domain note.

The current 3GPP standard for SMS uses the MSISDN to send an SMS. Ongoing 3GPP work is geared toward relieving this constraint by allowing for PS-only subscriptions and sending SMSs to an international mobile subscriber identity (IMSI), that is, without the need for an MSISDN.

3.2.3.4 Conclusions of Early Deployment Issues

While there are several issues relating to the early M2M deployments, the most important ones to date are related to overload and congestion, as well as to the shortage of scarce numbering and addressing resources. For both issues, standards and regulatory bodies are working toward providing long-term fixes that will allow for a seamless and cost-effective growth in M2M. As far as addressing is concerned, while the current momentum for IPv6 should provide concrete answers for the foreseeable future, its wide-scale deployment often encounters the obstacle of end-user deployment.

As far as congestion and overload control is concerned, it becomes very clear that while standards development should provide long-term fixes, a solution is needed in the short term to allow for current M2M deployment growth without impacting the network stability. Elements of such a solution are provided in the next section.

3.2.4 Possible Optimization of M2M Deployments

This section provides an overview of the best current practices in terms of network planning and architecture to allow for M2M deployment growth, while new standards are being developed to allow for a longer-term fixes. In the rest of this section, the need for traffic identification is introduced. Traffic identification is a basic capability that allows for optimized architectures to be built for M2M.

3.2.4.1 Traffic Identification

Identifying M2M traffic as such is a basic enabler for:

- optimizing an operational network to better handle M2M. In the rest of this section, it is demonstrated how the specific core network devices can be dedicated and optimized for M2M subscriptions.
- providing M2M specific, operational administration and maintenance (OA&M) functions. For instance, in the event of congestion due to a malfunctioning set of devices, it is often desirable to completely disable the set of devices causing the congestion, for example, M2M devices relating to M2M service "ABC." Alternatively, the related subscriptions can be routed to another network device so as not to impact sensitive services. In so doing, the network administrator has the means to stop the collateral damage and operationally maintain other services. Once the root cause of the issue has been solved – for example, a new firmware version of an M2M device is causing over-frequent connections to the network – the traffic can be reinstated on the network.

3.2.4.2 Use of IMSI Range

Traffic identification can be performed based on the use of dedicated IMSI ranges. Those dedicated ranges will be organized by usage category and implemented on home location registrar (HLR)/ home subscriber server (HSS) and GSM mobile services switching center (GMSC) in order to define a dedicated routing for the M2M traffic. The IMSI is composed of 15 digits: the first five digits are used for the mobile country code and mobile network code; the next two can be used to route toward a specific HLR for M2M, while a subset of the remaining digits can be used to identify a specific end-user, for example, Utility ABC or a certain M2M market segment, for example, eHealth.

3.2.4.3 Dedicated Core Network Central Equipment

In order to manage the M2M use cases in a seamless manner, the importance of identifying the M2M traffic has been demonstrated. The paragraph advocates the use of a network

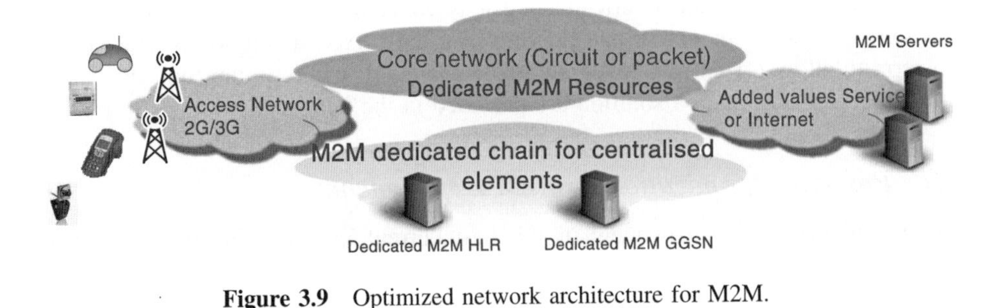

Figure 3.9 Optimized network architecture for M2M.

architecture where specific network devices are dedicated to M2M traffic. These are referred to as "*dedicated M2M chains*."

The Figure 3.9 shows two core network devices with a dedicated M2M chain composed of:

- **the HLR** – this is the central database that manages all subscriptions to the network. The use of specific IMSI ranges for M2M allows for all network registration requests to be routed to this particular HLR. In so doing, the network operator can avoid the congestion and overload resulting from a massive number of M2M devices wanting to register on the network at the same time, for example, roaming devices that lose coverage with Operator A and want to roam with Operator B. Additionally, by allocating a specific HLR to M2M, its design and maintenance can be optimized for M2M. For instance, the storage space in the HLR for an M2M subscription can be optimized through making use of the fact that a large amount of the subscription information is common to subscriptions pertaining to a certain device category. Also, as several M2M devices are stationary, the HLR could be optimized by reducing the frequency of mobility management procedures, thus allowing for high-capacity HLR devices.
- **the GGSN** – this is the network devices that provides, among other things, gateway function to IP-based networks, that is, companies or the Internet, where the M2M server is typically located. The engineering of this device could be fully customized for M2M. For instance, there have been cases where certain M2M applications require the use of persistent bearers but generate small amounts of traffic. The GGSN can then, in this example, be customized and engineered to process an infinitely larger number of PDP contexts with less memory for data transmission.

3.2.4.4 Specific Set-Up of Core Network Elements: GGSN APN-Specific Configuration for M2M

In addition to the use of a dedicated M2M chain for the core network, specific configuration and optimizations of the core network can ensure further efficiency in handling M2M. This section provides an example of the GGSN access point name (APN) configuration for M2M. Such a configuration of specific APN allows for *finer-grained* handling of M2M traffic. As an example, one can chose to use a specific APN for eHealth traffic because of the need to allow for low delay and jitter. Such a need will be translated into appropriate packet marking and higher priorities corresponding to such traffic.

The APN is a parameter of the GPRS that allows for specific traffic routing toward IP networks. Each subscription in the HLR can be assigned a specific APN that is activated upon establishment of a PDP context.

The APN can be configured with several parameters in order to allow for a specific form of behavior usually based on a common agreement with the end-user:

- **The connection mode** can be permanent and non-permanent. When the connection mode is non-permanent, the GGSN can be configured to release the PDP context after a certain time-out.
- **The IP addressing** can be configured to use public or private IP addressing.

Table 3.2 APN categorization and set-up

Connection mode	IP addressing plan	APN's name	APN's timer session time-out
Always on	Private	M2M_Al_On-Private.com	Off
	Public	M2M_Al-On-Public.com	
Non-permanent	Private	M2M_Non_Perm-Private.com	Daily
	Public	M2M_Non_Perm-Public.com	

- **The APN's timer session timeout** indicates a timeout after which a PDP context is released by the network. This mechanism is used to avoid the need to maintain context for bearers that are not used anymore.

In Table 3.2, by way of illustrating the various forms of categorization possible, four APNs are created and split into always-on connection or non-permanent, either with private or public IP addresses plan. The APN's parameter then could be configured according to the traffic they handle: as an example, the private APNs "M2M_Al_On-Private.com" and "M2M_Non_Perm-Private.com" are routed through NAT routers whereas the public APNs "M2M_Al_On-Public.com" and "M2M_Non_Perm-Public.com" are directly routed through the Internet. With this APN categorization method the M2M traffic can not only be separated via the use of dedicated M2M chain, but also assigned different priorities/traffic handling mechanisms depending on the customer/market segment needs.

3.3 Chapter Conclusion

While M2M applications are making their way in operational networks, it is essential that M2M is not simply considered as yet other personal communication service. A two-step approach is being advocated by the network operators and in particular the MNOs:

- Step 1 consists in optimizing currently deployed networks for M2M. These optimizations take into account some of the fundamental characteristics of M2M such as synchronization, burstiness, and stationary devices. A set of best current practices is progressively being built around:
  - **dedicated M2M chains** – which consist of allocating specific network elements for M2M handling. These elements are dimensioned according to M2M traffic needs, as opposed to the regular dimensioning pertaining to personal communications.
  - **traffic identification** – which consists of identifying the M2M traffic for both the CS domain and PS domain via the use of specific IMSI ranges and dedicated APNs.
  - **network equipment optimizations** – which consist of optimizing specific equipment for M2M. The example provided in this chapter is the use of an optimized HLR where, for example, smaller contexts per subscription are used and less frequent mobility management procedures are triggered for stationary M2M devices.

The list of optimizations that fall under Step 1 can go on. This chapter does not provide an overview of all of them, as these can vary from one operator to another and largely depend on the targeted M2M applications.

- Step 2, consists of providing a longer term fix of the deployed MNO networks based on the introduction of new features that are designed from the ground up for M2M. 3GPP TR 23.888 [1] provides an initial starting point for these optimizations. These are expected to pave the way for massive and more cost-optimized M2M deployment in the future and become operational through software upgrades or deployment of new equipment as needed. Examples of useful network optimization include:
  - PS only subscription/MSISDN-less: in this feature there is no need for the MNO to maintain a subscription context in the CS domain, or to assign an MSISDN for M2M devices that use only data services.
  - Online device triggering: consists of providing efficient mechanisms for triggering data bearer establishments.

Reference

1. 3GPP TR 23.888 (2011). System Improvements for Machine Type Communications.

Part Two

M2M Architecture and Protocols

4

M2M Requirements and High-Level Architectural Principles

Omar Elloumi and Franck Scholler
Alcatel-Lucent, Velizy, France

4.1 Introduction

The focus of this book is primarily on the horizontal enablers for M2M. In order to fully understand the motivation behind the work currently taking place in the different standards initiatives, as well as the most recent market developments, it is fundamental to become familiar with basic M2M requirements. Most standards organizations, including 3GPP, 3GPP2, and ETSI, have adopted a use-case-driven approach as a means with which to derive the set of requirements that further define the service architecture. ETSI, however, has adopted a more formal way of describing the use cases. In addition to the use-case-driven approach, it became very clear that all network optimization matters, both equipment features and the design or operational networks, also have to take into account fundamental characteristics of M2M-generated traffic and growth patterns. Both issues put forward new and particularly challenging requirements on the access and core network.

While this chapter focuses principally on M2M requirements for the services and network aspects, the aim is not to provide an exhaustive list of requirements which can be found in the different ETSI, 3GPP, and 3GPP2 standards. The objective is to show how

M2M Communications: A Systems Approach, First Edition.
Edited by David Boswarthick, Omar Elloumi and Olivier Hersent.

these requirements are derived and what new constraints M2M imposes on the underlying systems. The requirements relating to the role of IP in M2M communications are described in greater detail in Chapter 7.

The next section will mostly focus on the ETSI M2M use cases and requirements for the M2M service layer.

4.2 Use-Case-Driven Approach to M2M Requirements

4.2.1 What is a Use Case?

The term use case is very commonly used. However, a commonly agreed definition is more difficult to find. The Object Management Group (OMG) provides an excellent definition and approach regarding the description of use cases. According to the OMG, a use case describes the interactions between one or more actors on the one hand and the system under consideration on the other. These interactions are represented as a sequence of simple steps. Actors are something (a terminal) or someone (a person or group of people) which exist outside the system. Actors take part in a sequence of interactions with the system to achieve a given goal.

Use cases treat the system as a black box where interactions with the system, including responses, are perceived from outside the system. A use case will be described in an architecturally neutral manner that does not assume any particular physical architecture.

Use cases should not be confused with the functionalities, features, or requirements of the system under consideration. A use case may be related to one or more functionalities and/or requirements. A functionality or requirement may be related to one or more use cases.

A use case will also describe what the actor achieves by interacting with the system, that is, the goal.

4.2.2 ETSI M2M Work on Use Cases

The ETSI Technical Committee (TC) M2M was created in January 2009 and aims to provide an end-to-end view of M2M communication architecture. This end-to-end view does not mean that all elements of an M2M system are specified by ETSI TC M2M. The work fully endorses prior state-of-the-art work and ongoing standardization efforts taking place within other standards development organizations (SDOs) and forums. TC M2M focuses primarily on the service layer aspects and refers, where appropriate, to other standards activities. ETSI TC M2M made a conscious decision to base its work on a set of families of use cases where the use cases constitute the input toward requirements definition.

ETSI decided to base its use-case-driven approach on five main families of use cases:

- smart metering;
- eHealth;
- connected consumer;
- automotive;
- city automation.

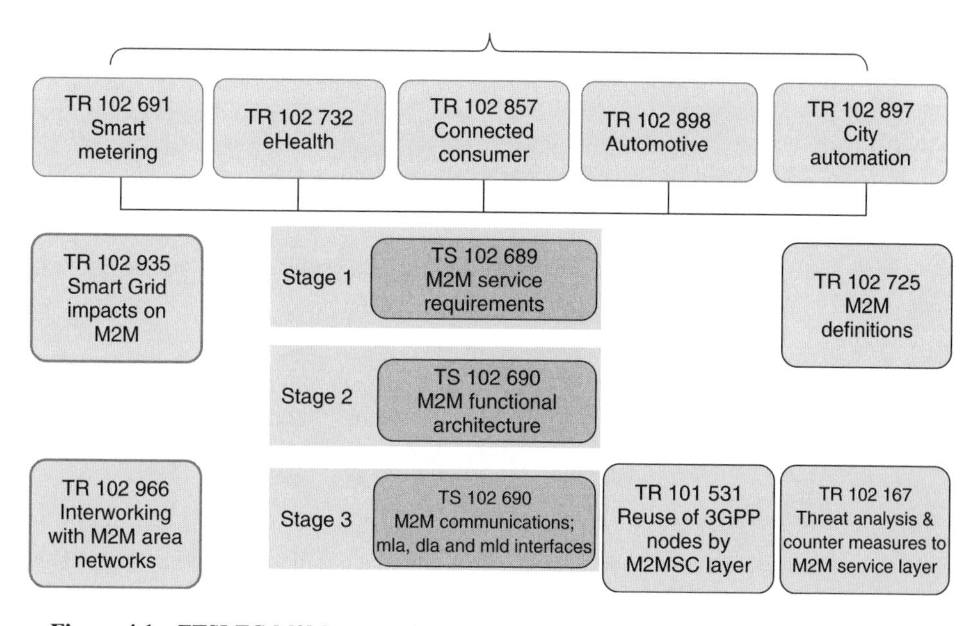

Figure 4.1 ETSI TC M2M approach, use-case-driven approach to service requirements.

Five technical reports have been developed by ETSI to capture these five families of use cases. From these technical reports, ETSI TC M2M is currently developing a set of specifications. Figure 4.1 depicts the relationship between the use case technical reports and the other deliverables.

The work on use cases is not meant to be exhaustive, since a systematic approach could become cumbersome and time-consuming without really providing much added value. Additionally, documenting all potential use cases might leave little room for innovation and differentiation among the actors. The objective is to cover enough use cases to ensure that all important requirements are captured so that the architecture work provides the foundation for a potentially large number of M2M applications.

4.2.3 Methodology for Developing Use Cases

Because of the multiplicity of possible use cases, it has become increasingly clear that a single methodology for developing use cases is an essential step in the process.

4.2.3.1 Use Case Template

In order to allow for a minimal formalism in describing use cases, the use of a single template is fundamental for achieving common understanding and a certain formalism when deriving the requirements.

The following template, as taken from [1], has been used by ETSI TC M2M:

Template

General Use Case Description

Objective/goals of this use case in high-level terms listing major issues that are highlighted.

Stakeholders

List of who or what the use case is referring to: consumer, network operator, database, billing entity etc.

Scenario Case Description

Descriptive text of the use case showing how the stakeholders use the system. This section may provide preconditions for the use case as well as the trigger for the use case.

Information Exchanges

Stepwise description of information flows, for example as registration, data retrieval, or data delivery implied by the use case.

Potential New Requirements

List of requirements derived from the use case.

Use Case Source

Reference to the document or the entity that developed the use case.

4.3 Smart Metering Approach in ETSI M2M

4.3.1 Introduction

Smart metering use cases were handled in ETSI M2M under a separate technical report [1]. The work was strongly related to and driven by the M/441 mandate of the European Commission on smart metering. Mandate M/441 has the following goal:

> The general objective of this mandate is to create European standards that will enable interoperability of utility meter (water, gas, electricity, heat), which can then improve the means by which customers' awareness of actual consumption can be raised in order to allow timely adaptation to their demands (commonly referred to as "smart metering").

Smart metering mainly targets the improvement of energy end-use efficiency, thus contributing primarily to the reduction of energy consumption and to the mitigation of CO_2 and other greenhouse-gas emissions.

Smart meters are utility meters (electricity, gas, water, heat meters) which eliminate the need for estimated bills and human meter readings and provide customers, energy distributors, and suppliers with accurate and timely information on the amount of the consumable utility (e.g., electricity, gas) being used. Smart meters may also provide other services such as the capability for real-time control of energy consumption.

A report [2] of the Smart Metering Coordination Group (SMCG) – a joint group of CEN, CENELEC, and ETSI created to coordinate standards actions in the area of smart metering – documents a list of functionalities that are suggested to be provided by smart metering information systems. This list, grouped into six categories, represents the additional functionalities proposed on top of those typically offered by conventional meters. The six categories of functionalities are described below [1]:

- **Remote reading of meteorological register(s) and provision to designated market organization(s):** Metering system capability to remotely provide the designated market organization(s) with the value of the meter register(s) through a standard interface at a predefined time schedule or upon request.
- **Two-way communication between the metering system and designated market organization(s):** Capability of the metering system to remotely retrieve data on, for example, usage, network and supply quality, events, network or meter status, and non-meteorological data and to make this data available to the designated market organization(s). Ability of the designated market organization(s) to remotely configure the metering system and to carry out firmware/software upgrades. Ability of the metering system to receive information, for example, information sent from the energy services provider (and/or via relevant third parties, e.g., distribution system operator or metering operator) to the end-user customer.
- **Meter supporting advanced tariffing and payment systems:** Capability of the metering system to allow the customer to prepay for usage by suitable means of payment, to connect to a supply and disconnect it after a predetermined consumption or a certain duration of time. Support for tariffing: metering system provided with multiple rate registers for consumption (and where applicable) injection to allow, for example, for time of use tariffs, critical peak, real-time pricing or combinations of these.
- **Meter allowing remote disablement and enablement of supply:** Capability of the metering system to remotely allow the designated market organization(s) to safely control or configure supply limitation (not applicable to gas meters), enable and disable supply through configurable parameters set at the meter.
- **Secure communication enabling the smart meter to export meteorological data for display and potential analysis to the end-consumer or a third party designated by the end-consumer.**
- **Meter providing information via portal/gateway to an in-home/building display or auxiliary equipment:** Capability of the metering system to provide information on total usage and other meteorological and non-metrological data for external visual display.

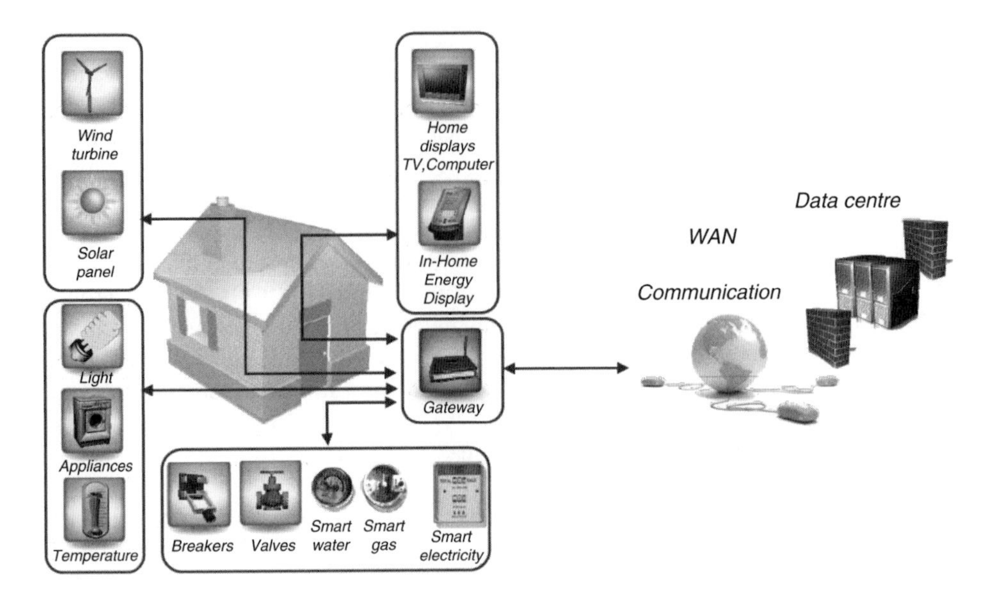

Figure 4.2 Typical deployment scenario of smart metering systems. Reproduced by permission of ETSI.

During use case development, the "additional functionalities" have been used as a general framework/scope for the development of use cases. As such, each use case is attached to one or more functionalities.

4.3.2 Typical Smart Metering Deployment Scenario

Figure 4.2 provides a typical equipment deployment scenario for a smart metering system. This scenario shows smart metering devices (e.g., valves, electricity meter, gas meter, and water meter) which are connected to data centers via a communications gateway.

The data center collects data from the smart metering devices and is able to control the relevant smart meter devices remotely via the communications gateway.

In this scenario, the gateway provides an interface to home automation devices such as sensors (e.g., temperature sensors), displays, and appliances and to electricity micro-generators deployed in residential environments.

Smart metering use case example: Pre-payment functionality. This use case is taken from [1]

General Use Case Description

A smart metering information system is configured to support the pre-payment functionality as defined by the billing entity. The use case describes occasions where the pre-payment functionality is managed locally on the smart metering information system as well as instances where the management is performed remotely by the billing entity.

Stakeholders

Billing Entity

Organization responsible for billing the consumer(s).

Consumer

Organization or person consuming the electricity, gas, heat, or water at the premise location. The consumer may also be the organization or person contracted with the billing entity to pay the bill.

Asset Entity

Organization responsible for the installation, configuration, operation, and maintenance of the smart metering information system assets (e.g., meters communication devices, smart metering gateway).

Scenario Case Description

Pre-conditions: The smart meter system is installed and configured to act as a pre-payment system. The smart metering information system recognizes the billing entity and/or asset entity and has an address for them.

Post-conditions: The smart metering information system confirms to the billing entity and/or asset entity that a prepayment-triggered action has been completed.

The billing entity and/or asset entity is/are made aware in the event of a failure.

Trigger: The billing entity or the consumer decides that there is a need to carry out a prepayment action on the smart metering information system.

Information Exchanges

Basic Flow

1. The asset entity or billing entity sends a message to the smart metering information system to change the payment mode.
2. The smart metering information system validates the request.
3. The smart metering information system changes the payment mode.
4. The asset entity or billing entity receives confirmation of the completion of payment mode change.
5. The smart metering information system displays preconfigured information from the asset entity and/or billing entity.
6. The consumer reads this information via a display.
7. The consumer is able to act upon this information.
8. The consumer undertakes an action as a result of this information, for example it adds credit to the smart metering information system or releases the emergency credit facility.

9. The smart metering information system sends confirmation messages back to the asset entity and/or billing entity when actions have been completed.
10. The smart metering information system provides status information at predetermined intervals or upon request from the billing entity.

Alternative Flow 1

Fails at Step 2: smart metering information system fails to validate a request:

1. The asset entity or billing entity sends a message to the smart metering information system to change the payment mode.
2. The smart metering information system deems the request invalid, and is unable to change payment mode or display the preconfigured information.
3. The smart metering information system records the event (error) along with date/time stamp.
4. The smart metering information system notifies the asset entity and/or billing entity that a failure has occurred.
5. END.
   (...)

Alternative Flow 3

Fails at Step 5: smart metering information system fails to display preconfigured information from asset entity and/or billing entity:

1. The asset entity or billing entity sends a message to the smart metering information system to change the payment mode.
2. The smart metering information system validates the request.
3. The smart metering information system changes the payment mode.
4. The asset entity or billing entity receives confirmation of the completion of payment mode change.
5. The smart metering information system fails to display preconfigured information from the asset entity and/or billing entity.
6. The smart metering information system deems the request invalid, and is unable to display the preconfigured information.
7. The smart metering information system records the event (error) along with date/time stamp.
8. The smart metering information system notifies the asset entity and/or billing entity that a failure has occurred.
9. END.

(Not all alternative flows are reproduced, the complete list is provided in [1])
New requirements derived from the use case:

- The M2M system should support auto-configuration functions for M2M area networks.
- The M2M system should support accurate and secure time synchronization. M2M devices and M2M gateways may support time synchronization or secure-time synchronization.
- The M2M system should support the capability to remotely change the state of an M2M device, for example to enable or disable.
- The M2M system should support transaction handling between cooperating objects capable of handling this functionality.
- The M2M system should support the following mechanisms for receiving information from M2M devices and M2M gateways:
  - receiving unsolicited information (passive retrieval)
  - receiving scheduled information
- End points of the M2M system should be able to verify the integrity of the data exchanged.

Use Case Source

ESMIG (European Smart Metering Industry Group), Document: Smart Metering Functionality Use Cases, ESMCR003-002-1.0, October 2009.

This smart metering use case contributed by the European industry association, ESMIG, allows for several requirements for the M2M system to be derived. Take for example the need to support transactions. The use case clearly shows that the steps of all the prepayment flows are atomic, in the sense that the service is only effective when all intended actions have been correctly executed. If this is not the case, then a roll-back to the original state is mandated. Transactions are defined to be a set of operations that are indivisible (also referred to as atomic), that is, if one operation fails, then the effect of the other operations must be cancelled. Transaction management represents an important requirement for M2M systems, and since it is required by several M2M applications, one could argue that this functionality could benefit from being developed and tested once, and then offered as a capability to the different applications through open APIs (application programing interfaces).

Consider now the fact that a smart meter will be connected to a communication infrastructure that could be owned by either a utility or a network operator. Bringing network connectivity to electricity meters makes the system vulnerable to malicious and fraudulent usage and imposes several security requirements, such as the need to perform mutual authentication between the smart meter and back-end utility applications or a platform (acting on behalf of the application), as well as the authentication of all the messages exchanged. Likewise, there should be a means to validate the integrity of the software running on the smart meter, which ensures that the equipment has not been tampered with or its software image changed by a malicious user. Another example of fraudulent use is when the SIM card is used in another device, to browse the Internet, for example.

Finally, it is often the case that smart meters are requested to periodically report meter data to the back-end applications (e.g., every hour). However, several use cases show the need for allowing utility back-end application to either send commands at any point in time (in this use case, prepayment-related commands) or request a non-solicited meter reading, for instance during peak demand periods. When the technology allows for the meter to benefit from always-on connectivity and be constantly reachable (e.g., a meter connected via PLC, power line communication), utility back-end application-triggered requests do not impose specific constraints. However, if the meter is connected via a cellular system (e.g., 3GPP or 3GPP2) where typically connectivity is requested by the terminal[1] (in this case, the smart meter), a new requirement is put forward, that is, for the network to allow network application-initiated connectivity to a particular terminal. This is typically done nowadays by making use of specific SMS (triggering the terminal to establish data connectivity). However, using SMS turns out to be costly in this particular scenario because it forces the network operator to maintain a circuit subscription (needed only for the purpose of sending an SMS) in addition to the data subscription used to exchange application data. In future, M2M will increasingly impose new requirements on the network to allow for network-initiated connectivity establishment that is requested by a network application or an entity on its behalf, that is, a horizontal service platform. This particular requirement is now being handled by 3GPP under the following technical report [3] and is known as device triggering.

4.4 eHealth Approach in ETSI M2M

4.4.1 Introduction

eHealth applications and related appliances are being increasingly deployed for applications such as:

- **Remote patient monitoring (RPM)**: Enables healthcare providers to monitor and diagnose health conditions by remotely collecting, storing, retrieving, and analyzing patient health-related information. RPM devices allow healthcare providers to treat patients before their conditions become more acute, thus avoiding unnecessary trips to emergency departments and readmissions to the hospital. Ultimately, RPM allows for hospital stays to be greatly minimized, resulting in the reduction of expenditure related to healthcare delivery. Typically, one or more sensors are used to monitor the patient's vital signs such as blood pressure. The monitoring information is usually reported using a single piece of equipment (known as the gateway), often connected to a cellular network.
- **Disease management**: A common use of M2M applications for eHealth is to support the remote management of patient illnesses, a process referred to as disease management, dealing with conditions such as diabetes or cardiac arrhythmias. For some disease management applications, an alarm function is needed to trigger an alarm to get the attention of a doctor or the patient in order to react to a critical health condition.
- **Ageing independently**: M2M applications for eHealth can enable the elderly to live an independent life and remain in their homes in cases where assistance would usually be needed. It consists of monitoring a patient's vital signs, such as pulse, temperature, weight, or blood pressure, ensuring that patients are taking the medication and that their activity level is being tracked.

[1] Except in Long Term Evolution, where a default data bearer provides the means for always-on connectivity.

- **Personal fitness and health improvement**: M2M applications for eHealth can be used to record health and fitness indicators, such as heart and breathing rates, energy consumption, and rate of fat burn, during exercise sessions. They can also be used to log the frequency and duration of workouts, the intensity of exercise, running distances, etc. When this information is uploaded to a back-end server, it can be used by the user's physician as part of their health profile, and by the user's personal trainer to provide the user with feedback on the progress of their exercise program. It allows for exercise programs or physiotherapy to be more precisely and more rapidly adapted to the needs of the patient/user.

ETSI TR 102 732 [4] provides the following explanation pertaining to the set of devices as well as the network aspect for eHealth:

"In order to acquire the information on a patient's health or fitness, appropriate sensors have to be used. For this reason, the patient or monitored person typically wears one or more sensor devices that record health and fitness indicators such as blood pressure, body temperature, heart rate, weight, etc. Because these sensors typically have to cope with severe limitations on form factor and battery consumption, it is expected in most cases that they need to forward the collected data via some form of short-range technology to a device that can act as an aggregator of the collected information and a gateway towards a back-end entity that is supposed to store and possibly react to the collected data. It is also possible that the sensors used to monitor parameters related to the health condition of the patient are located somewhere in the immediate vicinity of the patient."

Examples of sensors used for eHealth applications are shown in Figure 4.3. When multiple sensors are used, a gateway (typically with WAN connectivity) is often deployed.

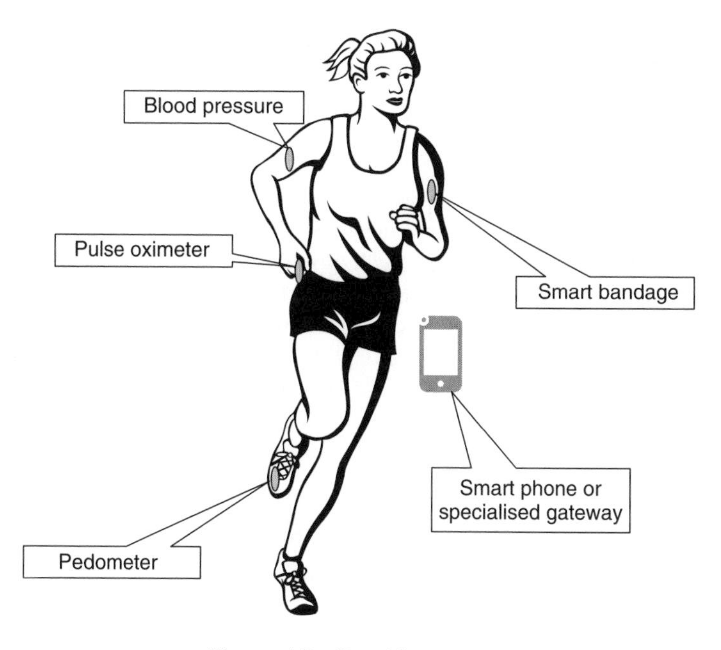

Figure 4.3 Portable sensors.

The gateway aggregates and forwards collected sensor data to a central application that processes the information and triggers actions when needed. Portable sensors commonly use short-range wireless interfaces to the gateway device with WAN connectivity, allowing for the data to be reported. While such a gateway could be a cell phone or a residential gateway at a home, often a dedicated device with a subscription to a cellular network is used.

The eHealth use cases introduce similar requirements to other M2M applications in terms of registration, authentication and data exchange which need to be both terminal- and network-initiated. The following provides a list of requirements where eHealth puts forward very specific requirements:

- Location tracking: the M2M device and the supporting system (and application) must provide means for the tracking and reporting of the patient's location. This information may be critical in emergency situations in order to provide location information to an emergency medical technician (EMT) or ambulance services, proximity to nearby hospitals, etc.
- Support for time-critical message handling and delivery: some critical alarms must be sent to appropriate applications in a timely fashion, especially in the event of an emergency.
- Support for secure-time synchronization: reported data must provide accurate timestamps because this accuracy is crucial during the data analysis process.

4.4.1.1 eHealth use case Example: Interference Isolation

This section provides another use case example described in ETSI TC M2M TR 102 732. This use case was chosen because it shows very unique requirements put forward on M2M systems.

General Use Case Description

In a remote patient monitoring scenario where low voltage body signals need to be acquired for remote health monitoring purposes, the acquisition process could be disturbed by radio transmission activities, such as GSM/GPRS, that can take place on the nearest co-located radio parts of the same M2M device.

Where the health monitoring process is continuously applied to the patient, for example, discovering arrhythmias, the acquisition activity could be disrupted by typical cellular radio communication activities performed by the M2M device.

It becomes vitally important to avoid or reduce any possible interference on incoming body signals to achieve a reliable eHealth service, even if it is not characterized as a life-saving service.

One way to cope with unexpected disruptive radio transmissions is to concurrently sample a radio transmission indication signal in order to adjust the signal-sampling process aimed at reducing the interference effects, for example, by discarding or correcting those received samples associated with active radio transmission indications, or slightly shifting the time of the signal sampling. Another way could be the control of the radio transmitter by suspending the radio transmission during time-limited measurements, then resuming it at the end of the measurement session.

Stakeholders

Patient: The patient may be any individual or surrogate who could use a remote monitoring device (RMD) to gather measurements, data, or events. The patient measurements may be taken in various clinical settings, such as a hospital, or non-clinical settings, such as at home, at work, at school, while traveling, or in assisted-living facilities.
RMD: An electronic M2M device with a sensor, user interface and/or actuator, and an interface into the M2M network. The device collects patient information and communicates to the appropriate M2M service-capability provider and/or M2M application via the M2M network. An RMD may also communicate with these entities via an M2M gateway. Furthermore, the device may receive and/or act upon commands from the M2M service-capability provider and/or M2M application or provide information to the patient. Low-power, low-complexity protocols are likely to be required for these devices.

M2M service capability provider

A network entity that provides M2M communication services to the M2M application entities. These applications may support specific functional capabilities that assist in facilitating health information-exchange activities. Additionally, the M2M service-capability provider communicates with the RMD to collect data or send commands.

M2M application entity

A term created to bundle together, and treat as a single system element, stakeholders above the M2M scope. High-level applications such as free-standing or geographic health information exchanges, data analysis centers, integrated care-delivery networks, provider organizations, health record banks, or public health networks, and/or specialty networks are all examples of M2M application entities. The term M2M application entity also includes the following typical RPM stakeholders.

Information Exchanges

The use case is applicable to most eHealth applications such as RPM. As such, the description of the information exchange is not relevant for deriving the related requirements. See [4] for information exchanges.

New Requirements Derived from the Use Case

- **Non-interference with electro-medical devices**: The M2M system, or parts of it, should avoid disrupting the detection and measurement of very low voltage signals to be acquired and used by the M2M Application. For example, in the case of eHealth applications, where ECG body signals are continuously measured by a wireless sensor, the body network can be seriously disrupted by the nearest GSM/GPRS transmitters belonging to the M2M gateway of the electro-medical device.

- **Radio transmission activity indication**: Depending on the type of M2M service, all the radio transmitting parts (e.g., GSM/GPRS) of the M2M device (or Gateway) shall provide a real-time indication of radio transmission activity to the application on the M2M device/Gateway.
- **Radio transmission activity control**: Depending on the type of M2M service, all the radio transmitting parts (e.g., GSM/GPRS) of the M2M device (or gateway) may be instructed in real time by the application on the M2M device/gateway to suspend/resume the radio transmission activity.

Use Case Source

Telecom Italia's contribution to ETSI TC M2M.

This use case provides another challenging set of requirements: if GSM/GPRS equipment is used to transmit measurement data, then it must not interfere with body signals so as to ensure accuracy of the measured data. In addition, a mechanism to allow the suspension and subsequent resumption of the transmission of data when critical sensing is being performed is required.

4.5 ETSI M2M Service Requirements: High-Level Summary and Applicability to Different Market Segments

Table 4.1 provides a high-level summary of ETSI M2M requirements as described in ETSI TS 102 689 [5]. This table is not meant to provide an exhaustive list of service requirements. It rather provides the reader with a summary of the basic requirements as well as their mapping to the different family of use cases.

The most important conclusion that can be drawn from this table is that several M2M applications have similar requirements which can be addressed through the use of stable and well-tested software building blocks, thus making a case for a thinner (more lightweight) application and the need for open APIs exposing M2M service capabilities.

A further conclusion is that M2M imposes new service requirements in comparison to personal communications. For instance, consider the case of smart meters. Utility providers expect these items of equipment to be deployed for a period of at least 20 years. Such a long deployment phase mandates the need to upgrade the security algorithms and key lengths during the life cycle of the item of equipment. The rationale for this requirement stems from the fact that, over time, security experts and cryptographers often discover new attacks on systems. This, coupled with the constant improvements in computing capabilities (deployed by hackers), will force security managers to upgrade key lengths and modify security policies and algorithms, in particular for devices that remain operational for long period of time (such as smart meters). In some instances,

Table 4.1 Mapping of M2M service requirements to ETSI M2M use cases

Requirement	Applicability to market segments				
	Smart metering	eHealth	Automotive	Connected consumer	City automation
Communication mediation					
Support of different bearer technologies (SMS, GPRS, and IP)	✓	✓	✓	✓	✓
Message delivery toward sleeping device (through storing and delivery when the device becomes available on the network)	✓	✓	✓	✓	✓
Communication path selection (select the appropriate bearer based on configured policies)	✓	✓	✓	✓	✓
Communication with devices behind a gateway	✓	✓	✓	✓	✓
Communication failure notification	✓	✓	✓	✓	✓
Confirmation of message delivery	✓	✓	✓	✓	✓
Accurate and secure time synchronization and time stamping	✓	✓	✓	–	–
Location support					
Report M2M device/gateway location to M2M applications	–	✓	✓	✓	✓
Enforce privacy pertaining to location	–	✓	✓	✓	–
Data collection and reporting					
Periodic reporting	✓	✓	–	–	✓
On-demand reporting	✓	✓	✓	✓	✓
Event-based reporting	–	✓	✓	✓	✓
Use of QoS provided by underlying networks	–	✓	✓	✓	✓
Path diversity (primary + backup path)	–	✓	–	–	✓
M2M devices interacting with multiple back-end applications	✓	✓	✓	✓	✓
Security					
Mutual authentication between communicating entities	✓	✓	✓	✓	✓
Data transfer confidentiality (encryption)	✓	✓	✓	–	✓
Data transfer integrity verification (authentication of the exchanged data)	✓	–	–	✓	–

(*continued overleaf*)

Table 4.1 *(continued)*

Requirement	Applicability to market segments				
	Smart metering	eHealth	Automotive	Connected consumer	City automation
Avoid an M2M communication module usage for generic communication purposes	✓	✓	✓	–	✓
Device/gateway integrity validation	✓	✓	–	–	–
Security credential and software upgrade at the application level	✓	✓	✓	✓	✓
Naming and addressing					
Device names rather than network addresses are to be used by the application	✓	✓	✓	✓	✓
Support of multiple addressing schemes such as IP, E.164, etc.	✓	✓	✓		✓
Remote management					
Fault management (proactive monitoring, fault discovery, connectivity verification, fault recovery, etc.)	✓	✓	✓	✓	✓
Configuration management (configuration of network and service parameters, installation of new application versions)	✓	✓	✓	✓	✓
Software upgrades, firmware upgrades, and patches	✓	✓	✓	✓	✓

there may be a need to upgrade algorithms. In other instances, there may be a need to distribute security patches to address vulnerabilities in protocols and applications that were unknown at the time of installation. The same logic equally applies to the software and firmware that may need to be constantly upgraded in order to deploy new features or to allow remediation for discovered functionality bugs. Device life cycle management (e.g., software and firmware upgrades) is therefore a critical enabler of M2M communications.

4.6 Traffic Models-/Characteristics-Approach to M2M Requirements and Considerations for Network Architecture Design

This section introduces nine different M2M market segments (also referred to as M2M applications) which are believed to have a significant impact both on:

- network features evolutions (these evolutions constitute a major piece of work undertaken by 3GPP and 3GPP2);
- network planning and dimensioning aspects.

While this list of market segments is not exhaustive, it provides in our view an excellent characterization of M2M traffic. This section will describe, in some detail, the different market segments, as well as the high-level characterization of the related traffic characteristics. A detailed set of service-level features will then be provided for each market segment, which will then be used to derive a traffic model taking into account:

- service features and related data payload;
- M2M module (2G, 3G, and 4G) sales predictions for the North American (NA) market (taken as a relevant example provided the level of market interest in M2M in the NA market) for each market segment (Source: ABI research [6]);
- an average concentration of M2M modules per site (base station) in a mobile network;
- the estimated application throughput versus total generated throughput (taking into account network overhead and retransmission estimates).

While the model contains several approximations and is based on average estimates (as opposed to worst-case scenarios), it provides excellent characterization for M2M traffic patterns and allows for the support of the ongoing 3GPP and 3GPP2 network optimization work, as well as network planning strategies in operator networks.

4.6.1 Why Focus on Wireless Networks?

While M2M applies to both wireline and wireless, the traffic characterization is more important for wireless access network because the impact is greater for several reasons.

- Wireless infrastructure provides seamless coverage and ease of installation. For example, vending machines that are increasingly being connected to a public communication network, bringing copper or fiber wires to this equipment have both CAPEX and OPEX structures that may not justify the operational efficiency gained through wiring the equipment. However, wireless will immediately be available virtually anywhere, including for indoor applications. The cost for an M2M communication module (the modem that allows data and SMS connectivity through a cellular network) is as low as $25 (roughly €20) for 2G technology, making the initial investment fairly reasonable.
- While several residential and in-business applications will rely on wireline infrastructure (mainly x digital subscriber line (xDSL) and fibre to the home (FTTH)), M2M devices deployed within residential environments may not be visible (and not incur significant extra traffic) at all to the access network because these will be typically hidden by means of a network address translator (NAT). The additional generated traffic is often negligible compared to bandwidth-hungry triple-play applications such as video. Some

of the residential deployments such as home surveillance will have cellular as primary or backup access. Besides IP addressing aspects, several M2M applications do not have strong QoS requirements, placing very little burden on wireline networks.

As a conclusion (and going beyond wireline versus wireless comparison aspects), it is generally agreed that M2M imposes a whole range of network optimization requirements on wireless networks. Therefore, this chapter focuses primarily on cellular network aspects when it comes to network optimization.

4.7 Description of M2M Market Segments/Applications

Based on the work done by ETSI M2M on the use cases and the research study by ABI Research [6], the following segments are considered for the purpose of M2M traffic characterization.

4.7.1 Automotive

Automotive M2M segment refers to items of equipment installed in vehicles and related network applications that provide any combination of the following services.

- **Car accident emergency call service**: provides automated assistance to motorists following a vehicle accident. The most notable deployment examples to date include General Motor OnStar service in the USA, and PSA, BMW, and Volvo in Europe. In the case of Europe, current services provide a prestandard implementation of the European Commission eCall project and related standards (developed by 3GPP and endorsed by ETSI). Over time, it is commonly agreed that car accident emergency call services will be mandated through regulation in different countries.
- **Tracking and interactive services**: provide vehicle tracking services that can be used by multiple applications, including:
  - tracking and interactive stop of stolen cars;
  - "pay-as-you-drive" car insurance;
  - rental-area limitation or zoning area-based payment;
  - fleet management;
  - tourist information push;
  - location-based advertising and commercial services.
- **Interactive commercial and infotainment applications**: provide commercial (e.g., advertisement) and entertainment services to vehicle passengers. These include:
  - on-demand video download;
  - interactive gaming;
  - interactive payment;
  - Internet access.

It is likely that future automotive deployments will be dominated by 3G and 4G connectivity, which will be especially needed if interactive and infotainment applications are targeted.

The traffic characteristics/network impacts of automotive applications can be described as:

- high mobility;
- high data transfer rates, in particular for commercial and infotainment applications;
- multibearer connectivity.

4.7.2 Smart Telemetry

Smart Grids and smart metering applications are being deployed by the energy sector to achieve a higher degree of energy usage efficiency and reliability. Utility applications typically interact with smart meters deployed at customer premises and a variety of other sensors or M2M devices that allow for the provision of a grid-wide monitoring infrastructure, the collection of consumption information and the sending of information about tariffs or incentives. Smart meters and Smart Grid sensors will be connected via a variety of wired and wireless technologies such as PLC, meshed RF, cellular access, etc. These provide network connectivity between smart meters, sensors or M2M devices and the utility's back-end applications, databases, and management systems. In several western European countries, and in North America, a combined connectivity model is emerging where PLC is used from the electricity meter to the data concentrators located at the distribution service operator (DSO) network, then cellular connectivity is used from the concentrators to utility back-end applications (see Figure 4.4). However, this model is only economically viable when the number of meters per concentrator is above a certain threshold (typically 25 meters). For rural areas, bringing cellular all the way to the smart meter could complement the PLC/cellular deployment.

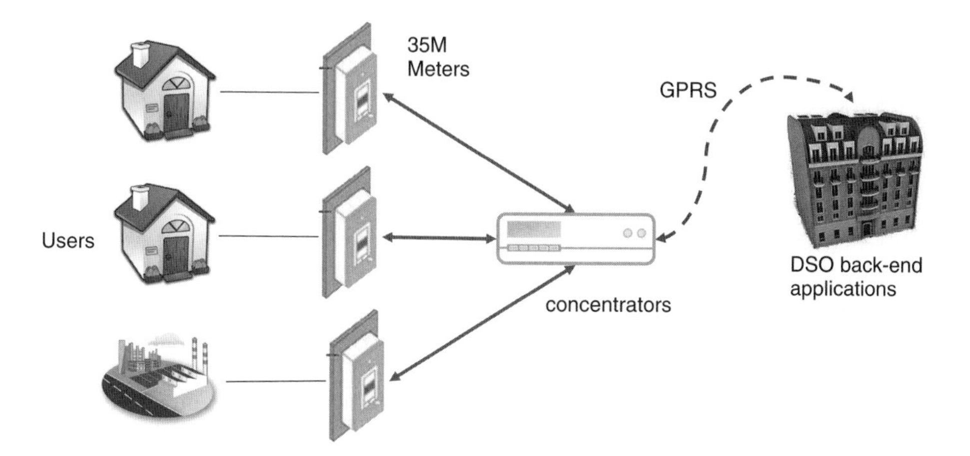

Figure 4.4 Typical PLC/cellular smart metering deployment scenario.

The traffic characteristics/network impact of smart telemetry applications can be described as:

- no mobility for stationary devices;
- low data transfer rate;
- predictable behavior (typically, smart telemetry devices are configured to report data periodically);
- delay tolerance: while smart meters are configured to report data every period (e.g., every 30 minutes), utility applications could (in particular during non-peak hours) tolerate reporting of the data at any time during the period;
- synchronization effect (typically, devices are configured by the utilities to simultaneously establish connectivity and report data).

4.7.3 *Surveillance and Security*

Surveillance and security devices are mainly deployed in residential and small business premises to provide picture and video surveillance information to security alarm applications. Security devices generally use cellular networks as primary or backup access to provide connectivity to security monitoring applications or to the owners of the premises.

The information exchanged is mostly composed of alarm information and occasionally low-to-medium-resolution video signals.

The traffic characteristics/network impact of surveillance and security applications can be described as:

- no mobility;
- cyclical exchange of information but unpredictable bursty exchanges of information when an alarm procedure is triggered;
- low transfer data with occasional peak traffic using picture/video transfer data;
- multibearer connectivity.

4.7.4 *Point of Sale (PoS)*

PoS terminals (including automated teller machines (ATMs)) provide services such as cash dispensing, payments, and all tasks of store-checkout counters, such as payments by bank or credit cards, transaction verifications, sales reports, and coordination of inventory data.

PoS terminals are connected either to a local area or a wide area network in order to exchange information with a multitude of applications. The use of cellular connectivity in PoS is enjoying rapid growth thanks to both the wide coverage provided by cellular operators and the ease of installation. In some countries, jurisdictions require secure and redundant connectivity using both wired and wireless technology.

The traffic characteristics/network impacts of PoS/digital signage applications can be described as:

- limited mobility;
- low data transfer rate;
- persistent bearers.

4.7.5 Vending Machines

Vending machines are terminals that dispense goods to consumers in exchange for a fee. Vending machine terminals are increasingly being connected to a network in order to exchange information relating to logistics and payments, such as to inform back-end application that a certain stock threshold has been reached. The main objective of connecting vending machines to a communications network is to improve the operational efficiency through better knowledge and control of demand and supply. Cellular access technology is the dominant network connectivity for vending machines thanks to its wide coverage, as well as ease of deployment.

The traffic characteristics/network impact of vending machines can be described as:

- no mobility;
- low data transfer rate;
- persistent bearers;
- synchronization effect;
- delay tolerance.

4.7.6 eHealth

Some eHealth applications include the following (more information is provided under the eHealth use case):

- **RPM**: enables healthcare providers to monitor and diagnose health conditions by remotely collecting, storing, retrieving, and analyzing patient-health-related information. RPM devices allow healthcare providers to treat patients before their conditions become more acute, thus avoiding unnecessary trips to the emergency departments and readmissions to hospital. Ultimately, RPM allows for hospital stays to be significantly minimized, resulting in a reduction in the cost of healthcare delivery. Typically, one or more sensors are used to monitor the patient's vital signs (e.g., blood pressure). The monitoring information is usually reported using a single piece of equipment often connected to a cellular network.
- **Personal emergency response system (PERS)**: provides the patient with access to a 24/7 call center for assistance. Deployed devices are worn on a neck pendant or wristband and allow immediate access to assistance services through the push of a single button.

- **MPM (mobile personal monitoring)**: being very similar to RPM devices, MPM enables preventive health and fitness monitoring through the reporting of monitored information to central applications.

The traffic characteristics/network impact of telehealth/medical applications can be described as:

- low mobility;
- low to medium data transfer rate;
- multibearer applications;
- synchronization effect.

In addition to these traffic characteristics, telehealth and medical applications often require network services offering high reliability and availability; both could be the subject of a service level agreement between the service provider (e.g., healthcare provider) and the network operators.

4.7.7 Live Video

Live video is often used for supervision of applications, in particular in the field of transport infrastructures (roads, railways, etc.) or city surveillance. Live video cameras are deployed by public safety agencies (e.g., police departments) or another security-related operating center to monitor and record any events relating to the operations of monitored areas.

Because of coverage, ease of deployment or mobility needs, cellular networks constitute the technology of choice for this market segment with a slight preference to 3G connectivity (and 4G in the future) to cope with bandwidth needs of live video.

Three types of video resolutions could be considered:

- low-resolution (bit rate <128 kb/s);
- medium-resolution (bit rate <768 kb/s);
- high-resolution (bit rate <10 Mb/s).

Another feature that could be implemented is the ability of the device to record an HD scene and transfer the related file upon demand.

The traffic characteristics/network impact of live video applications can be described as:

- no mobility for stationary devices;
- high mobility for embarked devices;
- emergency connectivity;
- medium to high data transfer rates.

4.7.8 Building Automation

Building automation applications are based on devices used to provide services in commercial buildings. Building automation systems aim to provide a safe, comfortable and productive facility for its occupants. The focus is very often centered on operational efficiency by reducing energy consumption and operations staff levels.

Building automation is characterized by an important set of sensors, switches, monitoring and telemetry devices, all connected first to a private and second to a public wired or wireless public network using a gateway.

Examples of applications include cooling equipment maintenance and monitoring, lift maintenance and monitoring, power supply monitoring, etc.

The traffic characteristics/network impact of building automation applications can be described as:

- low or no mobility,
- low data transfer rate.

4.7.9 M2M Industrial Automation

Industrial automation applications and related devices such as PDAs (personal digital assistants) are used in commercial and industrial areas to support a set of workflows and clearly defined business processes. Industrial automation applications may additionally collect data from telemetry devices.

The traffic characteristics/network impacts of industrial automation application can be described as:

- low or no mobility,
- low data transfer.

4.8 M2M Traffic Characterization

The following table provides the different features that generate traffic for M2M devices. A description of individual service features is provided in Annex A. Each service feature is considered in the traffic model as a source for data traffic. The next section will focus on the impact of smart metering on a typical site for the North American market. We will then provide global traffic model estimates without detailing all the M2M applications.

4.8.1 Detailed Traffic Characterization for Smart Metering

In this section, we will provide an overview of the methodology used to characterize the M2M traffic by detailing a specific use case of smart metering. We assume the following typical deployment architecture.

Table 4.2 Typical features generating traffic for M2M devices

Automotive	Smart telemetry	Surveillance and security	Point of sale	Vending machines
Connect/ disconnect[a]	Connect/ disconnect	Connect/ disconnect	Connect/ disconnect	Connect/ disconnect
Voice communication	Daily meter readings (single phase)	Daily presence control	User account creation	Sales-related information exchange
Alert management	Daily meter readings (three phases)	Alert setting	User information retrieve	Alert settings
GPS position transfer	Individual reads[b]	Presence detection	Online payment	Reading settings
Maintenance data transfer	Load control and emergency[c]	Fire detection	Alert settings	Reading event logs, individual
Firmware and software updates	Alert settings	Power shut down detection	Internet access	Reading event logs, all
Internet access	Reading settings	Reading event logs, individual	Reading event logs, individual	Security alarms
E-mail	Reading event logs, individual[d]	Reading event logs, all (within a certain area)	Reading event logs, all	Firmware and software upgrade
Game/movie download	Reading event logs, all	Presence picture/video upload	Database update	Registration, authorization, etc.
Alarm management	Control signals to home area network appliances	Firmware and software upgrade	Firmware and software upgrade	NTP
Reading event logs	Firmware and software upgrade	Registration, authorization, etc.	Registration, authorization, etc.	–
Registration, authorization, etc.[e]	Registration, authorization, etc.	NTP	NTP	–
NTP[f]	NTP	–	–	

eHealth	Live video	Building automation	Industrial automation
Connect/disconnect	Connect/disconnect	Connect/disconnect	Connect/disconnect
Single parameter reading	Monitoring, QCIF video	Daily meter reading	Daily meter reading
Complex parameters reading[g]	Emergency video transfer QVGA	Individual reads	Individual reads

Table 4.2 (*continued*)

Automotive	Smart telemetry	Surveillance and security	Point of sale	Vending machines
Video transfer (QVGA)	Investigation VGA	Alert settings	Alert settings	
Bulk data transfer	Alert settings	Reading settings	Reading settings	
Data + voice + video communication	Reading settings	Reading event logs, individual	Reading event logs	
Device remote configuration and control	Reading video logs, individual	Reading event logs, all	Firmware and software upgrade	
Alert settings	Reading video logs, all	Firmware and software upgrade	Registration/ authorization, etc.	
Reading settings	Firmware and software upgrade	Registration/ authorization, etc.	NTP	
Reading event logs, individual	Registration/ authorization, etc.	NTP		
Reading event logs, all	NTP			
Firmware and software updates				
Registration/ authorization, etc.				
NTP				

[a]Covers all procedures managing the connection and disconnection of the devices to the network.
[b]Individual access reading for one device as opposed to a request to read a massive group of devices.
[c]Specific urgent request to control the energy demand. Measures taken can be as drastic as powering down some equipment.
[d]Individual request to access events log (example: bill investigation for reconciliation) as opposed to an access to the logs of a group of devices.
[e]Procedures to allow service-level authentication, authorization and registration. These procedures are the necessary steps before the data exchange phase can start.
[f]NTP: network time protocol used for global clock synchronization.
[g]Transfer of critical data requiring a high level of network reliability/availability.

In Figure 4.4, smart meters are connected via PLC to concentrators located on the DSO power network. The concentrator is then connected to the DSO back-end application via the use of cellular network connectivity. Figure 4.4 reflects a widely adopted and accepted model for combined PLC/cellular Smart Metering deployments. This model has been assumed for the smart metering traffic model.

Table 4.3 Smart metering traffic matrix

Function	Transaction volume assumptions	Hour of day when activity occurs		To back-end application (UL)					From back-end application (UL)				
		From	To	Payload size per message (bytes)	Packet size	Number of messages per transaction	Average bitrate- (kbit/s)	Average bitrate (kbit/s) during busy hours	Payload size per message (bytes)	Packet size	Number of messages per transaction	Average bitrate (kbit/s)	Average bitrate during busy hours
Daily meter readings single phase	100% of meters over 30 min	0	1	5000	1024	1	2.7	0.0	2000	1024	1	1.1	0.0
Daily meter readings three phase	100% of meters over 30 min	0	1	5000	1024	2	1.3	0.0	2000	1024	2	2.2	0.0
Individual reads	7/8 of 2% of meters	0	1	5000	1024	2	1.3	0.0	2000	1024	2	2.2	0.0
Individual reads (busy hour)	1/8 of 2% of meters in busy hour	7	8	5000	1024	7	0.4	0.0	2000	1024	7	0.0	0.0
Connect-disconnect	All concentrators over 24 h	0	24	5000	1024	1	1.0	1.0	200	1024	1	0.0	0.0
Load control and emergency	Maximum of 200 concentrators	7	8	2000	1024	1	48.0	0.0	1000	512	1	1.3	0.0
Alter settings	7/8 of 2% of meters	8	21	200	64	1	0.0	0.0	500	512	1	0.0	0.0
Alter settings (busy hour)	1/5 of 2% of meters in busy hour	7	8	200	64	1	0.0	0.0	500	512	1	0.0	0.0
Reading settings	7/8 of 2% of meter in 13 h	8	21	1000	512	1	0.0	0.0	200	64	1	0.0	0.0
Reading settings (busy hour)	1/8 of 2% of meters in busy hour	7	8	1000	512	1	0.0	0.0	200	64	1	0.0	0.0
Reading events logs (individual)	2% of meters in 30 min period during busy hour	7	8	1000	512	1	0.2	0.0	200	64	1	0.0	0.0
Reading events log (all)	100% of meters in 7 d	1	24	2000	1024	1	0.1	0.1	200	64	1	0.0	0.0
Sending HAN (Home Area Network) messages	100% of meters in 3 h	7	10	100	64	1	0.2	0.0	500	512	1	0.0	0.0
Firmware and software upgrade	All meters within 30 d	1	24	50000	1024	1	0.0	0.0	5000000	1024	1	2.0	2.0
Registration, authorization etc.	All meters in 24 h	0	24	1000	512	1	0.2	0.2	1000	512	1	0.0	0.0
NTP	All meters evenly over 24 h	0	24	500	512	1	0.1	0.1	500	512	1	0.0	0.0

Source: Alcatel-Lucent.

Applying an estimate for each software feature within a smart meter or a concentrator that generates traffic (as per Table 4.2), Table 4.3 provides an estimate of average bitrate per 2G site for each of the features. In order to derive the values in Table 4.3, the main assumptions are:

- number of concentrators per site: 15;
- total number of smart meters concentrator: 100;
- protocol overhead including network signaling and TCP/IP: 50%;
- percentage of single-phase meters: 80% (the remaining meters account for 20%).

Table 4.3 is specific to smart metering. Similar studies have been performed for each of the major market segments in order to provide aggregate traffic model per site and on an average basis.

Each device feature generating network traffic is considered separately in terms of the uplink and downlink generated traffic.

4.8.2 Global Traffic Characterization

In this section, we will provide the global traffic characterization for 2G, 3G, and 4G traffic. Based on the following considerations:

- global worldwide wireless operator deployment characterization in 2G, 3G, and 4G (including market share, technology deployed...);
- global worldwide geographical characterization in term of population density, city/rural knowledge;
- global worldwide economics profitability in term of global revenue and operator income;
- local regulatory body recommendation.

A database has been built to characterize the split of devices and applications per country and operators. The traffic prediction provided in the remainder of this chapter assumes an NA suburban city of 4 million inhabitants where the M2M terminal mix is provided in the following table (in 2010 the percentage of 3G and 4G modules is relatively low):

	2012			2014		
	2G (%)	3G (%)	4G (%)	2G (%)	3G (%)	4G (%)
Smart telemetry	13	4	0	8	3	1
Live video	0	1	1	0	1	1
Vending machines	1	1	0	1	1	0
eHealth	1	1	0	1	2	0
Surveillance and security	7	4	0	5	5	1
Point of sale	3	2	0	2	2	1
Industrial automation	1	4	0	1	4	1
Building automation	10	3	0	7	4	1
Automotive	20	21	2	13	30	6
Total	**56**	**40**	**4**	**37**	**51**	**13**

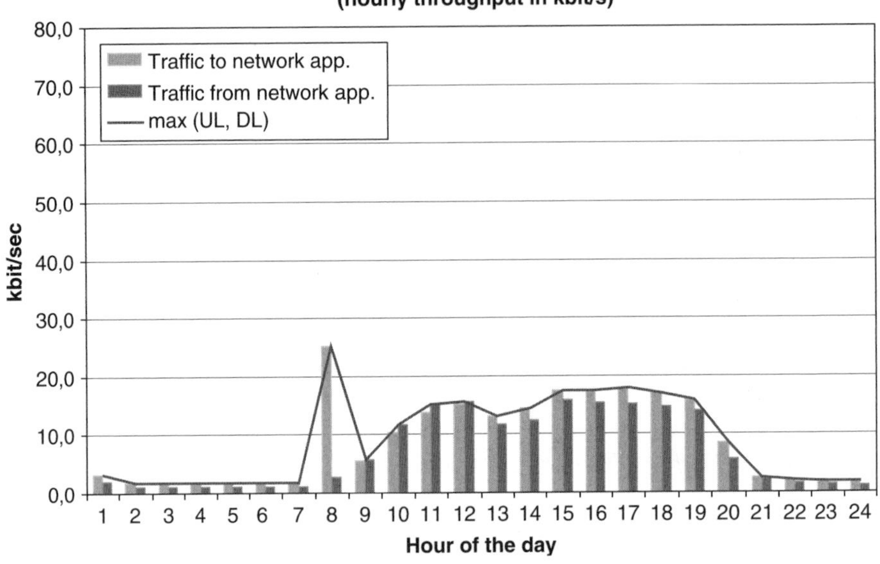

Figure 4.5 Average M2M traffic per site, 2G, NA, 2010.

4.8.2.1 2G Traffic

Figures 4.5–4.7 provide, on an average basis, the bitrate (calculated over a period of 1 hour) generated by M2M applications interacting with 2G M2M devices. These figures are an estimate for the NA market based on projected 2G module sales for each of the market segments under consideration. Based on these figures, the following conclusions can be drawn:

- 2G M2M traffic will roughly double every couple of years between 2010 and 2014.
- Global 2G module traffic tends to be symmetric. While we have shown in a previous table that 2G smart metering traffic mostly exhibits asymmetric traffic (with on average ratio of 5/100 between upstream and downstream data), other 2G M2M traffic is more downstream-intensive, particularly in the case of automotive where considerable amounts data are being downloaded (maps, Internet browsing, etc.).
- The average bitrate on an hourly basis is fairly low. The peak rate could be significantly larger, but harder to show because it depends on several parameters, such as link capacity and site load. Application developers very often tend to build their applications in a service-friendly way. For example, in the case of smart metering, the meters will generally be programmed to report data at the start of every hour, generating a huge synchronization effect and a heavy burden on the network. This is further depicted in Figure 4.8, which provides daily patterns of RADIUS usage (protocol used for authentication and accounting) taken from an operational network running M2M applications. This figure shows the synchronization effect that takes place every 90 minutes. The result of this synchronization effect is that M2M traffic generally generates traffic bursts, which are, on average, 20 times higher than the hourly estimated bitrates shown in Figures 4.5–4.7.

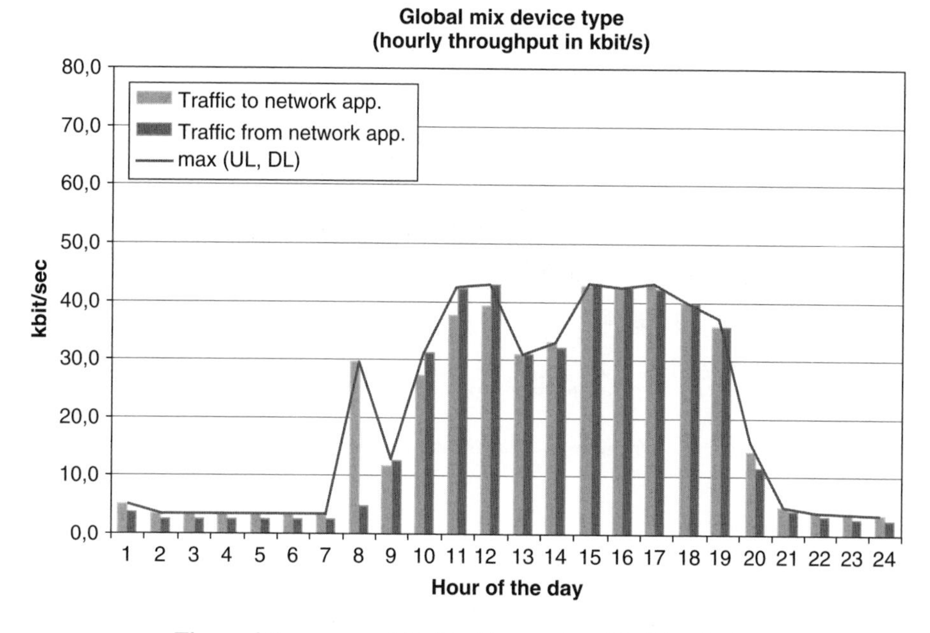

Figure 4.6 Average M2M traffic per site, 2G, NA, 2012.

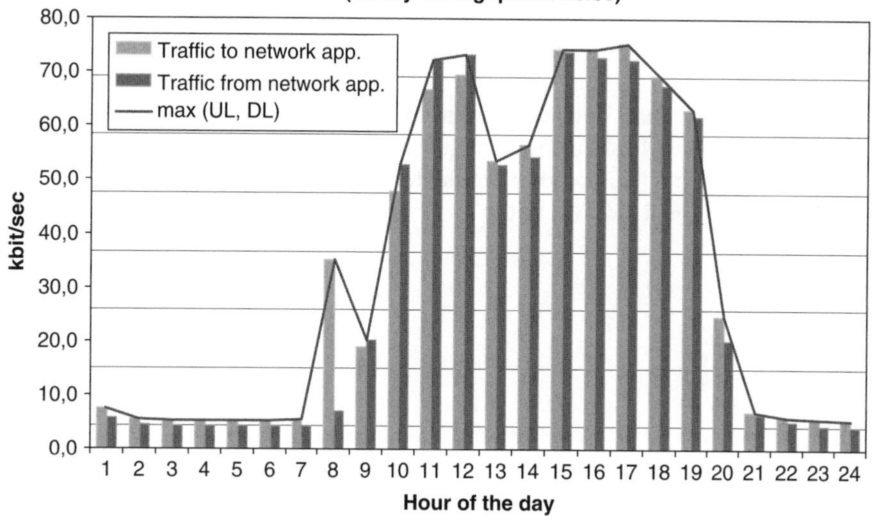

Figure 4.7 Average M2M traffic per site, 2G, NA, 2014.

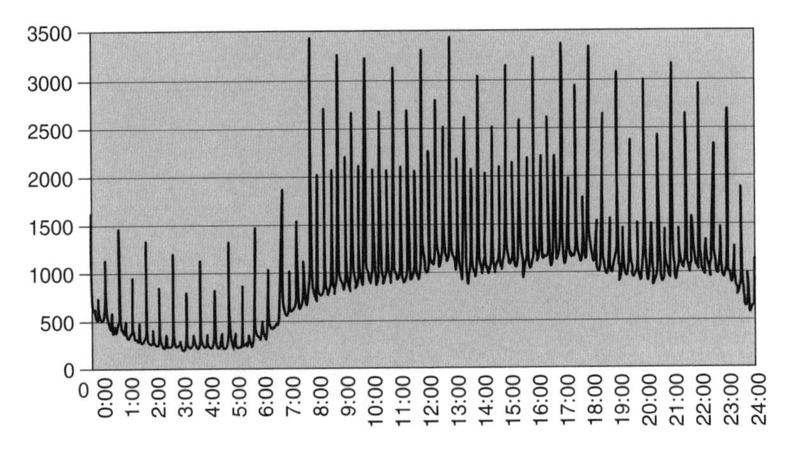

Figure 4.8 Example of synchronized data access from M2M applications (load on a RADIUS server).

4.8.2.2 3G Traffic

Figures 4.9 and 4.10 provide, on an average basis, the bitrate (calculated over a period of 1 hour) generated by M2M applications interacting with 3G devices. These figures are estimates for the NA market based on projected 3G module sales for each of the market segments under consideration. Based on these figures, the following conclusions can be drawn:

- Because of module prices, which remain relatively expensive, 3G M2M traffic will only begin to have a significant impact on the network in 2012.
- As opposed to 2G, 3G global traffic is very asymmetric where uplink traffic is five to six times higher than downlink traffic. This is mainly due to the fact that some bandwidth-hungry applications, such as live video, will de facto use 3G and 4G (when the corresponding modules become available).
- By 2014, 3G global M2M traffic will be 10 times higher compared to 2G M2M traffic.

4.8.2.3 4G Traffic

Figures 4.11 and 4.12 provide, on an average basis, the bitrate (calculated over a period of 1 hour) generated by M2M applications interacting with 4G devices. These figures are an estimate for the NA market based on projected 4G module sales for each of the market segments under consideration. Based on these figures, the following conclusions can be drawn:

- Because of the price point of 4G modules, 3G-generated traffic will be dominant compared to 2G and 4G.
- All other conclusions derived for 3G are also valid for 4G traffic.

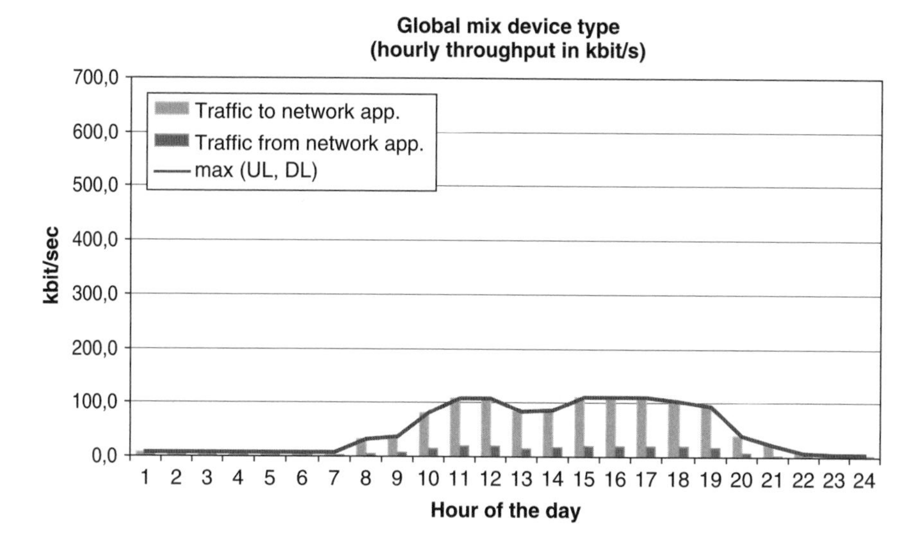

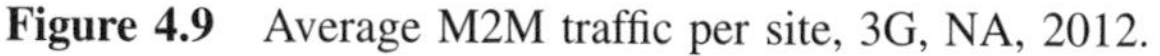

Figure 4.9 Average M2M traffic per site, 3G, NA, 2012.

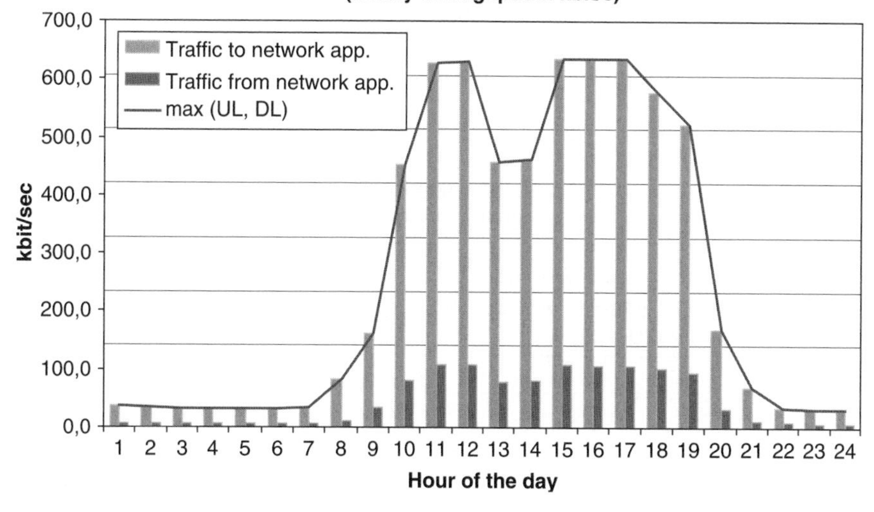

Figure 4.10 Average M2M traffic per site, 3G, NA, 2014.

We can drawn the following conclusions as regards M2M global traffic:

- As further depicted in Figures 4.13 and 4.14, the global 3G and 4G M2M traffic is asymmetric with more upstream traffic than downstream traffic. This introduces a new challenge for operational deployments where operators are used to the opposite pattern for personal communications. Both radio and network design must take this characteristic into account. Current communication networks, as well as related standards, have been designed and optimized for downlink-intensive traffic. Certain M2M applications deployed over 3G and 4G will change this paradigm and impose new challenges on the network design.

- M2M traffic is bursty, since M2M applications have not been developed with network load shaping in mind. Burstiness is the result of the synchronization effect resulting from several M2M devices simultaneously reporting data.
- Overall, peak hours for the M2M traffic mix are the same as for personal communications, even if some M2M applications have some disjointed peak hours. Network operators cannot therefore benefit from disjointed peak hours for personal and M2M communications. Therefore, other mechanisms, such as overload control or delay tolerance, must be considered to ensure network stability and cost effectiveness.

As a result of the above conclusions, it becomes clear that not only equipment features but also operational practices have to take advantage of all the characteristics of M2M traffic, particularly in terms of flexibility, in order to provide cost-effective M2M communications, master the traffic load, and avoid disrupting other communications where stricter service level agreement guarantees are required.

The following table provides a summary of the characteristics of M2M traffic that can be taken advantage of in terms of "system improvements for machine type communications", the umbrella term used in 3GPP for its work on M2M. This table is based on a previous section providing M2M segment characteristics.

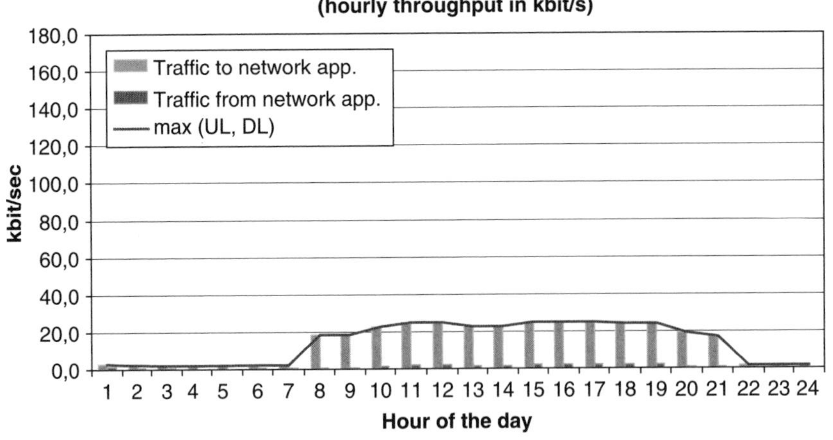

Figure 4.11 Average M2M traffic per site, 4G, NA, 2012.

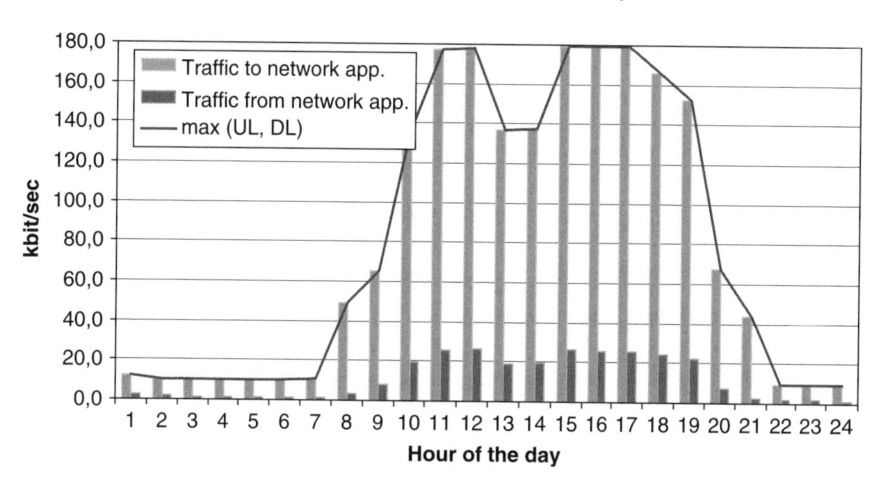

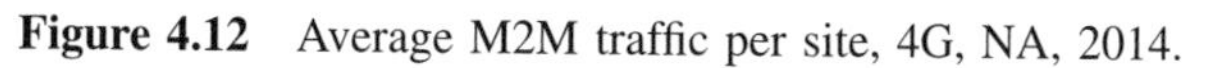

Figure 4.12 Average M2M traffic per site, 4G, NA, 2014.

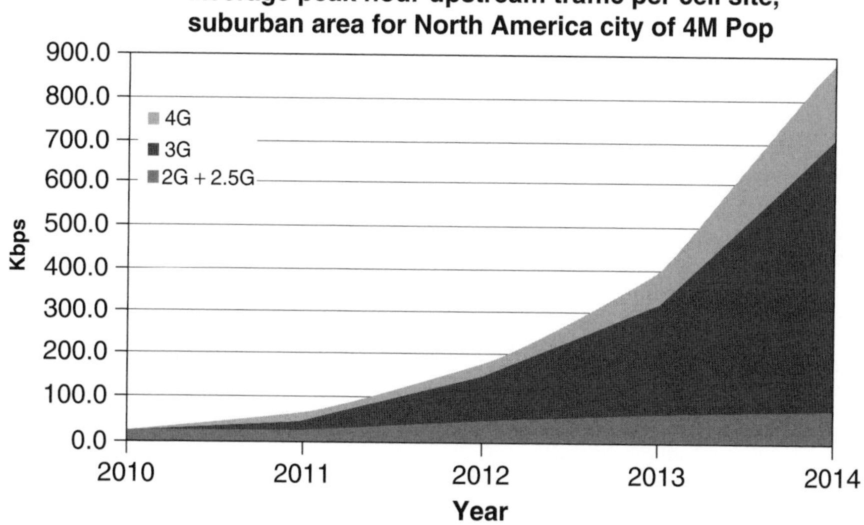

Figure 4.13 Average peak hour traffic per cell site (upstream).

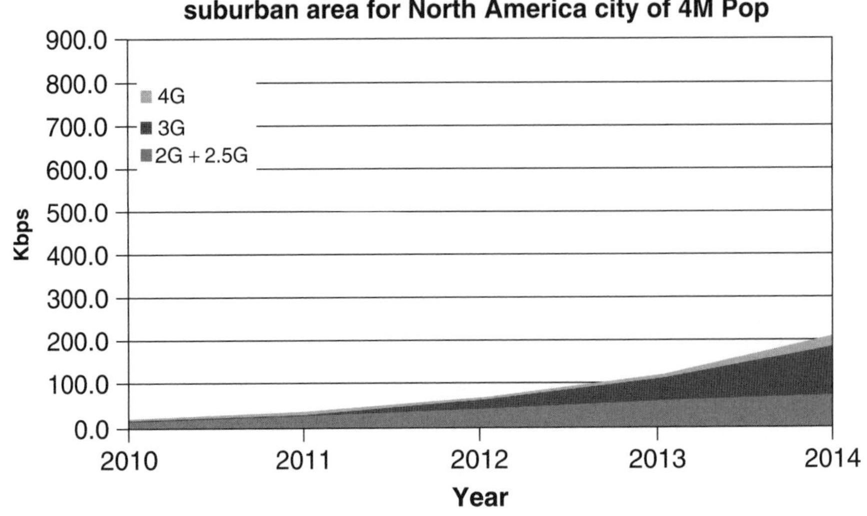

Figure 4.14 Average peak hour traffic per cell site (downstream).

Traffic characteristics	Market segment								
	Automotive	Smart telemetry	Security	PoS	Vending machine	eHealth	Live video	Building automation	Industrial automation
No mobility	–	✓	✓	✓	✓	–	[3]	✓	✓
Low mobility	–	–	–	–	–	✓	–	✓	✓
High mobility	✓	–	–	–	–	–	[3]	–	–
Low data rate	–	✓	✓	✓	✓	[2]	–	✓	✓
High data rate	✓	–	[1]	–	–	–	✓	–	–
Persistent bearers	–	–	–	✓	–	–	–	–	–
Synchronization effect	–	✓	–	–	✓	–	–	–	–
Multibearer connectivity	✓	–	–	–	–	✓	–	–	–
Delay tolerance	–	✓	–	–	✓	✓	–	✓	–
Predictable behavior	–	✓	✓	–	–	✓	✓	✓	–

[1]In the case of an alarm, a security device might use a high data rate bearer in order to send a video stream relating to the alarm.
[2]Occasionally, the data rate could be medium, when there is a need to bulk transfer statistics taken over a long period of time.
[3]High mobility for on-board live video devices and otherwise no mobility.

4.9 High-Level Architecture Principles for M2M Communications

Several lessons have been learned from early M2M deployments. First, vertically integrated applications are hard to develop, deploy and test. Not only do these need to deal with the application business logic, but they must also include a large number of

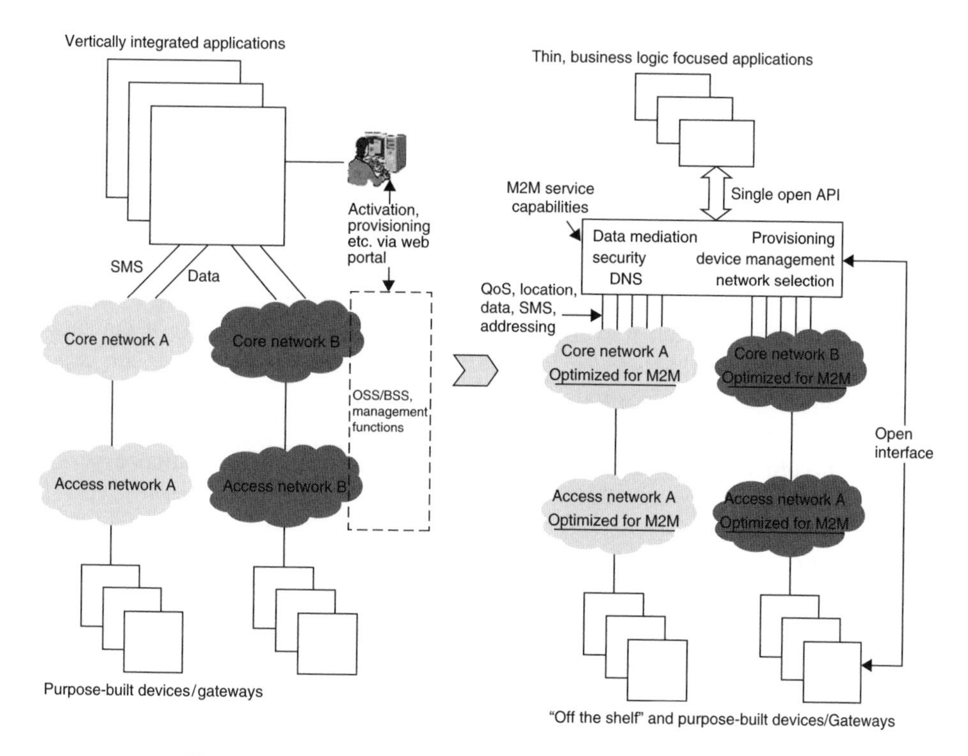

Figure 4.15 M2M from initial to mainstream deployments.

network-aware functionalities. Looking at the M2M applications requirements, it is clear that they all rely on a set of similar building blocks, particularly when it comes to all aspects related to communications and underlying networks.

Second, M2M is all about cost-effectiveness. M2M is known to generate a fraction of the ARPU (average revenue per user) of other consumer handsets such as smart phones. However, all analysts agree that the number of deployed M2M devices will be an order of magnitude higher than personal communication devices. As a result, the network has to become M2M-aware and take advantage of all traffic characteristics of M2M in order to provide the level of scaling and cost that matches projected M2M devices and ARPU structure.

Figure 4.15 provides at a high level the architectural principles of the work taking place in different standards organizations. It also shows the migration plan from current deployments toward new horizontal platform-based deployments.

Figure 4.15 shows the following key principles for future M2M deployments:

- Applications will increasingly focus on the application-business logic and outsource functions such as data mediation, security, DNS, device management, etc. to a horizontal platform. The access to the functionalities of this horizontal platform will be provided through a set of open, standardized, and IT-friendly APIs.
- As opposed to today's vertically integrated applications that make use of a limited set of operators interfaces (mostly data and SMS), making use of a horizontal platform

provides access to a larger set of core network interfaces, such as location, QoS, and addressing, without the burden of implementing protocols to access them, since the complexity is hidden through the use of a single open API.
- The access and core network must become M2M-aware. In contrast to the work being done in ETSI on the horizontal platform, which mandates the need for new architecture work, the work on network optimization is not aimed at new architectures. It aims to reuse existing 2G, 3G, and 4G architecture (for wireless) and making the network optimized for M2M traffic.
- Devices implement an open interface toward the horizontal platform and may come equipped with off-the-shelf functionality implementing the counter part of the network horizontal platform: M2M service capabilities. These service capabilities are exposed to device client applications similar to the network side.

In order to provide a high-level overview of the network optimization aspect, the following provides examples of functionalities that are being standardized by 3GPP [3]:

- **Low mobility**: intended for use with MTC (machine-type communications) devices that do not move, move infrequently, or move only within a certain region. The optimization consists of less frequently (based on operators policies) triggering mobility management functions that consume network resources.
- **Time controlled**: intended for MTC applications that can tolerate sending or receiving data only during defined time intervals. The network operator may allow such MTC applications to send/receive data and signaling outside of these defined time intervals but charge differently for such traffic.
- **Time tolerant**: intended for use with MTC devices that can delay their data transfer during network overload phases in exchange for better rates.
- **Packet-switched only**: intended for use with MTC devices that only require packet-switched services, meaning that there is no need to access SMS or circuit voice services.
- **Small online data transmissions**: intended for use with MTC devices that send or receive small amounts of data. The objective will be to dramatically reduce the signaling overload when limited amounts of data are being sent.
- **Priority alarm message (PAM)**: intended for use with MTC devices that issue a priority alarm in the event of, for example, theft, vandalism, or other instances requiring immediate attention.

4.10 Chapter Conclusions

This chapter provided an overview of the M2M requirements relating to network evolution, as well as the service architecture. It clearly shows that M2M imposes very specific requirements on the network in order to achieve lower investment and operational costs that match ARPU constraints imposed by M2M communications. Based on the extensive M2M use case analysis, this chapter makes the case for the clear need to deploy a horizontal architecture as a means of reducing development, deployment, and test cost and time for new M2M applications. Such a horizontal platform also provides easier access to network enablers, such as location, QoS, and addressing, which would otherwise be difficult to access because of the multiplicity of interfaces and the complexity involved in building business relationships with multiple operators.

The following chapters provide a deeper focus on the network optimization work being carried out in 3GPP, as well as the ETSI M2M Platform architecture work.

References

1. TR 102 691. ETSI Technical Report, Machine-to-Machine Communications (M2M); Smart Metering use cases (05-2010).
2. Smart Meters Coordination Group Standardisation Mandate to CEN, CENELEC and ETSI in the Field of Measuring Instruments for the Development of an Open Architecture for Utility Meters Involving Communication Protocols Enabling Interoperability M/441, final report (Version 0.7, 10-12-2009).
3. TR 23.888. 3GPP System Improvements for Machine-Type Communications (Release 11 – to be published 2012).
4. TR 102 732. ETSI Technical Report, Machine-to-Machine Communications (M2M); eHealth use cases (to be published 2012).
5. ETSI TC M2M Service Requirements for Release 1, published 08-2010.
6. ABI Research Study (2009) Cellular M2M Markets.

5

ETSI M2M Services Architecture

Omar Elloumi[1] and Claudio Forlivesi[2]
[1]*Alcatel-Lucent, Velizy, France*
[2]*Alcatel-Lucent, Antwerp, Belgium*

5.1 Introduction

As with every vision, the M2M promise has taken time to materialize. The early phases have been dedicated to refining the vision, testing new business models, developing standalone solutions to test concept feasibility, and also overcoming the limitations of insufficient interoperability. The use of horizontal service platforms, as a means for an operator to provide value-added services, or for an application provider to have a modular and future-proof application enablement strategy, is an integral part of the M2M vision. A recent report from the Yankee Group on the US carriers' perspective on M2M, *A Closer Look at M2M Carrier Strategy*, very clearly depicts the fact that providing value-added services based on horizontal platforms is becoming a business imperative for telecom operators:

> Carriers view value-added services as new differentiators. As all four major carriers have increased their coverage across the U.S., network availability is diminishing as a key differentiator. Overlapping coverage areas increase price competition, as solution providers have more carrier options in various geographies and can shop around for the cheapest price on a per-contract basis. The trend among all four carriers is to promote flexible, easy channels and robust device management services in order to attract business.
>
> (Source: Yankee Group)

M2M Communications: A Systems Approach, First Edition.
Edited by David Boswarthick, Omar Elloumi and Olivier Hersent.

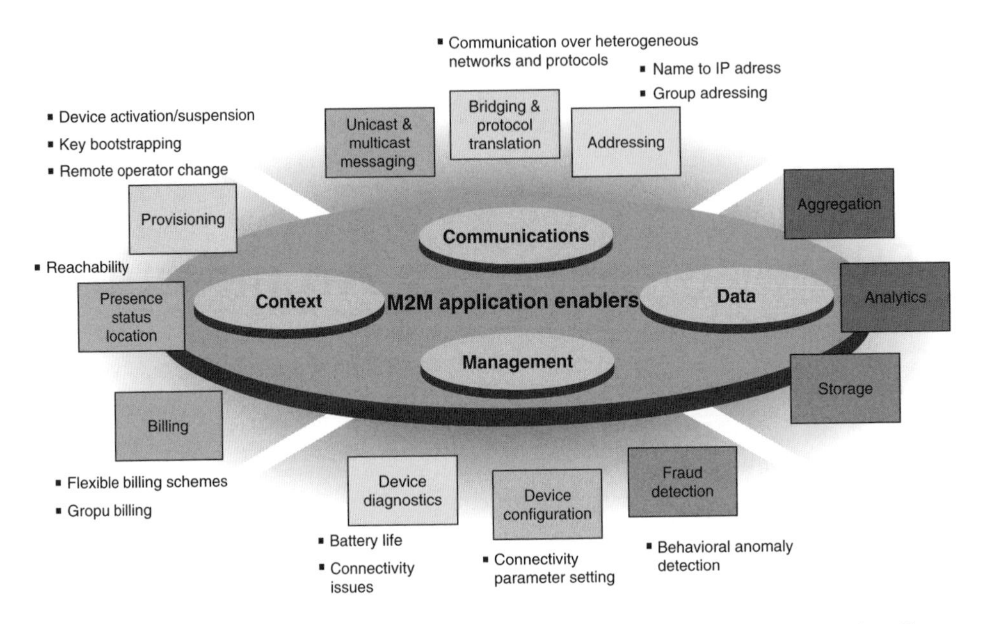

Figure 5.1 Typical components of a horizontal M2M service platform. Source: Yankee Group.

The components of such a horizontal platform may vary based on the targeted M2M applications. For instance, the support of location services is needed for fleet-tracking applications but not for smartmetering. Additionally, a stepwise approach is being adopted by the operators who are in the process of deploying horizontal M2M platforms. The original focus primarily targets connectivity and activation, basic data mediation, device management, and security. Figure 5.1 provides an overview of the M2M basic horizontal service platform components. These can be grouped into the following categories.

- **Data mediation functions**: Allow for basic data collection, storage, and subscription/notification on events pertaining to data availability. Additionally, further complex data functions, such as data aggregation, and analytics, can be provided.
- **Communications**: Hides network transport specificities as well as communication protocols from the applications. Communication functions include name-to-network address translation, bearer selection (SMS, data bearers) and establishment, protocol translation, etc.
- **Management functions**: Provides applications with configuration management (CM), fault, and performance management (PM) (such as battery level monitoring) as well as firmware and application software upgrade. Management functions are an important aspect of M2M, especially when considering the relatively long lifetime of M2M devices (more than 20 years for smart meters).
- **Context**: Provides access to other functions such as location, provisioning (device activation, security key bootstrapping, etc.), and billing functions.

This chapter will focus on the standardization work taking place in the ETSI Technical Committee Machine to Machine (ETSI TC M2M). This technical committee aims to provide a standards-based foundation for the deployment of such horizontal platforms.

The chapter is structured as follows. First, it provides the high-level system architecture and explains the generic framework used to build the horizontal platform service capabilities (SC). This framework is then further examined to show individual service capabilities. A motivational and a functional description are also provided for the principal capabilities. Second, an introduction to representational state transfer (REST) and the motivation underpinning the use of REST as an architecture style are provided. The rest of the chapter focuses on resource structure, as well as interface procedures, which are explained using an example.

5.2 High-Level System Architecture

High-level system architecture provides an overview of the components of a system, as well as the relationship between the individual components. It provides the starting point for a stepwise approach to the description of the functional architecture. This second presentation of the architecture provides a more formal description of the interactions between functions within the system, where each function can be mapped onto components of the high-level system architecture.

ETSI TC M2M has adopted the following high-level system architecture which allows for a common understanding of the system that is under standardization (see Figure 5.2). This high-level architecture fully endorses the need for M2M service capabilities that are exposed toward applications, be it in the network, the device or in the gateway. One important aspect of this high-level system view is that it provides an end-to-end representation of an M2M system. However, not all elements of this architecture are targeted as standardization work in ETSI TC M2M. That group fully recognizes that M2M communications will maximize the usage of already deployed access and core networks as well any other form of local and personal area networks. Additionally, the ongoing work on mobile network improvements for M2M currently underway in 3GPP and 3GPP2 is considered as a means to *improve* the delivery of M2M services as opposed to *must-have* features. As such, ETSI TC M2M, at least in Release 1 of its set of specifications, does not rely on the ongoing 3GPP and 3GPP2 network improvement for M2M, because the implementations must be based on existing network deployments. However, it is expected that subsequent M2M standards releases will increasingly seek to optimize these network improvements.

The M2M high-level system architecture (Figure 5.2) includes the concept of the M2M device domain, as well as the network and applications domain.

The M2M device domain is composed of the following elements:

- **M2M device**: a device that runs M2M application(s), referred to in the rest of this chapter as device applications (DAs), using M2M service capabilities and network domain functions. M2M devices can connect to the M2M core in the following manner:
  - **Scenario 1 "direct connectivity"**: The M2M device is equipped with a WAN communication module and accesses directly the operator access network. Examples of such devices include a smart meter that is directly connected to a GSM/GPRS infrastructure. In this case, the M2M device performs procedures such as registration, authentication, authorization, management, and provisioning with the network and applications domain.

- **Scenario 2 "gateway as a network proxy"**: The M2M device connects to the network and applications domain via an M2M gateway. M2M devices connect to the M2M gateway using the M2M area network. This case is applicable for "low-cost" devices that only run applications and make use of M2M SC available in the M2M gateway in order to perform the same procedures as in Scenario 1.

- **M2M area network**: a generic term referring to any network technology providing physical and MAC layer connectivity between different M2M devices connected to the same M2M area network or allowing an M2M device to gain access to a public network via a router (not shown in Figure 5.2) or a gateway. Examples of M2M area networks include wireless personal area networks, such as IEEE 802.15.x, Zigbee, Bluetooth, etc., or local networks such as PLC (powerline communications) or WiFi. While several M2M area networks are based on wireless RF (radio frequency) technologies, other wireline-based technologies are also being considered. The most notable example, beyond PLC, is the G.hn family of standards [G.hn] which has been designed with the objective of providing multiple profiles adapted for both multimedia/bandwidth-hungry applications and low-complexity terminals with lower bandwidth requirements. This latter option applies naturally to M2M applications such as home energy management.
- **M2M gateway**: of equipment implementing M2M SC to ensure the interworking and interconnection of M2M devices to the network and application domain. The M2M gateway may also run M2M applications. It is typically a piece of equipment that has at least a WAN communication module (e.g., GSM/GPRS) in addition to one or several communication modules that allow access to the M2M area network (e.g., Zigbee, PLC, etc.). An M2M gateway may implement local intelligence in order, for instance, to activate automation processes resulting from the collection and treatment of various sources of information, such as from sensors and contextual parameter).

The network and applications domain is composed of the following elements.

- **Access network** is a network which allows the M2M device domain to communicate with the Core Network. Examples of access networks include xDSL, HFC, satellite, GERAN, UTRAN, eUTRAN, W-LAN, and WiMAX.
- **Transport network** is a network allowing transport of data within the network and applications domain.
- **M2M core** regroups the various M2M SC that are exposed toward network applications (NAs) as well as operator core networks.
  - **Core network** provides functions relating to connectivity (including IP), service, and network control functions (such as 3GPP PCRF,[1] SMC-SC, etc.) and interconnection functions with other networks.
  - **M2M service capabilities (SCs)** provide M2M functions that are exposed to NAs through a set of open interfaces. M2M SCs typically use core network functionalities through known and standardized interfaces such as 3GPP Gi (used for exchanging

[1] Policy and Charging Rules Function (PCRF) is a server that provides policy decisions for network subscribers. Such policies are generally related to QoS levels and charging rules.

of IP data traffic) or 3GPP Rx (used to access QoS control functions). M2M SCs interface at the service level with peer M2M SCs residing in the M2M devices or M2M gateways.

- **M2M applications** run the service logic and use M2M SCs accessible via an open interface. Examples of M2M applications in the network and application domain include utility back-end applications responsible for collecting and analyzing smart meter data.
- **Network management functions** are all the functions required to manage the access, transport, and core networks. These include (but are not limited to) provisioning, supervision, and fault management (FM).
- **M2M management functions** are all the functions required to manage M2M applications and M2M SCs in the network and applications domain. The management of the M2M devices and gateways may use M2M SCs.

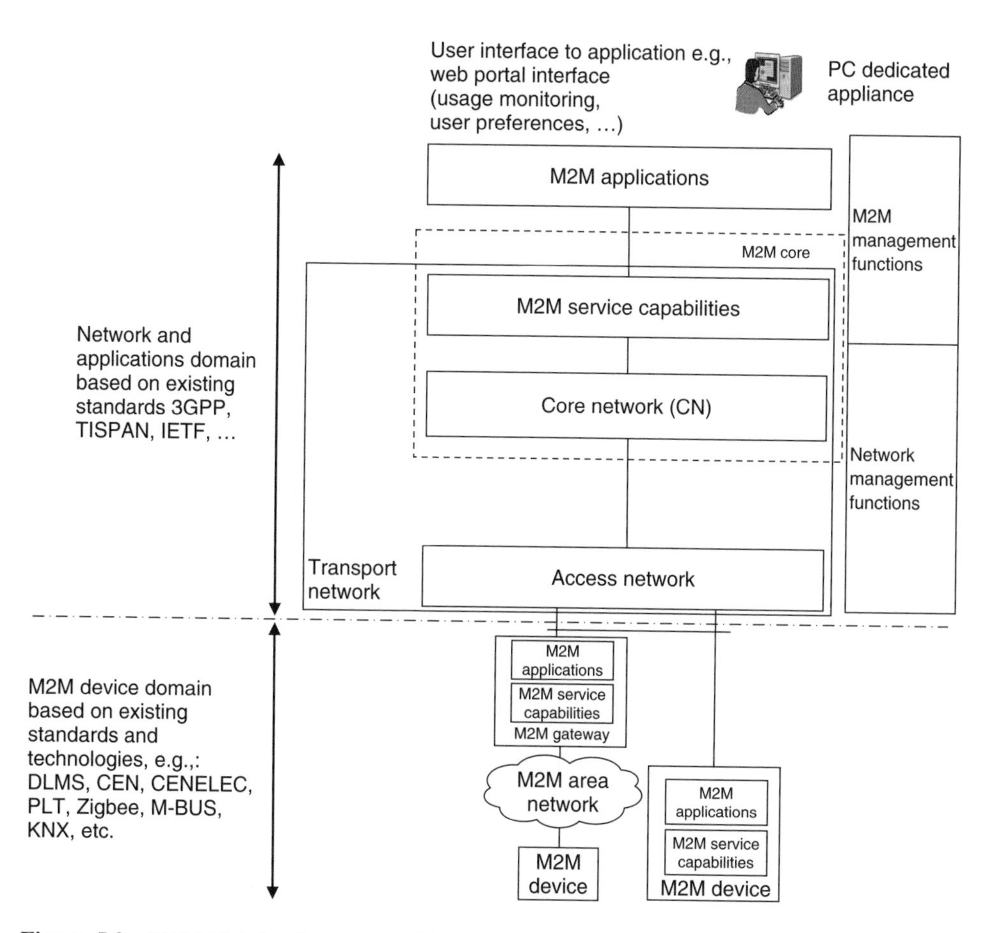

Figure 5.2 M2M high-level system architecture overview. Reproduced by permission of ETSI.

M2M SCs in the network and applications domain and in the device domain are the cornerstone of the ETSI TC M2M work. The main objectives of the standardization work are to:

- *Identify the functionalities that are to be exposed to the applications, be it NAs, gateway applications, or DAs.*
- *Standardize vertical interfaces (also referred to as APIs) that allow applications to make use of M2M SCs.*
- *Standardize horizontal interfaces at the service level between M2M SCs.*
- *Identify how SCs in the network and applications domain make use of core network capabilities (but this item has not been fully addressed within the first release of the ETSI TC M2M specification).*

In order to allow for an extendable and flexible structure of the M2M SCs, ETSI TC M2M has made the choice to develop an SC framework. An overview of this framework is provided in the next section.

5.3 ETSI TC M2M Service Capabilities Framework

A framework is a toolbox structured according to an architecture or a set of design patterns which can be instantiated to achieve a specific purpose. In the ETSI TC M2M context, the framework is the skeleton for SCs, and a set of reference points between the different entities of the system. The skeleton is then instantiated with a set of specific SCs, whereby the reference points constitute initial placeholders for the protocols used to interact between the entities of the M2M system. This framework was created with extensibility in mind since not all the SCs are known at the time of the standards specification, nor are all capabilities mandatory for an operational deployment. The ability to plug in SCs and make them discoverable by the application is of prime importance for future-proof standards and operational flexibility.

Figure 5.3 provides the framework that is used to build the ETSI TC M2M architecture. It shows a service layer architecture that is articulated around a set of SCs, either in the network and applications domain or in the device domain. The framework shows a set of reference points that are classified as:

- vertical reference points toward applications, be it in the device/gateway or in the network.
- a *single* horizontal reference point between SCs in the device or gateway on one side and the SCs in the network on the other – this reference point is at the service level, meaning that it uses network connectivity available in the M2M area network as well as the access and core network.
- a set of vertical interfaces toward the Core Network, which are required in order to make use of operator core network functionality – examples of these interfaces include 3GPP Gi/SGi for data or 3GPP Rx to access QoS offered by the core network [3GPP TS 23.060, 3GPP TS 23.401].

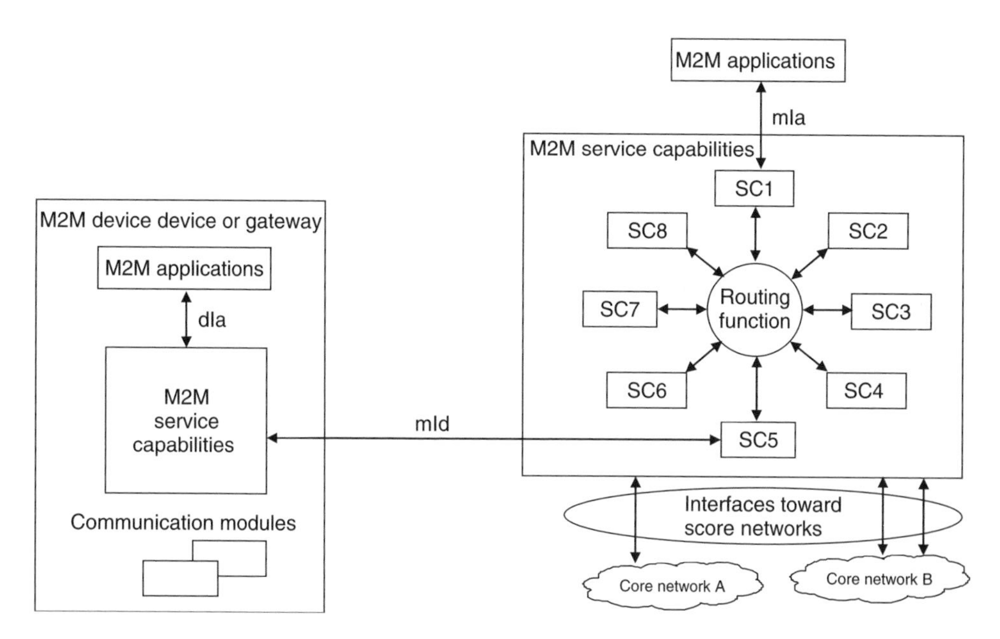

Figure 5.3 M2M service capabilities functional architecture framework. Reproduced by permission of ETSI.

The service capabilities (SCs) provide functions that are to be shared by different applications; these are software modules that provide specific services, such as location of a particular device. Note that the architecture defined in ETSI is a functional architecture, as opposed to box architecture, which means one or more SCs can be grouped. The ETSI TC M2M architecture does not impose a specific implementation of the SCs as long as the SCs behave externally according to the reference point specifications.

The SCs in the network and applications domain may interface with multiple core networks.

ETSI TC M2M identifies the following capabilities for the M2M Release 1 specification set (though not all the SCs have been exposed via the standardized API in Release 1 of the specification set):

- **Application enablement (xAE)** provides a single API interface to applications.
- **Generic communication (xGC)** manages all aspects pertaining to secure-transport session establishment and teardown, as well as interfacing with bearer services provided by the core network.
- **Reachability, addressing, and repository (xRAR)** provides a storage capability for state associated to applications, devices, and gateways and handles subscriptions to data changes.
- **Communication selection (xCS)** provides network and network bearer selection for devices or gateways that are reachable via multiple networks or multiple connectivity bearers, for example, WiFi or GPRS.

- **Remote entity management (xREM)** provides functions pertaining to device/gateway life cycle management, such as software and firmware upgrade and fault and performance management.
- **SECurity (xSEC)** implements bootstrapping, authentication, authorization, and key management. Interfaces with a M2M authentication server (MAS) – for example via Diameter – to obtain authentication data.
- **History and data retention (xHDR)** (optional) stores records pertaining to the usage of the M2M SCs. xHDR may be used for law enforcement purposes such as privacy.
- **Transaction management (xTM)** (optional) manages transactions.
- **Compensation broker (xCB)** (optional) manages compensation transactions on behalf of applications.
- **Telco operator exposure (xTOE)** (optional) provides access, via the same API used to access the SCs, to *traditional* network operator services, such as SMS, MMS, USSD, and location.
- **Interworking proxy (xIP)** allows a non-ETSI-compliant device to interwork with the ETSI standard.

where x, in each of the above, is:

- N for network.
- G for gateway.
- D for device.

ETSI TS 102 690 adopted the following terms to refer to SCs in the network, device, and gateway.

- **NSCL**: network service capabilities layer refers to M2M SCs in the network and applications domain.
- **GSCL**: gateway service capabilities layer refers to M2M SCs in the M2M gateway.
- **DSCL**: device service capabilities layer refers to M2M SCs in the M2M device.
- **SCL**: service capabilities layer, refers to any of the following: NSCL, GSCL, DSCL.

- **mIa reference point** allows an application to access the M2M SCs in the network and applications domains.
- **dIa reference point**:
  - allows an application residing in an M2M device to access the different M2M SCs in the same M2M device or in an M2M gateway
  - allows an application residing in an M2M gateway to access the different M2M SCs in the same M2M gateway.
- **mId reference point** allows an M2M device or M2M gateway to communicate with the M2M SCs in the network and applications domain and vice versa. mId uses core network connectivity functions as an underlying transport.

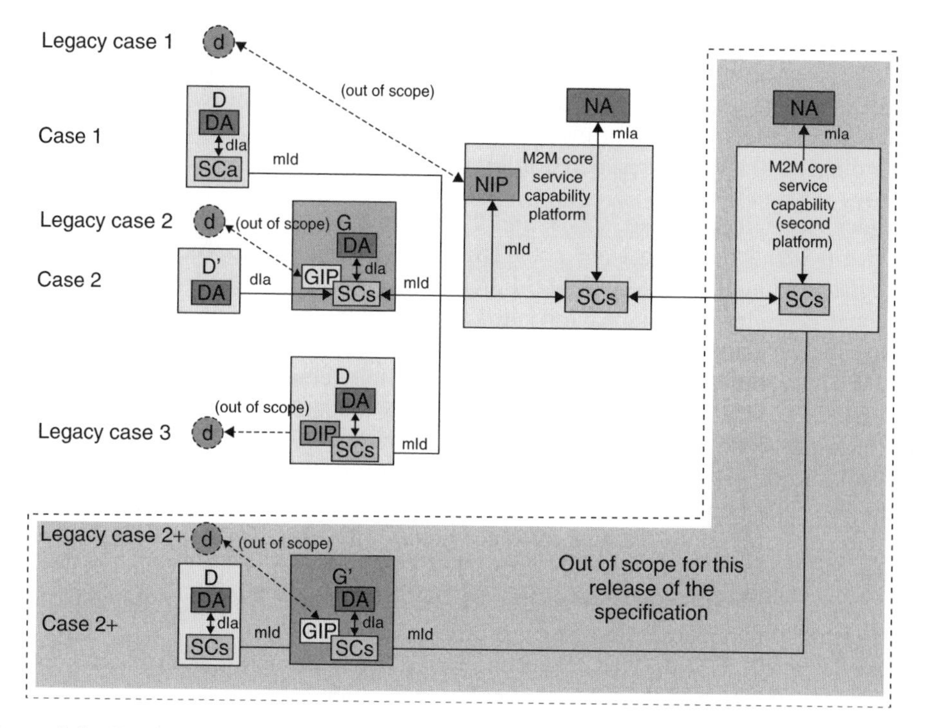

Figure 5.4 Service capabilities functional architecture framework. Reproduced by permission of ETSI.

5.4 ETSI TC M2M Release 1 Scenarios

This section provides a more in-depth look at the scenarios that are presently considered in ETSI TC M2M, as well as explaining their use of the specified reference points. These scenarios are depicted in Figure 5.4.

The following device types are considered in Figure 5.4:

- **Gateway (G)** is an ETSI M2M device specialized to directly manage M2M area networks of ETSI M2M devices; the gateway directly communicates with an ETSI M2M core network via the mId reference point.
- **Device (D)** is an ETSI M2M device that can directly communicate with an ETSI M2M core network or with an ETSI M2M gateway.
- **Device' (D')** is an ETSI M2M device that does not implement ETSI M2M SCs. It interacts indirectly with the M2M core via the use of SCs in the M2M gateway (G).
- **Gateway' (G')** is an ETSI M2M gateway that provides connectivity to both D and D' devices. Note that G' is not handled within the ETSI TC M2M Release 1 specification.

Additionally, there is a non-ETSI M2M-compliant device (d) belonging to an M2M area network managed by an ETSI M2M compliant device or gateway entities. The (d) type devices cannot access M2M SCs directly. However, interworking with ETSI-compliant entities is possible via the xIP (interworking proxy).

The different scenarios considered in Figure 5.4 are depicted below:

- **Legacy cases (legacy case 1, 2, 3, and 2+)** relate to legacy devices, referred to as "d" device (lower case d), that are not ETSI TC M2M-compliant. These devices can be integrated within the ETSI TC M2M architecture via an xIP service capability where x could be N for network, G for gateway, or D for device. xIP (NIP, GIP, DIP) is a specific capability that makes a non-compliant device looks like an ETSI-compliant M2M device through the implementation of the mId reference point. Note, however, that the ETSI TC M2M specification does not say how the interworking is done. The details of such interworking are not standardized and are left for individual implementations.
- **Case 1** shows a D device which is defined as a device that implements M2M SCs and runs DAs. The D device interfaces with the M2M SCs in the network and applications domain, making use of the mId reference point.
- **Case 2** shows a D' device which is an ETSI TC M2M-compliant device but that does not implement M2M SCs. Instead, it relies on SCs in an M2M gateway, referred to as "G", via the dIa external (in this case) reference point.
- **Case 2+** relates to D devices connecting to the M2M SCs in the network and applications domain via an M2M gateway, G. The motivation for such a case is to allow a D device to make use of WAN connectivity but also when needed – for example, in the case of mobile devices – reuse existing WAN connectivity provided by a gateway.

5.5 ETSI M2M Service Capabilities

This section provides the rationale and description of the most important M2M SCs. The intention is not to provide a detailed description, but rather to focus on the most important features, as well as the philosophy behind their introduction. This section also describes how these capabilities differ when implemented in the network, gateway, or device, respectively.

5.5.1 Reachability, Addressing, and Repository Capability (xRAR)

xRAR is the cornerstone of the ETSI TC M2M SC work. It has resulted from the merger of two other capabilities introduced in the early stages of the ETSI TC M2M work:

- **Device application repository**: originally introduced to maintain a record of registered DAs.
- **Naming addressing and reachability**: originally introduced to provide address translation functions and to keep track of reachability status of devices or applications running on those devices.

The merging of these two capabilities, resulting in xRAR, simplifies the structure and avoids the need for unnecessary frequent exchanges of messages between the two capabilities.

The features of the new defined xRAR revolve around the maintenance of registration information of SCs, applications (NA, DA, and GA) and, most importantly, the facilitation of data exchanges between M2M applications in accordance with access rights and permission.

The need to store data and make it available to other applications eventually became the most important feature of xRAR. The motivation for such a need is as follows:

If an M2M device (and consequently the hosted application) is not always on, it can become unreachable. However, there are situations where its last-known sampling should be rapidly returned. If the DA makes the data samples available in a timely fashion as a living document stored on a web server then the latest status of the DA is always accessible. xRAR is the SC that allows data mediation without necessarily the need to manage complex mechanisms maintaining wake-up periods.

Building on this mechanism, it also becomes possible for network applications to subscribe and be triggered when relevant data (matching a certain subscription criteria) is produced.

There now follows a list of xRAR functionalities:

- provides a mapping between a name (of a device, gateway, or application) to a network address, for example, an IP address, to allow the connectivity of the entity corresponding to the name.
- maintains the reachability status of an M2M device or M2M gateway, as well as the next planned wake-up time and duration (applicable to NRAR – the instantiation of xRAR in the network domain) – this feature is useful when the device or gateway is configured to accept incoming connections.
- provides a mechanism to allow applications or M2M SCs to register and be notified when a particular event happens, for example, data becomes available.
- allows the creation of a group of M2M devices, applications, or SCs to allow simpler interactions with the users of NRAR – for example, it is easier for a NA to send a message to a group than to the individual members of the group.
- allows applications to store application data and share with other applications in accordance with access rights and authorizations.
- allows local applications to register, a precondition for initiating a data exchange via M2M SCs. Local applications refer to the NA for network service capabilities (NSC), DA for device service capabilities (DSC) or gateway service capabilities (GSC) for DAs running on D' devices, and finally GA for GSC.
- allows mutual registration of SCs as precondition for exchanging data on behalf of applications.

5.5.2 *Remote Entity Management Capability (x REM)*

xREM was introduced in ETSI M2M to provide the means of managing device and gateway life cycles. As several M2M devices and gateways are installed for long periods of time, it is simply not possible to assume that the software they run is stable and includes all the functionality needed for the foreseeable future. In addition, M2M devices or gateways might run multiple applications, while some of them are installed well after the deployment of the devices. The US Smart Grid initiative, for instance, recognized the need for upgrading the software of deployed smart meters, and assigned a dedicated PAP (priority action plan) aimed at driving the corresponding necessary standards. The PAP mentions the following as part of its scope:

> Advanced Metering Infrastructure (AMI) and smart meter investments now as a precursor or enabler to additional Smart Grid, energy management, and consumer participation initiatives.
>
> One of the critical issues facing these electric utilities and their regulators is the need to ensure that technologies or solutions that are selected by utilities will be interoperable and comply with the yet-to-be-established national standards.
>
> (Source: http://collaborate.nist.gov/twiki-sggrid/bin/view/Smart grid/PAP00MeterUpgradability)

Device life cycle management involves several aspects including:

- **software and firmware upgrade**, where, in this context, the software is generally applicable to the operating systems, software modules implementing the M2M SCs and related APIs, while the firmware is the particular piece of software, often compact, that controls various electronic parts of the device.
- **application life cycle management** – the ability to install/remove applications or to upgrade to the new version of an existing application.
- **fault and performance management**, which applies to detecting faults pertaining to all components of the device (be it software or hardware), as well as monitoring performance indicators. Ultimately the goal of fault and performance management is to take corrective measures to ensure, wherever possible, service continuity. For instance, if an M2M device is battery-operated, it is useful to monitor the status of the battery and ensure that the process that allows battery changeover is triggered on time in order to prevent service interruption. Other parameters that may be monitored include the CPU or memory usage.
- **configuration management (CM)** applies to setting up different parameters of the device to allow its proper operation. In the context of ETSI TC M2M, CM applies to any parameters related to device components, such as a USB port or a camera, etc.

5.5.2.1 Motivations for Device Management

- Device management encompasses complex operations that are largely orthogonal to the application business logic. Take, for example, the case where a new version of software is not correctly functioning. In this case, device management procedures may need to perform a roll-back to a previous version of the software. Roll-back is a complex procedure that requires special expertise and possibly complex state machines to be maintained in the device and on the network.

- Device management can benefit from a better knowledge of network specificities. The software upgrade of a large number of devices can be scheduled over a long period of time. Therefore, upgrades could profit from network incentives if scheduled during non-busy hours. Additionally, if provided by network operator, it could take into account the actual network load so as not to disrupt more QoS-sensitive operations.
- The application does not need to know the details of the device management protocols that are used. In most current deployments, device management uses mature technologies such as BBF TR069 [BBF TR069] or OMA device management [OMA DM]. Both protocols have their own specificities. For instance, most current OMA DM deployments request the device to establish a device management session through the sending a special SMS to the device. On the other hand, in the case of BBF TR069, mostly used in wireline environments where there is no equivalent of SMS, triggering the device to establish a management session requires other mechanisms.
- Simplified fault and performance management assumes that the device management functions include monitoring the status of the battery. Conveying all battery parameters (in particular exhaustion level) will not help the NA. Typically, the NA will only be interested in being notified whenever a certain threshold is reached so as to schedule and plan for maintenance cycles. Also, the NA might be interested in being notified if the battery is being exhausted too quickly so as to remedy this issue, wherever possible.

The ETSI TC M2M specification of NREM lists the following functionalities to be offered to NAs:

- provides configuration management (CM) functions, such as the configuration of a device peripheral.
- collects and stores performance management (PM) and fault management (FM) data. Informs the NA when preconfigured events occur, for example, battery level reaches a certain threshold.
- hides connection (bearer and transport) establishment and teardown from M2M applications.
- performs software and firmware upgrades of M2M devices or M2M gateways.

In addition to NREM, which acts as a device management server, GREM and DREM provide the following functionality.

- GREM acts as a device management client vis-à-vis NREM, but also as a device management proxy vis-à-vis D' devices or other devices. A network management proxy allows GREM to perform device management function on behalf of NREM.
- DREM acts as a device management client to NREM.

In order to illustrate at a high-level the use case for NREM, Figure 5.4 provides a high-level message flow showing how NREM can be used for device management.

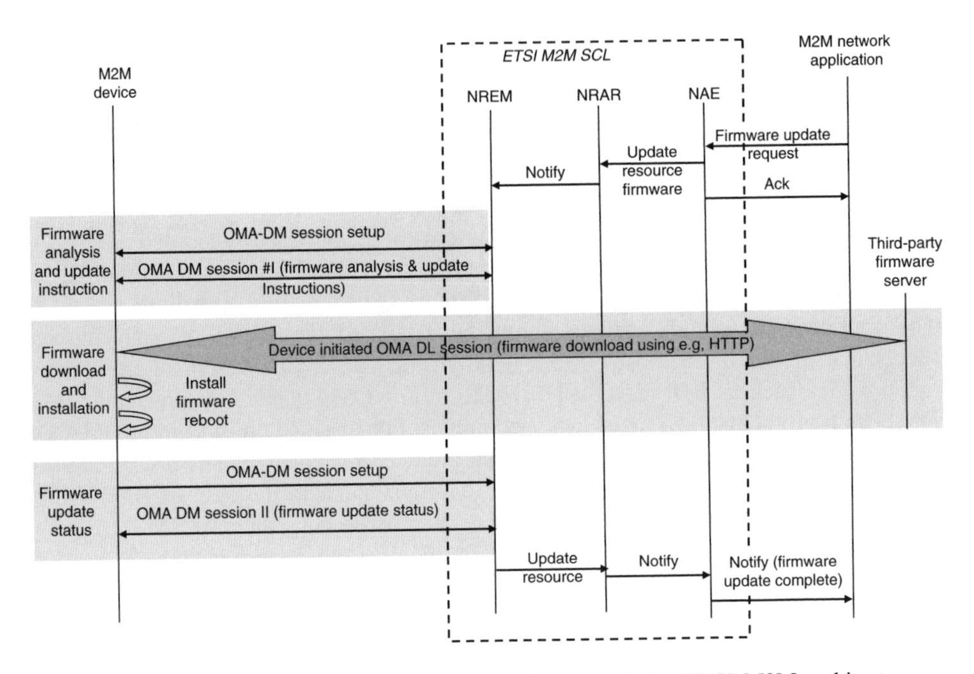

Figure 5.5 Use of NREM to upgrade a device firmware in the ETSI M2M architecture.

Figure 5.5 shows how the NREM service capability performs a firmware upgrade for a device at the request of an M2M NA. Initially, the M2M NA requests a firmware update. The parameters for such a request must include an identifier of the device subscription as well as the URL of the firmware image stored on a server ("Third-party firmware server" in Figure 5.5). The request will result in the creation of a special resource in NRAR containing a particular state for firmware upgrade operations. NREM is then notified about the need to perform a firmware upgrade. In this scenario, it is assumed that OMA device management is the management protocol used for managing the device life cycle. The OMA DM server (NREM) requests the establishment of an OMA DM session which may be performed by sending a specific SMS message or through HTTP push. This session will be used to convey to the M2M device DREM service capability a set of instructions corresponding to the firmware upgrade. These contain among others, the URL of the firmware image along with the credentials needed to access that image: user name and password. DREM will then download the firmware image, install it, and initiate a reboot. Then DREM establishes a session with the DM server in order to report about the status of the performed operations, in particular as to whether these were successful or not. If successful, a confirmation is propagated all the way to the NA via NRAR and NAE – the instantiation of xAE in the network domain.

5.5.3 Security Capability (xSEC)

This allows for service-level registration – the necessary step to allow a device or gateway to begin operations with the network service capability. Service registration includes mutual authentication, authorization, and bootstrapping. The high-level description of these functions is provided below. However, the reader may refer to Chapter 5 for a more in-depth overview of M2M security.

- **Mutual authentication**: In order for a device or gateway to be permitted to access certain resources that are managed by the M2M operator, authentication needs to first be performed. On the NSCL side, NSEC is the service capability that interfaces with the MAS, that is, the server that has access to user subscription information. In particular, whenever a device (DSCL) or gateway (GSCL) wishes to register with the M2M operator's core infrastructure (NSCL), NSEC obtains authentication material from the MAS and challenges the device (or gateway), while at the same time providing sufficient authentication information via which the NSCL is also authenticated to DSCL or GSCL. As such, NSEC can be considered as the "authenticator" in authentication, authorization and accounting (AAA) terminology. Authentication material is generated from a permanent, shared key credential called "root key" (see Chapter 8). As a result of such a mutual authentication procedure, a symmetric session key is mutually agreed and is stored within xSEC (NSEC on the one side and DSEC or GSEC on the other). The session key (or so-called "service key") is used for further deriving key material per M2M application for the purposes of setting up secure data connections over the mId interface, as well as for application-level authorization. Note that different M2M operators may be equipped with different types of authentication servers. This is particularly the case when such operators wish to reuse their existing authentication service infrastructures (such as HSS or AAA servers), which they may also use for other types of services, such as network access. Given this, the implementation of NSEC needs to take into account the most commonly deployed authentication technologies in order to facilitate different authentication protocols. Often this is an easy task if certain authentication frameworks are considered, such as the extensible authentication protocol (EAP).
- **Authorization**: Upon successful mutual authentication, MAS provides authorization information to NSEC. Such information specifies which resources a particular application in a particular device is supposed to access. Based on such information and in conjunction with the session key, NSEC generates keying credentials for each application residing in a specific user's device. Such material is populated by NSEC to NGC (the instantiation of xGC in the network domain) and is used to determine whether certain HTTP REST commands over the mId interface can be authorized. Note that not all operations performed by NSEC correspond to operations performed by GSEC and DSEC. For example, only NSEC interfaces with the MAS, which DSEC/GSEC interfaces with a secured environment in the device/gateway, within which all sensitive security operations are performed, such as key derivation functions that make use of the root key. On the other hand, there are certain functionalities that are similar, such as the derivation of key material from session keys which is used for establishing M2M secure data sessions over the mId reference point.
- **Bootstrapping**: In certain cases, NSEC may be used even before the activation of a user subscription. As discussed in Chapter 8, service bootstrapping allows a device/gateway to be permanently associated with a particular M2M operator's infrastructure, via the bootstrap of a permanent shared root key along with other service parameters and

configurations. In this context, whenever NSCL is supposed to facilitate automated bootstrapping of root keys, the device/gateway communicates with the M2M service bootstrap function (MSBF) using the NSCL infrastructure. Given that MSBF may be entity-owned and operated by the M2M service provider, the device/gateway needs to be first authenticated in order to be authorized to reach MSBF. Such a (mutual) authentication is again performed by NSEC. An example of an automated bootstrapping procedure that leverages the xSEC functionality is discussed in Chapter 8.

5.6 Introducing REST Architectural Style for M2M

5.6.1 Introduction to REST

REST (representational state transfer) is an architectural style defined by Roy T. Fielding in 2000 [Fielding]. It is a set of principles that enables distributed systems to achieve greater scalability and allow distributed applications to grow and change over time, thanks to the loose coupling of components and stateless interactions.

The main concept of REST is that a distributed application is composed of resources, which are stateful pieces of information residing on one or more servers. Regardless of their content, in REST it is possible to manipulate resources through a uniform interface that is composed of four basic interactions: CREATE, UPDATE, DELETE, and READ. Each of these operations is composed of request and response messages, and, with the exception of CREATE, they are idempotent, meaning that the end result of each operation is unchanged regardless of how many times the operation itself is repeated. In other words, these operations do not have side effects, meaning that it is possible to distribute resources and to use proxy functions. The lack of side effects allows more efficient use of caching and greater scalability.

More importantly, however, since the same set of operations can manipulate the most diverse kind of resources, it is not necessary to develop dedicated clients or infrastructures whenever the application domain changes. Therefore, the same underlying architecture can be reused for multiple applications.

The most common implementation of REST is HTTP, whereby the REST operations are mapped into the HTTP methods: CREATE is mapped on HTTP POST, READ on HTTP GET, UPDATE on HTTP PUT, and DELETE on HTTP DELETE.

5.6.2 Why REST for M2M?

Today, many of the devices that surround us contain an increasing number of sensors. Examples include the GPS antenna of many mobile phones, the accelerometers contained in several devices, weight sensors (as used in elevator or games consoles), and more traditional devices, such as smoke, humidity, or temperature sensors.

Sensors are typically characterized as being devices that convert one or more physical measurements into an analog or digital signal for purposes such as data processing and/or provide input to actuators. The latter are devices that take some form of input and use it to perform some form of physical action, such as closing a valve or lighting a lamp.

Sensors and actuators are very often coupled together, usually using some form of computing device to transform the output of the sensor into the input of the actuator. This

Figure 5.6 Sun SPOT M2M device.

means that both sensors and actuators often contain very little computing power and are highly dedicated devices that limit themselves to their specific measurement activities. Over time, microcontrollers – equipped with computing capability and standard-based communication interfaces, such as WiFi, USB, or ZigBee – have emerged. They may also offer some onboard sensing capability, such as GPS modules or accelerometers. This has allowed some greater degree of intelligence in performing measurements and controlling actuators. However, their computing capabilities may vary greatly, especially because many of these devices are battery-powered, which means that they must prevent the CPU from constantly draining the batteries. Below are some examples of M2M devices with sensing capabilities:

- **Sun™ SPOT** (see Figure 5.6) is a 180 MHz, 32-bit ARM920T core with 512 KB RAM, 4 MB Flash, and 2.4 GHz IEEE 802.15.4 radio with integrated antenna. It provides an AT91 timer chip, a USB interface, and several sensors, including temperature, light, and a three-axis accelerometer. It can support external analog input, and can be programmed in Java.
- **Sentilla™ JCreate** is an AAA battery-powered microcontroller, equipped with a TI MSP430 16-bit microprocessor and a TI/Chipcon™ CC2420 wireless transceiver. It hosts an onboard three-axis accelerometer and eight LEDs, and it can run Java/J2ME software from third parties.
- **Crossbow™ TelosB** is a TI MSP430 16-bit microcontroller with 10 KB RAM, IEEE 802.15.4, and USB interfaces, supporting the Contiki, TinyOS, SOS, and MantisOS operating systems. It can be accessed/programmed via USB and offers optional onboard temperature, light, and humidity sensors.
- **Nano-RK FireFly™** is an Atmel™ ATmega1281 eight-bit microcontroller with 8 KB of RAM and 128 KB of ROM with Chipcon™ CC2420 IEEE 802.15.4-standard-compliant radio transceiver. Optional expansion cards provide light, temperature, audio, passive infrared motion, dual-axis acceleration, and voltage sensing.

Applications that use these examples of devices are very much focused on low-level aspects of sensor communications, such as real-time interaction, energy consumption, error correction, and management of hardware faults.

On the other hand, the typical approach in the development of distributed systems is to create a consistent abstraction on top of the lower layers of communication (for instance using TCP/IP) and to define interaction models on top of this. Over time, several architectures have appeared in the world of distributed systems, like Java RMI, CORBA, DCOM, or SOAP. Most of these are based on the assumption that communication is like the invocation of a remote service whose nature can be ignored, and can therefore be considered in terms of remote procedure, remote function, or remote methods. This approach is called a remote procedure call (RPC) and assumes that the communication paradigm always involves an abstract entity that issues a request and another one that sends back a response, both of which are transmitted over a reliable channel.

Traditionally, distributed-computing experts have approached sensor applications using the same RPC paradigm. However, the above assumptions that characterize RPC are not always well-suited for M2M-type applications.

Indeed, M2M is more about tangible states, and a technology that models them would be more appropriate than the usual procedural approach used when dealing with real states. REST is the ideal method for M2M modeling. The concept behind REST is that every physical and/or logical entity is a RESOURCE that has a particular STATE that can be "manipulated." This concept maps naturally to the world of sensors and M2M devices in general, a sensor being a device that can be read or configured, and can even stream data to the outside world.

Of course, there is nothing to prevent the modeling of a sensor or M2M device as a web service and using the WS-* standards to their full extent. However, such a solution is highly complex, and complexity entails a cost that could exclude a large class of resources-constrained devices, especially the ones that could have the largest penetration into the market.

REST-based architecture offers several advantages that may be useful to sensor and M2M application developers, such as the possibility to visualize and manipulate the sensor data and calibration parameters through a web browser, or even to create web mash-ups using one or more sensors or M2M device as data sources.

A beneficial consequence of this is that the state transferred to the client can be cached by browsers and HTTP proxies, allowing a greater scalability than *any RPC-based approach, where each request must flow end-to-end*.

Scalability is also increased by the fact that every communication is stateless, and each request can be handled independently from the others.

Sensors, M2M devices in general, and M2M applications can benefit greatly from this, since it means that a simple HTTP proxy with caching can shield M2M devices (including sensors) from most of the network load. This is important, given the fact that most sensor devices are transient entities which have very limited processing power.

Finally, the effort needed to develop applications is greatly reduced, since REST adopts a much lighter tool chain than most service oriented architecture (SOA) technologies. Simple HTML and Javascript is all that is needed to create fully fledged M2M applications.

Another side effect of adopting HTTP is that the sensors can be addressed as web links, which can be posted on the Internet, inside RSS feeds or sent via email.

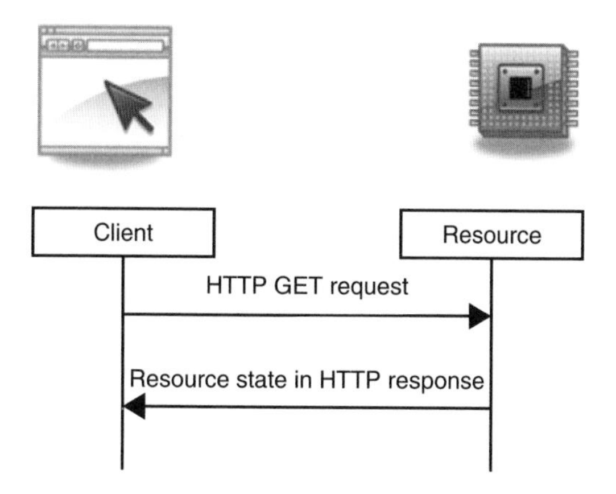

Figure 5.7 Addressing a REST resource.

5.6.3 REST Basics

REST is an architectural style that relies heavily on HTTP, and conceptualizes the idea of web-accessible resources. A resource, in REST terms, could be any entity that can be addressed using a HTTP URI (see Figure 5.7).

The host, port, and path parts of the URI are used to locate the resource inside a web server, and the HTTP method used in the HTTP request determines what action to take.

The HTTP GET method reads the "state" of the resource and returns it in the HTTP response. In REST, the GET method is considered to be free of side effects, which means that repeating the same GET request several times on the same resource will not alter its state.

This also means that any parameters passed in the query string of a GET request can only affect the content of the current response being returned (for filtering out some parts or altering the format). It does not affect the remote state of the resource.

The response of the HTTP GET can also be cached by the client or by any intermediary proxy (as shown in Figure 5.8), reducing the load on the resource.

The HTTP PUT method is used to update the state of a resource. Using this mechanism to update the state has several advantages, including the fact that the state is set in a single action. This makes the writing of the state an atomic action, which can only succeed or fail, avoiding the need for a stateful interaction between the client and the server.

Moreover, the intermediate proxies are able to recognize the URI and subsequently clear their caches accordingly.

There are also two other HTTP methods, POST and DELETE, which are both non-idempotent operations and are typically used to create and destroy resources, respectively. They work in the same way as PUT, and are also atomic.

Thanks to atomicity, this approach promotes a way of "talking" between a web client and a resource that limits the number of side effects, since only resource states, and not actions, are transferred to and from the resource. This is what is required when dealing with sensors.

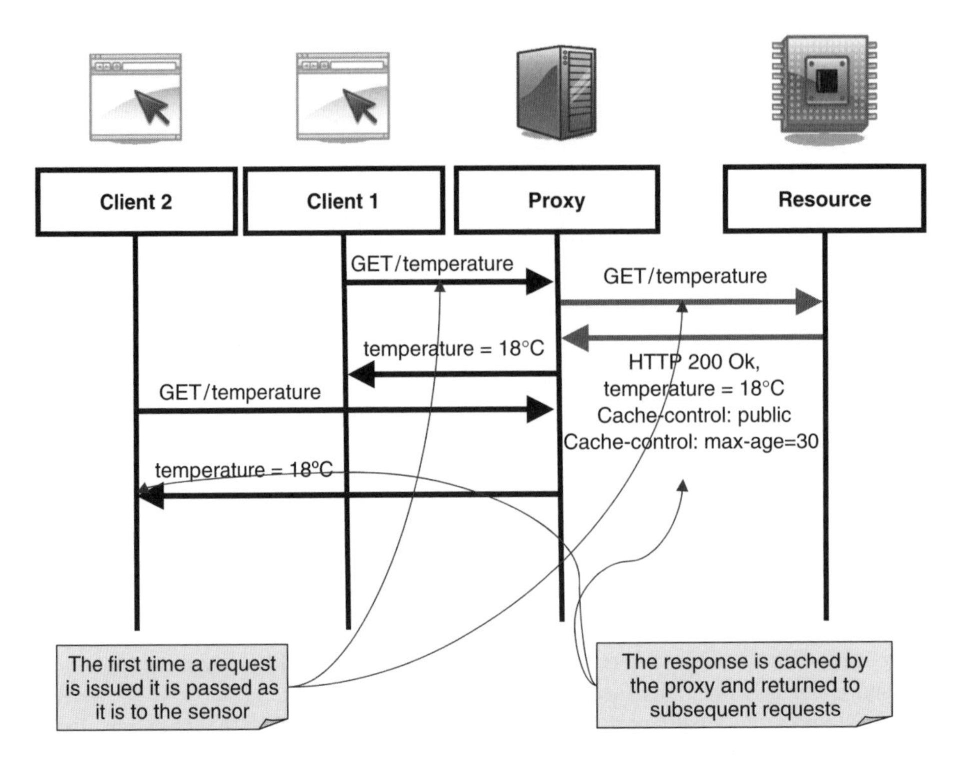

Figure 5.8 Proxy caching the temperature value.

5.6.4 Applying REST to M2M

How can REST be used when communicating with M2M devices?

There are multiple approaches that can be used to achieve effective communication, based on the fact that a physical device is seen as a resource that is accessible via a HTTP URI, whose samples can be read using GET and can be calibrated using PUT.

POST and DELETE may look strange when used in conjunction with physical devices, since these can be neither created nor destroyed. However, they can be used to configure more complex entities, such as programmable filters. POST can be used to "add" a filter to an existing set of filters that will be applied by a computing device that is processing sensor data.

A sensor can be directly reachable using HTTP requests, and will return the sampled data in the response message. This works well for those devices which are always on, but exposes them to potentially huge amounts of traffic. In order to prevent this, a caching proxy could be used between the clients and the sensors.

If a sensor is not always on, it can become unreachable. However, there are situations where the last known sampling status should be returned.

In this case, the resource will not be the sensor, but simply a document living inside a web server, which is updated with the latest sample data when needed by the sensor using PUT (see Figure 5.9).

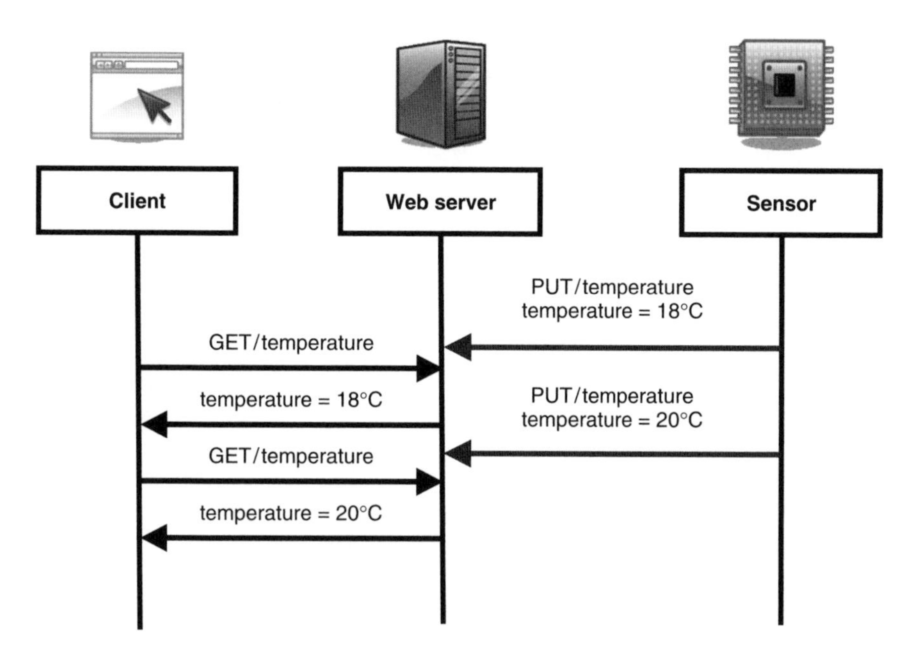

Figure 5.9 Sensor updating its samples using HTTP PUT.

If this is not possible – for example if a sensor does not have an IP interface – it might be necessary to have an intermediate entity that is able to communicate with the sensor using its native interface, as well as communicating with the web server. This can also be contained in the web server itself, as many servers can host third-party applications, which may implement the sensor native protocols (see Figure 5.10).

5.6.5 Additional Functionalities

5.6.5.1 Event Handling

Some browsers, and almost every available proxy, support the "multipart/x-mixed-replace" MIME type. This MIME type allows responses to be sent in "chunks" back to the client, each chunk replacing the previous one. This technique is known as "server push," and is used to force a HTTP server to enforce changes on the web pages displayed by connected clients. This technique is usually exploited by webcams to update the picture they are generating.

A sensor can use this technique to provide samples to connected clients in a streaming fashion. Whenever a sensor produces a new sample, each connected client will receive it as a chunk of the multipart response. This allows asynchronous events to be delivered to clients (see Figure 5.11).

One major disadvantage of server push is that one TCP connection needs to be kept open for the entire duration of the client subscription to sample events, severely affecting scalability. However, HTTP proxies can be used to mitigate the problem to a great extent.

It is possible to reduce the number of direct connections to each sensor to one, made by the proxy whenever a client subscribes to the events flow. Instead of connecting

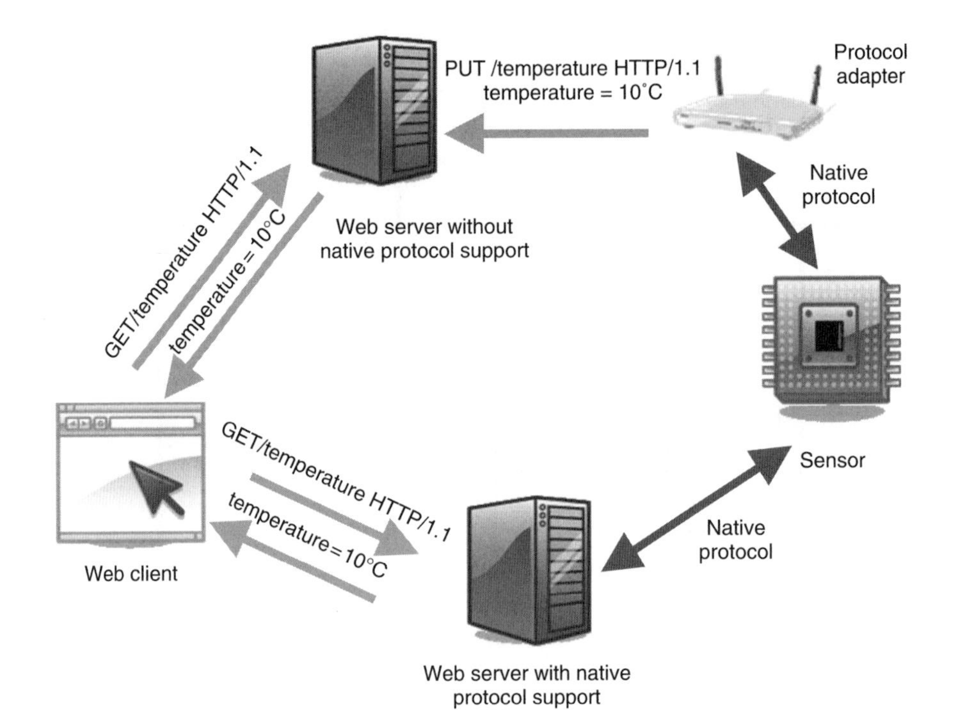

Figure 5.10 Connecting sensors using native protocols.

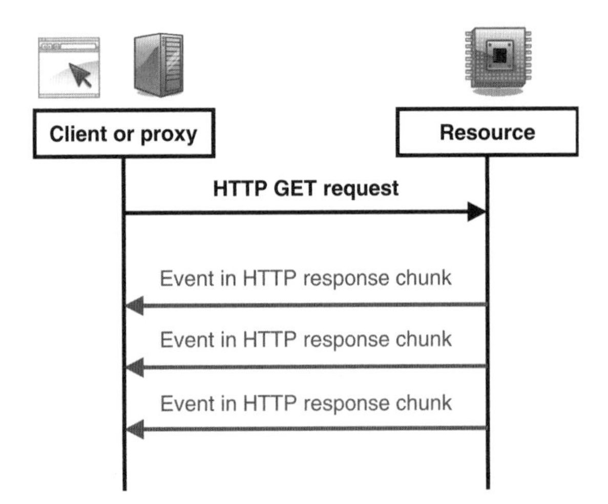

Figure 5.11 Multiple events streamed in the same HTTP response.

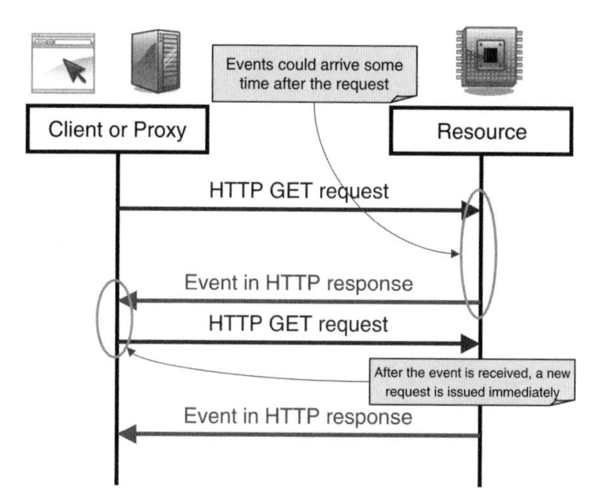

Figure 5.12 Events are sent as HTTP responses and new requests are issued to receive the next event.

to the sensor, clients will connect to the proxy that will be able to serve multiple connections.

Another HTTP technique that enables sensors to send events is HTTP long polling (see Figure 5.12). According to this technique, a client connects to the resource and sends its request on a TCP connection, and then it waits for the response to arrive. The sensor being queried in the HTTP request will make the client wait on the TCP connection until a new sample is available, and then it will send it back as an HTTP response. Upon receiving the sample, the client will issue a new HTTP request, while waiting for the next sample to be produced.

In this case, a connection is kept open for quite long periods of time, meaning that the resource consumption is very similar to the previous mechanism (Figure 5.11). However, this technique is more browser- and proxy-friendly, since there is no need to support special MIME types, and the sensor/M2M device is just simulating a very slow web server. However, the optimization considerations of the multipart case still apply. In fact, it is possible to use one or more proxies in order to multiplex the HTTP response toward several connected clients using a single connection to the actual sensor/M2M device.

5.6.5.2 Improving the Efficiency through the Use of Caching Proxies

Caching of the sensor sampling data can increase the scalability of the overall system in a transparent way. Transparency is ensured by the fact that caching requires only the introduction of common HTTP proxies, with no perceived changes from the point of view of both the sensors and their clients. This modified topology can serve a far greater number of requests. Cache control occurs in the same way as with normal web browsing. One way is to use the "If-Modified-Since" HTTP header in requests. In fact, a sensor can return the actual time of its last sampling using the HTTP "Last-Modified" header in responses to GET requests. The client and/or every intermediate proxy retains this value and can place it in the "If-Modified-Since" header of every subsequent request.

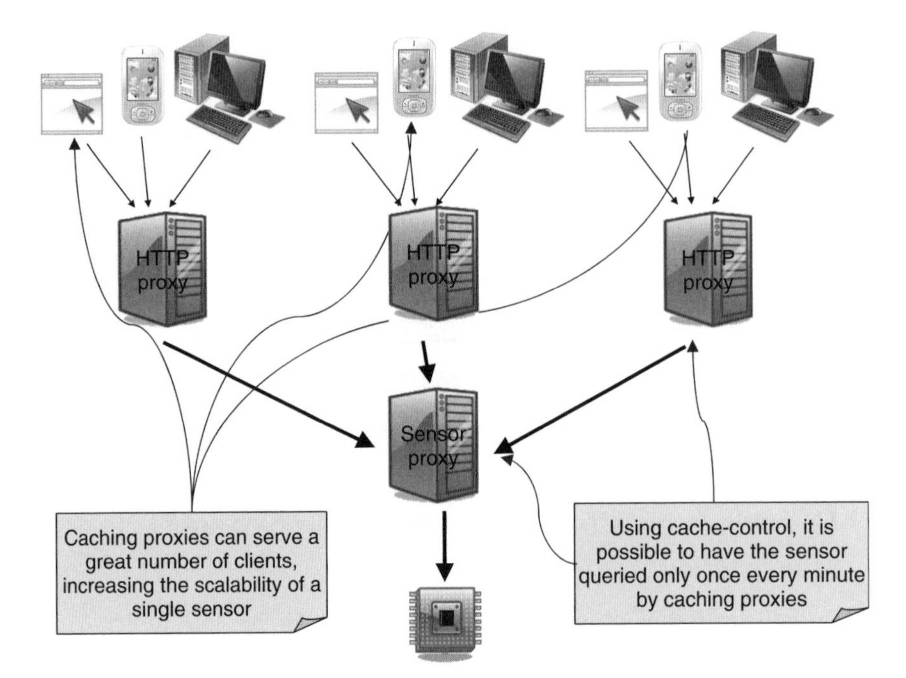

Figure 5.13 Scalability increased using caching proxies.

If there was no new sample, the sensor simply returns an HTTP Code 304 (Not Modified) to these requests, with the "Last-Modified" header containing the time of the last sample.

A proxy in between the client and the server inspects all the passing traffic, and is aware of the latest sample data of several sensors. Whenever a GET request arrives from one of the clients, it will pass it on to the sensor, and if a 304 Code is returned, the proxy could return the last known sample instead, even if the client never specified an "If-Modified-Since" header in its request.

Moreover, using the "Cache-Control" header in requests and responses, sensors, clients, and proxies can control the caching of resources in a more sophisticated way (see Figure 5.13). It is also possible to specify new caching directives, in addition to those already foreseen by the HTTP protocol. It is possible, for instance, for a sensor to inform the recipients of a response (either clients or proxies) of the time validity of the sample being returned and whether revalidation is needed before the expiration of the sample.

This mechanism can be used by proxies to return the last known sample to clients until it expires, without having to contact the sensor before expiration.

Caching greatly reduces the amount of traffic on the sensor, since only one request will be sent to it after the current sample expires, and no further requests will then be sent until the expiration of the new one.

5.6.5.3 Device Configuration Example

Devices do more than just produce data from measurements. They also need to be configured and calibrated. For instance, a laser meter needs to be recalibrated from time to time, in order to align its internal hardware with measurement parameters. When calibration takes place, some reference information about a physical dimension are sent to the sensor while it is performing sample measurements, and the difference between the sample and the reference is then used by the sensor to align its further samplings. Configuration, on the other hand, can be also done when the sensor is not performing samples (in some cases, this is mandatory) and it involves sending the sensor the necessary information concerning how to operate and/or to communicate.

In all these cases, it is necessary to send the sensor some calibration or configuration data. In REST, this is essentially possible in three ways.

The first method allows sensor configuration data to be sent directly to the sensor using HTTP PUT or POST methods. However, this is only possible if the sensor itself can be addressed by HTTP, which is not the case for not always-on devices or when a device does not support IP connectivity.

In this case, there must be an HTTP proxy between the client application and the sensor that is capable of converting the data received from HTTP POST requests into a set of commands for the sensor. These commands are then immediately sent to the sensor as a consequence of the POST request (see Figure 5.14).

However, when the sensor is not always on, it is necessary that the configuration data being posted by an application is stored by the HTTP proxy inside a cache or repository and then retrieved by the sensor when it goes online, as in Figure 5.15. Finally, the

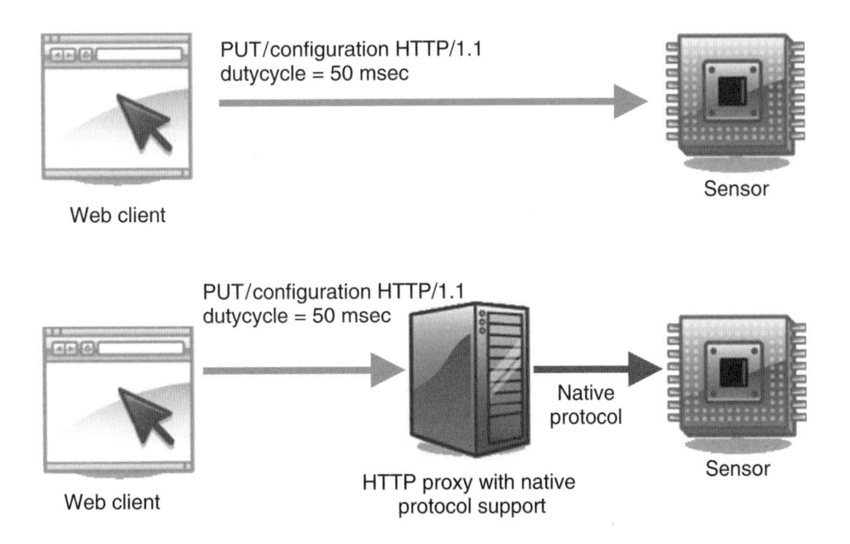

Figure 5.14 Sensor configured directly or with a proxy.

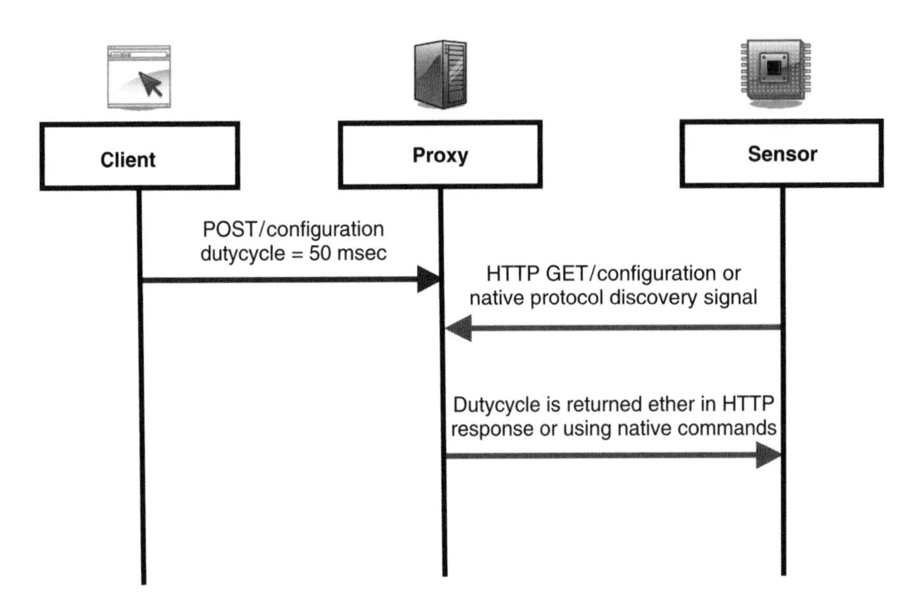

Figure 5.15 Asynchronous configuration using an HTTP proxy.

retrieval of this information can take place in two ways: the sensor may issue a HTTP GET request to the proxy or use its native protocol if IP connectivity is not supported. In the latter case, the proxy must also support this protocol.

5.6.5.4 REST Additional Features

Another feature of REST is that it is possible for each device to choose the URL that it may want to use in order to expose its state. This can be done using a HTTP URL where the state could be addressed using the context-path part of it. It is also possible, for complex systems, to use different URLs in order to store the state of the different subsystems.

For instance, the status of the parts of an "intelligent" car could be posted in several URLs:

- http://carmaker.com/numberplates/00001/wheels.
- http://carmaker.com/numberplates/00001/engine.
- http://carmaker.com/numberplates/00001/battery.

It should be noted that the URLs of resources are sometimes generated dynamically when a resource is created. This typically happens when resources are created using the POST-Redirect-GET paradigm. Under this model, a device exposes its state for the first time by creating a resource using HTTP POST to a predefined URL. The resource is then created within another URL that is returned to the device inside the HTTP response (see Figure 5.16).

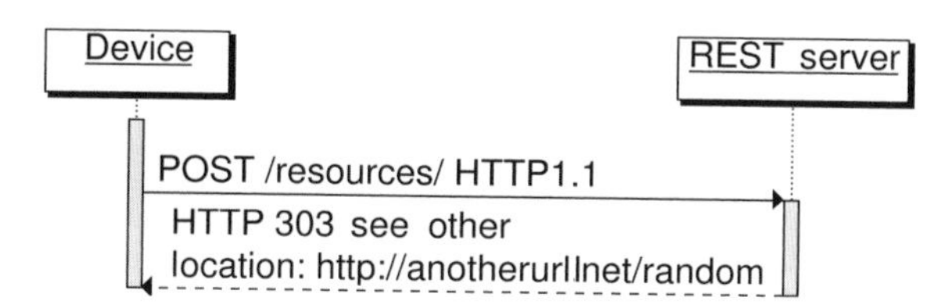

Figure 5.16 Resource creation on behalf of a device.

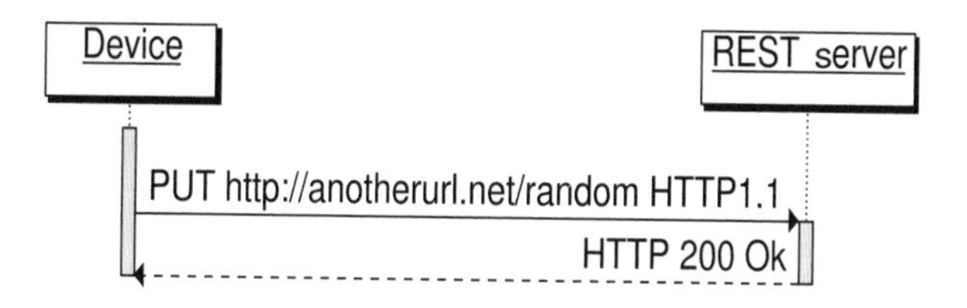

Figure 5.17 Device updates a resource with its current status.

The newly created URL can then be used by the device to update its state using HTTP PUT (see Figure 5.17).

Other devices can use that URL in order to retrieve the state of a device even if it is offline. In order for these devices to retrieve the dynamically generated URL, it can be stored inside a discovery service, for example, with DNS-SD in a DNS TXT record. Since the publication is long-lived (it can exist until the underlying M2M device is disposed of), the time-to-live of the entry can be longer than the typical wake-up of the device (see Figure 5.18).

With this approach, the access control to information becomes easy, as device manufactures or administrators do not have to consider how their devices authenticate other devices and users, only with connected devices and the REST/HTTP server. This can be done, for example, with DNSSEC and HTTPS if authentication of the server is required. The REST/HTTP server administrator must authenticate devices, so only certain devices can write to certain addresses. This can be done also with TLS (transport layer security).

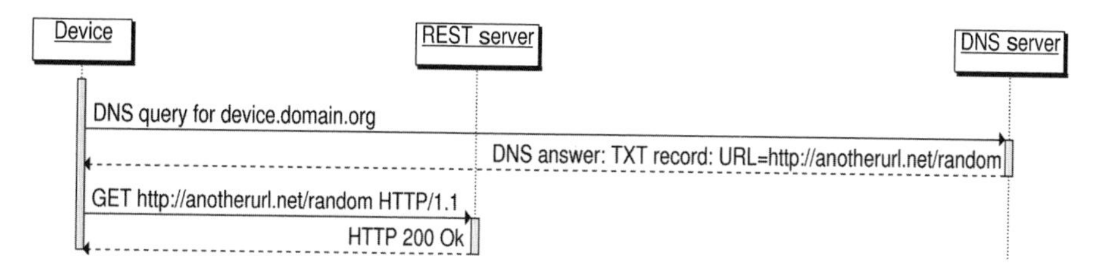

Figure 5.18 DNS query to retrieve the URL of a device's state.

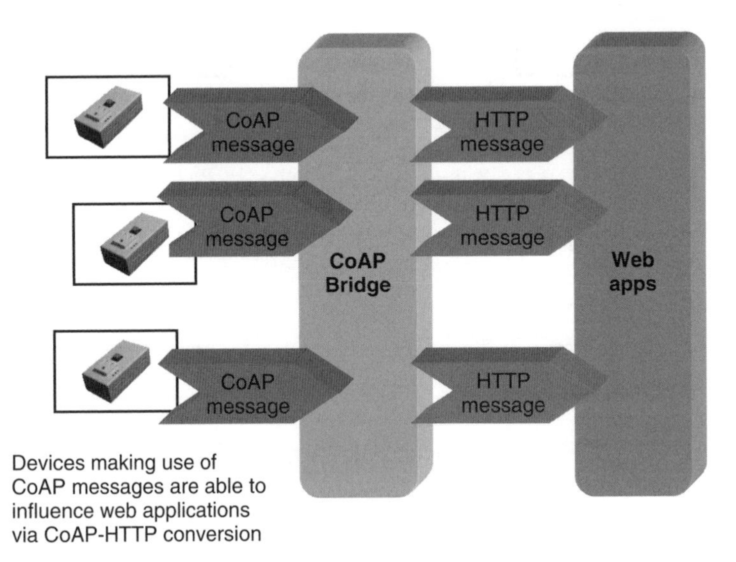

Figure 5.19 CoAP to HTTP protocol bridge allows for seamless integration between devices and web applications.

5.6.5.5 Relationship with Other Standards

IETF CoAP

As an extension of the 6LoWPAN (IPv6 over lossy and low-power PAN) concept, CoAP[2] is another IETF initiative revolving around the idea of porting the features of a REST-based architecture to a wireless sensor network, especially to networks supporting 6LoWPAN.

REST-based architectures define essentially four basic operations to become available when dealing with resources (devices in this case). These operations are Create, Read, Update, and Delete. All REST resources may be manipulated using these operations. Commonly, REST servers use HTTP as a transport, and the operations above are mapped into the HTTP methods POST, GET, PUT, and DELETE, respectively.

What CoAP proposes is a non-HTTP approach to REST. The reason for doing this is mainly due to the fact that it is difficult to implement HTTP in small, constrained devices like those of 6LoWPAN. Therefore, CoAP defines some primitives that allow REST on top of TCP and/or UDP.

In order to simplify the message format, the HTTP headers have been simplified to fixed-position, fixed-size binary fields inside a CoAP payload, and only a limited set of MIME types are available.

However, even with these limitations, it is possible to create bridges in the network that make a proper conversion between CoAP and HTTP/REST (see Figure 5.19). This allows a much more seamless integration of limited sensor devices and web applications.

[2] CoAP (constrained application protocol) is a protocol that is being specified by the IETF. CoAP is a Restfull protocol that runs on top of UDP (user datagram protocol) and that is particularly adapted to constrained devices.

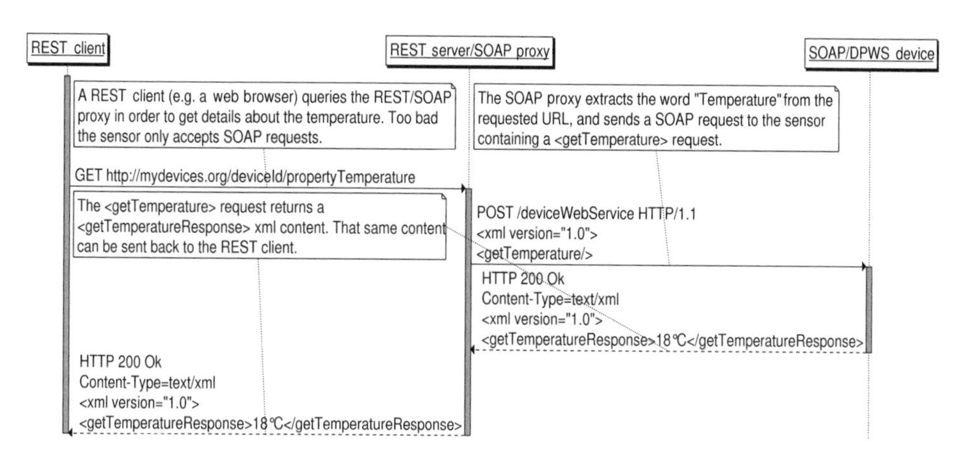

Figure 5.20 REST to SOAP conversion.

SOAP

Since REST is an architectural style, it does not impose HTTP as a protocol. If the underlying protocols can be re-adapted into resource-style manipulation mechanics, it is then possible to use them instead of HTTP. This is the case of RESTful web services, for instance. In this case, a device can expose a SOAP[3] interface containing appropriate GetXXX and SetXXX methods that can be used to manipulate its internal state as a resource (see Figure 5.20).

A similar approach enables gateways to support multiple protocols without much re-adaptation. In fact, it now becomes possible to convert GetXXX methods into HTTP GET requests.

What happens in practice is that a SOAP request can be easily synthesized using the last part of the URI, which indicates the part of the state of the resource we are interested in. The SOAP response could be passed as is in the HTTP response as a piece of XML.

This mechanism could be employed when exposing devices profile for web services (DPWS) devices and back-end web services as REST resources.

However, it should be noted that there is a fundamental difference between REST-based and SOAP-based systems.

In fact, REST principles dictate that the interface through which the system is manipulated must be *uniform*, meaning that the same HTTP methods described above are the same for each and every REST resource.

This uniformity is not present in SOAP and its related standards, since it is very possible for different web services (and for different SOAP-exposed devices) to exhibit interfaces that are clearly different.

In order to bridge the gap between different SOAP systems, it is necessary to adopt complex and heavy frameworks, such as J2EE,[4] where it is possible to generate

[3] SOAP (simple object access protocol) is a simple XML-based protocol that lets applications exchange information over HTTP using RPCs.

[4] J2EE (Java 2 enterprise edition) is a Java specification that is targeted at enterprise applications.

programming language binding for SOAP interfaces at the cost of extra programming whenever a new interface is added to the system.

This means that when the number of involved interfaces increases – which will happen in M2M as every different device type will provide its own interface – the complexity and development costs of the system will similarly increase.

In particular, it becomes unfeasible to scale the system up to serve new devices when they appear on the market, since their interface needs to be supported.

Consider the following example of a thermometer that reads the temperature of a room. Accessing it via SOAP requires the following steps:

1. The thermometer vendor should publish (usually in an UDDI[5] repository) the SOAP interface of the thermometer.
2. An application developer that wishes to include the thermometer in its application should download the interface description (usually a web service description language (WSDL) file) and compile it using a WSDL-enabled technology, for example, J2EE. This will create a dependency between the application and the technology used to compile the WSDL. Consequently, if different parts of an application are implemented on top of different technologies, it may be necessary to recompile the WSDL for every form of technology involved.
3. The generated code must be integrated into the application.

If, for any reason, the thermometer's interface changes at some point in the future, the entire application will need to be rebuilt.

Using REST instead makes things much simpler. In fact, reading a thermometer would always be implemented using an HTTP GET.

The thermometer client would specify the preferred MIME type (HTTP accept header), encoding (HTTP accept-encoding header), and charset (HTTP accept-charset header) and the M2M server would have to convert the data provided by the thermometer into the requested one.

This is much less problematic since most MIME types, encodings, and character sets are known and the code that implements the conversion can be reused across different applications.

For instance, the thermometer client may ask the M2M server to return the temperature sample as an audio/mp3 MIME type, whereas the thermometer returns its samples as plain text.

5.7 ETSI TC M2M Resource-Based M2M Communication and Procedures

5.7.1 *Introduction*

This section provides a more in-depth look into how the RESTful style is applied to the ETSI TC M2M architecture. In ETSI TC M2M, it is assumed that all three normative reference points defined in the M2M architecture, namely mIa, mId, and dIa, will use

[5] UDDI (universal description discovery and integration) is a specification for a directory service based on XML.

REST, although there may be some exceptions. For instance, on the mId reference point, device management is performed using existing protocols such as BBF TR069 and OMA DM that are RPC-based.

ETSI TC M2M did not assume that the HTTP protocol would be the protocol to be used for implementation, although HTTP is the natural choice until CoAP (more adapted for constrained devices) becomes widely deployed. To avoid the use of specific HTTP primitives, this section will mostly use the four primitives:

- **Create**: create a resource.
- **Retrieve**: read the content of the resource.
- **Update**: write the content of the resource.
- **Delete**: delete the resource.

These methods are referred to as the CRUD methods below. In addition to these basic methods, it is often also useful to define methods for subscribing to a change of a resource (S) and a notification about a change of a resource (N) included in a more general RESTful architecture. It is assumed in what follows that CRUD and SN are applicable to the resources used in the SCL (service capability layer).

Figure 5.21 (source ETSI TC M2M TS 102 690) provides an example of how data is exchanged between a device application and a network application using REST.

In Figure 5.21, Step 1 shows how a DA can request that the particular data be stored under the NSCL (network service capability layer). Since the DA does not speak directly

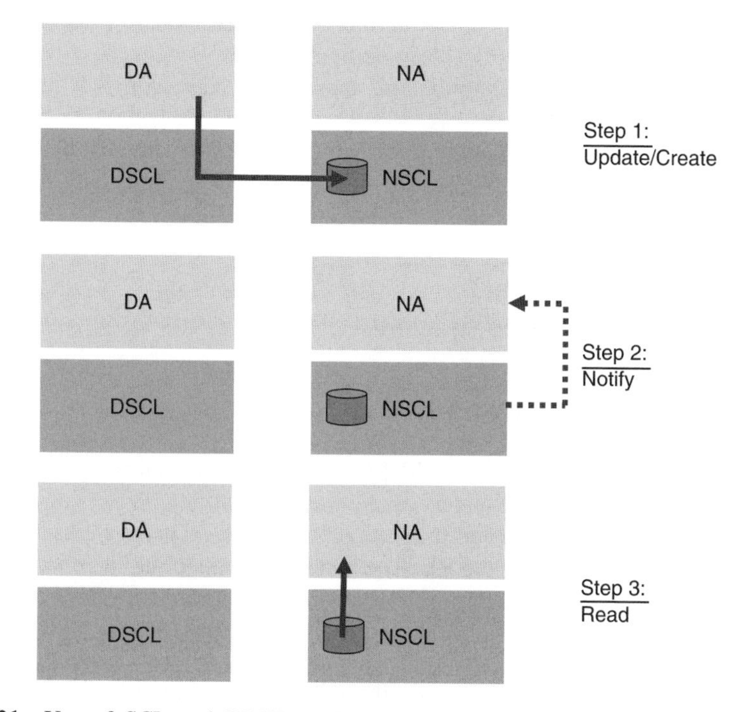

Figure 5.21 Use of SCL and REST architectural style for data exchange. Source: ETSI TS 102 690.

with the NSCL, storing this data under the NSCL is facilitated by the local SCL – the DSCL in this case. The DA can either use an update or a create primitive to this end.

In Step 2, which is optional, the NSCL notifies the NA that some data is available. This step assumes that the NA has already subscribed to be notified when data matching a certain criterion becomes available. Step 3 consists of reading the data received by the NA.

5.7.2 Definitions Used in this Section

Below are the definitions used in the remainder of this section. These definitions are aligned with the ones provided in ETSI TS 102 690.

- **Issuer**: the actor performing a request. An issuer can either be an application or an SCL.
- **Hosting SCL**: the SCL where the addressed (master/original) resource resides.
- **Local SCL**: the SCL to which an application or a SCL will register. It is the first SCL that receives the request from the original issuer of the request (either an application or an SCL).
- **Announced resource**: the content of this resource refers to a resource hosted by the hosting SCL (master/original resource). The announced resource is a resource which consists of only a limited set of attributes (generally stored information about the resource itself, such as lastModifiedTime), the most important being the search strings and the link to the original resource. The purpose of this resource is to facilitate a discovery of the original resource, so that the issuer of the discovery does not have to contact all the SCLs in order to locate specific resources.
- **Announced-to SCL**: an SCL that contains the announced resource. A resource could be announced to multiple SCLs.
- **Receiver**: the actor that receives a request from an issuer. A receiver is an SCL or an application.

5.7.3 Resource Structure

5.7.3.1 Why Define a Resource Structure?

M2M service capabilities aim to provide data mediation functions. ETSI TC M2M in TS 102 690 defined a resource structure with a tree representation as a means to:

- provide M2M applications, using M2M SCs for data-mediation functions, which is a meaningful way for addressing resources. However, great flexibility is offered in terms of where new resources are created (in particular for data exchanges). As such, the data structure is not meant to break the REST principle of decoupling the client and server, but rather to provide a minimal approach that simplifies addressing (URI construction), and also to facilitate maintenance and data-life-cycle-related actions. Take, for example, the case where DA data is stored on the network SCL, when the device is no longer registered. If all of its related data is easily identifiable, it is much easier to remove it from the network SCL.

- describe how the different types of resources relate to each other.
- improve the overall system performance through the use of minimally structured data. Assume, for instance, that there is a requirement to discover all applications running on a particular device – it is much easier to use a URI where there is a representation of all applications running on a given device.

5.7.3.2 Resource Types Used for the Resource Structure

Figure 5.22 provides an overview of the resource tree structure specified in TS 102 690. In order to provide an overview of this resource structure, the next section will provide the list of resource types used in the resource structure. A particular resource type is often used for different resources that are part of the resource tree structure.

- **sclBase** is the root for all other resources hosted by the M2M SCL. The sclBase resource is addressed by an absolute URI. All other resources under the sclBase will have a URI that is hierarchically derived from the URI of the sclBase. For example, a specific sclBase resource identifying a network SCL could be addressed through the URI: <protocol>://m2m.operator.com/. An example of a URI of a container resource (specific resource used for data exchange) hosted by the same SCL could be: <protocol>://m2m.operator.com /containers/meterDataSamples/.
- **scl resources** represent M2M SCLs managed on other entities. For an sclBase corresponding to an NSCL, scl resources allow the storage of information on DSCL and GSCL registered to interact with the NSCL. Registration – and therefore the existence of an appropriate scl resource – is a precondition for the SCLs to exchange data. Registered scl resources will contain other resources, for example, providing a partial representation of applications that are registered on the remote scl.
- **application resources** store information about the applications. Application resources are created as a result of successful registration of an application. Applications register to their local SCL only, but are known to the sclBase where the local SCL has registered.
- **container resources** are generic resources used to exchange data between applications and/or M2M SCLs by using the M2M SCL as a mediator that takes care of buffering the data.
- **accessRight resources** store a representation of permissions that govern which entity is entitled to access which part of the resource tree structure. accessRight resources are linked to other resources under the SCL.
- **group resources** store information about a group of other resources. The advantage of a group resource is that it allows an issuer, such as an application, to write once when it wants to send data to a set of receivers, thus simplifying the interactions on the API interfaces. The SCL may in this case need to replicate the data on the resources of the group members.
- **subscription resources** keep track of status of active subscriptions to its parent resource. Subscriptions allow subscribers to be notified when a particular event happens, for example, an update performed on a resource.
- **collection resources** are used in ETSI M2M to allow the grouping together of resources with certain commonalities. Grouping resources in collections provides a means for an

issuer to refer to a group of resources using a single URI. The ETSI M2M resource tree structure uses collections at different places such as scls, applications, or containers.

- **announced resource** – a particular resource that contains a limited set of attributes such as a searchString and a link (URI) to the original resource. Typically, announced resources are permitted to have a partial representation of a resource in a remote SCL so as to allow discovery (in particular making use of the searchString attribute) without having to query the hosting SCLs. An announced resource will only be visible when it is directly accessed via its full URI.
- **discovery resource** – a specific resource used to allow discovery. It is used to retrieve the list of URIs of resources matching a discovery filter criterion.

5.7.3.3 ETSI M2M Tree Structure Model

Figure 5.22 provides a tree structure for resources: a logical grouping of resources that allows simple addressing of resources, flexibility in exchanging application data, and simple APIs. The following notations are used:

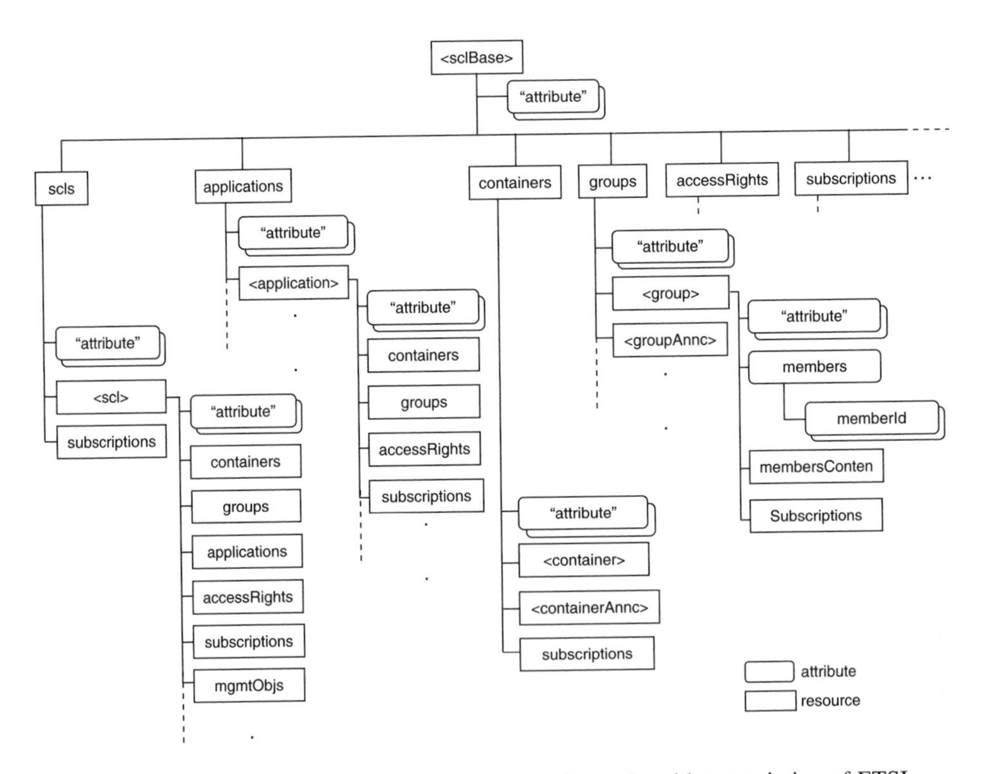

Figure 5.22 ETSI TC M2M resource structure. Reproduced by permission of ETSI.

- The notation <resourceName> means a placeholder for the name of a resource of a certain type. The actual name of the resource is not predetermined. In practice, it can either be dynamically allocated by the SCL or chosen by the issuer during the creation of the resource. Take, for instance, the example of the <application> resource under the applications collection resource, where its actual value could be: "utility1smartmeterApplication." In this example, the URI for such an application according to the resource structure in Figure 5.22 could be: <protocol>://m2m.operator.com/applications/utility1smartmeterApplication/ assuming the <sclBase> is <protocol>://m2m.operator.com/, where <protocol> is the actual protocol used to access the resource. ETSI M2M R1 has chosen HTTP and CoAP as the two possible protocols to be used for implementation.
- When the delimiters <and> or ".." are not used, the names appearing in square boxes are literals for fixed-resource names.
- Square boxes are used for resources and sub-resources.
- Rounded boxes are used for attributes. Attributes store information about the resource itself, such as lastModifiedTime. Attributes can be mandatory or optional and have read/write permissions. Attributes are addressed in the exact same manner as resources. For example, if an issuer wants to read information about when an sclBase was last accessed, it will need to issue a Read primitive using the URI: <protocol>://m2m.operator.com/lastModifiedTime.
- All collection resources are always plural (name ending with "s"). The cardinality between a collection resource and its child, delimited with <and>, is always plural.

Figure 5.22 provides a partial representation of the tree structure upon which the following observations can be made:

- **scls resource** is a collection of resources. For each registered SCL, an <scl> child resource is created to keep track of SCLs with which the local SCL has a relationship. The other child resource of scls is subscriptions, which is a collection providing a representation of subscriptions (e.g., issuer, delay tolerance, etc.) to the resource scls. <scl> contains multiple sub-resources such as applications which are a collection of all applications that are local to the subscribed SCL. <scl> also contains a containers collection, which is really the basic resource where applications can create containers to exchange data. Containers can also be created under <application>. The decision to use one or the other containers resource is a choice of the applications themselves, depending on their business logic but also the lifetime of the data to be stored under the containers resource.
  In order to simplify the big picture to the reader, here is an example of how the scls resources are used. Assume that a utility company back-end metering application uses the sclBase:<protocol>://m2m.operator.com/utility1/. An <scl> instance will be created under the scls collection for each registered meter. This instance will be addressed through the use of, for example, the URI: <protocol>://m2m.operator.com/utility1/scls/meter123456. Assume that a smart meter runs two applications – one for reporting

smart metering data and the other for home automation. For each application, an <application> resource instance will be created (the resource being an announced resource containing a limited representation of the original resource), and these can be addressed via: <protocol>://m2m.operator.com/utility1/scls/meter12345/applications/meteringApp12345/ and <protocol>://m2m.operator.com/utility1/scls/meter12345/applications/homeAutomationApp12345/. Assuming that the smart metering application running on the smart meter is configured to report the meter data under the NSCL, one way of reporting this data is to store it under a containers collection of the <application> announced resource: <protocol>://m2m.operator.com/utility1/scls/meter12345/meteringApp/applications/meteringApp12345/containers/meterSamples/ where an hourly value can be stored under: <protocol>://m2m.operator.com/utility1/scls/meter12345/meteringApp/applications/meteringApp/containers/meterSamples/contentInstances/Day1H1/.

- **applications collection resource** is used to store resources pertaining to locally registered applications (NA for NSCL, DA for DSCL or GSCL and GA for GSCL).
- **accessRights collection resource** allows <accessRight> instances to be grouped under a single resource. An <accessRight> resource allows access to store permissions, that is, "allowed" entities for certain access modes (permissionFlags: read only, read and write, write once, etc.). The resource structure for an accessRight is represented in Figure 5.23 where permission allows rights to be stored for the entities that can access the resource linking to this <accessRight> while selfPermission stores the rights for the entities that are allowed to change the accessRight resource itself. As an example, building on top of the previous metering example, assume the accessRight resource: <protocol>://m2m.operator.com/utility1/accessRights/smart metersRight/. It is possible to configure this accessRight to allow only the metering back-end application to access meter data. All meter application resources will need to link back to this accessRight resource in order to give only the utility back-end application access to this data.

Table 5.1 provides a list of some commonly used attributes, source ETSI TC M2M TS 102 690.

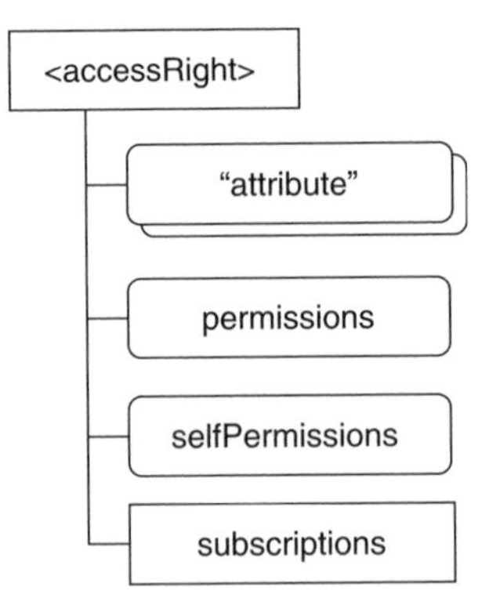

Figure 5.23 <accessRight> resource structure.

Table 5.1 M2M common attributes

Name	Description
accessRightID	URI of an access rights resource. The permissions defined in the accessRight resource that is referenced determine who is allowed to access the resource containing this attribute for a specific purpose (retrieve, update, delete, etc.)
creationTime	Time of creation of the resource
expirationTime	Absolute time after which the resource will be deleted by the hosting SCL
lastModifiedTime	Last modification time of a resource
searchStrings	Key words used for discovering resources

5.7.4 Interface Procedures

Interface procedures define the set of primitives and interactions on the mIa, dIa, and mId interfaces that allow resource manipulation. Ultimately, the objective of the interface procedures is to allow exchange of data between applications within the M2M system. However, data exchange can only take place when a certain number steps have been accomplished:

- **SCLs discovery** defines the procedures that allow a DSCL or GSCL to discover an NSCL so as to perform registration. SCL discovery procedures have not been defined at the time of the publication of this book. Initial deployments of the ETSI M2M architecture may rely on configuration.
- **SCLs registration** defines the procedures that allow a GSCL or a DSCL to register to a NSCL once discovery has been performed. SCL registration is a necessary procedure to allow the SCL to start resource management procedures. SCL registration assumes that SCLs have performed appropriate authentication and authorization mechanisms, as described in Chapter 8.
- **Applications registration** defines the set of procedures that allows an application (NA, DA, or GA) to register to its local SCL. Application registration is the necessary step for an application to become known and to start exchanging data using the resource-based procedures.

Once these procedures have been performed, SCLs can perform any of the following procedures upon the request of applications.

- **access-rights management** consists of manipulating (creation, deletion, update, retrieve) resources pertaining to access rights.
- **container management** consists of procedures that allow the exchange of application data through the use of specific resources, referred to as container resources.
- **group management** consists of manipulating groups of resources. Group resources allow for smoother interaction between applications and SCLs and among SCLs.
- **resource discovery** allows the discovery of resources stored on a specific SCL or resources announced on that SCL through the use of filter criteria. The filter

criteria may consist of a combination of attributes such as the creationTime and searchString.

- **collection management** defines the set of procedures to manage collection resources.
- **subscription management** defines the set of procedures allowing an application or an SCL to subscribe and be notified when specific subscription criteria are matched.
- **resource announce/de-announce** defines the set of procedure to allow resource to be announced and de-announced towards a remote SCL.

The interface procedures are better explained through the use of a concrete example. However, this example will not provide all possible procedures defined in TS 102 690.

5.7.4.1 Resource Management through an Example: Smart Metering

In this example, the following assumptions are made:

- The NA is a smart metering application that has already registered with the NSCL operated by an M2M service provider. The <sclBase>, as agreed between the two entities, is: <protocol>://smartmetering.utility1.com/.
- A particular smart meter has been installed and is operational. The smart metering device runs a metering application that will generate meter data measurements on an hourly basis. The smart meter device runs a DSCL that is configured to have the <sclBase>: <protocol>://meter12345.utility1.com/.

Figure 5.24 provides a graphical representation of the NSCL and DSCL along with their corresponding resource structure.

When the smart meter becomes operational, it will perform the following functions:

- **network bootstrap and network registration**: These procedures depend on the access network. Network bootstrap consists of configuring a device with all the necessary parameters to allow it to connect and register to a network. An example of network bootstrap includes bootstrap from universal integrated circuit card (UICC – often referred to as the SIM card). Once bootstrap has been performed, the device will have the necessary credentials to authenticate and to obtain authorization to access the network.
- **service bootstrap**: This involves the provisioning of permanent M2M service credentials (identities, security keys, etc.), which will be used for connecting and registering with the M2M service layer.

5.7.4.2 NA Subscribes for Registering Smart Meters

Assume the NA wants to be informed whenever a smart meter is deployed. One possible way of doing this is to subscribe to the NSCL to get notifications when DSCLs register under the scls resources of its <sclBase>. In the ETSI M2M architecture, subscriptions are performed via the creation of a specific resource, that is, a subscription resource. Since the NA wants to monitor the creation of an <scl> instance under <protocol>://smartmetering.utility1.com/scls/ the subscription resource will need to be

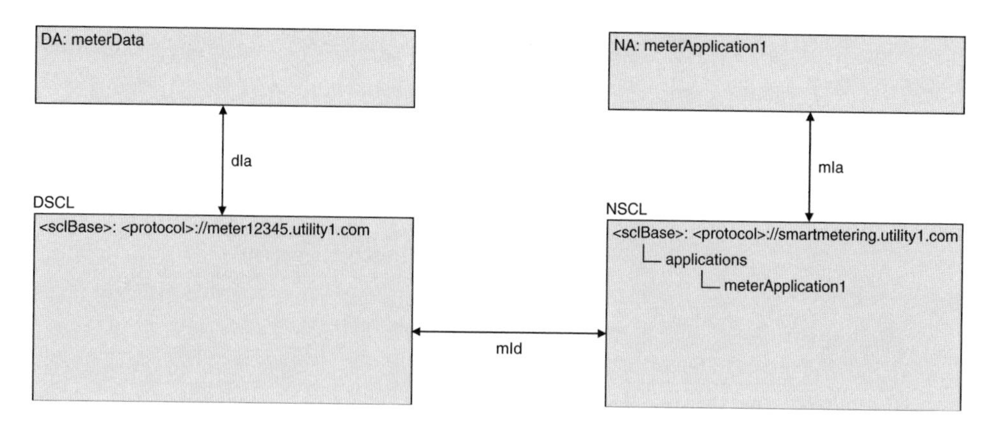

Figure 5.24 Graphical representation of the initial resource structure.

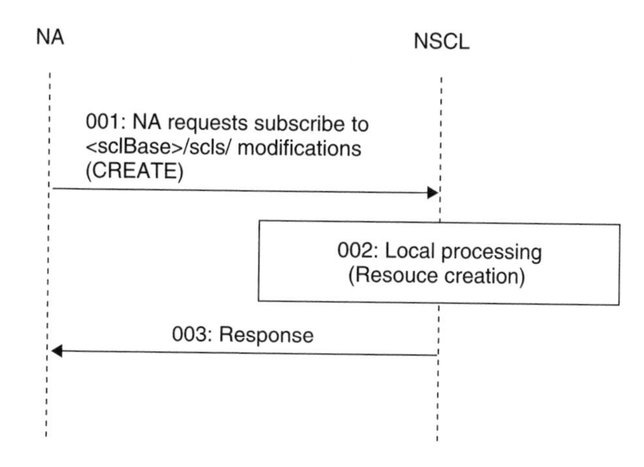

Figure 5.25 Subscription to the scls resource.

created under the scls subscriptions collection resource. Figure 5.25 provides the procedure flow for a subscription.

The details of the message flows are provided below:

- **001**: The NA requests the creation of a subscription instance under the resource <protocol>://smartmetering.utility1.com/scls/subscriptions/. The NA provides the name of the instance (not mandated in the standard): newMeters. The request may also provide other parameters, such as delay tolerance, which provides an indication of how quickly the NA will be notified.
- **002**: The NSCL validates the request and creates the subscription.
- **003**: The NSCL confirms the resource creation to the NA.

Notification will not take place until a smart meter registers to the NSCL, which will be translated into a DSCL registration procedure, as described in the next section.

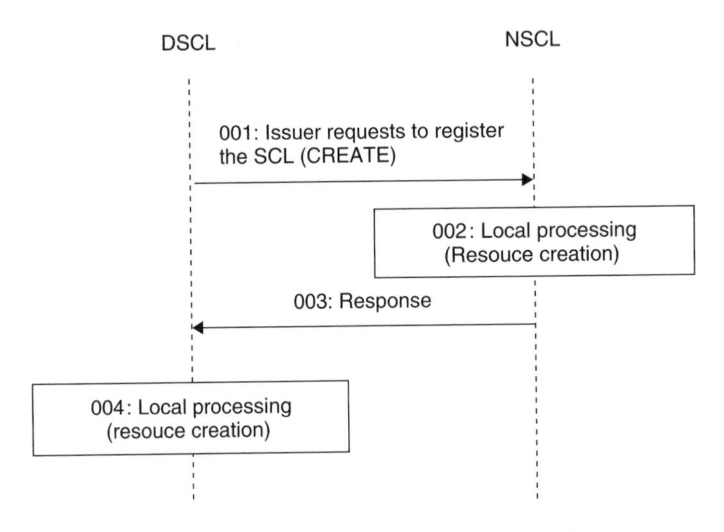

Figure 5.26 Registration procedures example.

5.7.4.3 The Smart Meter Registers to the NSCL

Once the bootstrap has been performed, the DSCL can perform NSCL discovery and SCL registration procedures. For the sake of simplicity, this example assumes that the necessary information, for example, FQDN,[6] port number, etc., that is needed to reach the NSCL is configured on the smart meter device. This means that SCL registration can immediately take place. Figure 5.26 provides the message flow for SCL registration.

The following steps are performed:

- **001**: The DSCL requests registration through the creation of a resource under the <protocol>://smartmetering.utility1.com/scls/ collection with the resource identifier meter12345, which is the unique identifier of the SCL of the smart meter.
- **002**: The NSCL ensures that the DSCL has properly authenticated and that the scl resource name, typically a configured identifier of the DSCL, does not already exist in the <sclBase>/scls collection. An scl resource is then created with the resource name <protocol>://smartmetering.utility1.com/scls/meter12345. Default attributes, such as the expiration time, are populated in the resource.
- **003**: The hosting SCL (NSCL in this case) responds positively to the request.
- **004**: The DSCL creates a resource representing the NSCL, which does not need to explicitly register to the DSCL, since only the DSCL and GSCL are requested to explicitly register to the NSCL.

[6] FQDN (fully qualified domain name) is a unique name that can be resolved into a network address using the DNS.

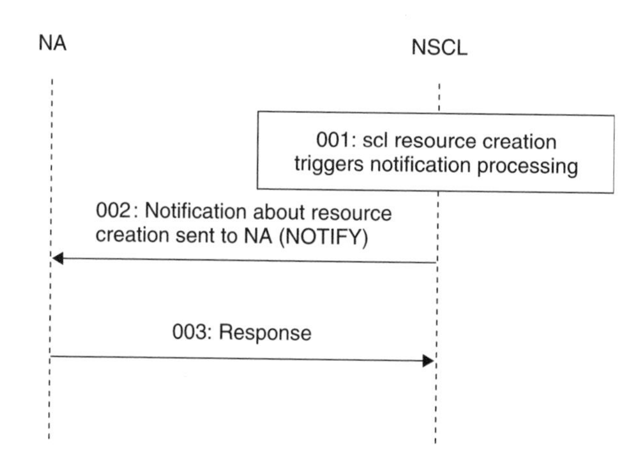

Figure 5.27 Notification procedures example.

5.7.4.4 Notifying the Network Application about a Registered Smart Meter

Once the DSCL registers to the NSCL, a natural procedure to be triggered by the NSCL will be to notify the NA about the registration. The corresponding message flow is provided in Figure 5.27.

The steps of the notification message flow are as follows:

- **001**: An <scl> resource instance is created under the scls collection resource. This corresponds to the registration of a smart meter.
- **002**: The NSCL notifies the NA about the registration of a new SCL, corresponding to the deployment of a new smart meter.
- **003**: The NA acknowledges the receipt of the notification.

5.7.4.5 Device Application Registration to the DSCL

The next step would be for the DA to register locally to the DSCL using the dIa interface. Application registration allows for the application resource to be created on the local SCL (see Figure 5.28). It is also a necessary condition to allow an application to manipulate resources on its local SCL or remote SCL in accordance with the access rights. Because DSCL has registered to the NSCL, there is no need for the DA to register to the NSCL.

- **001**: The DA requests registration through the creation of a resource under the <protocol>://meter12345.utility1.com/applications/ collection with the identifier meterData for the applications collection instance, addressable through the link <protocol>://meter12345.utility1.com/applications/meterData.
- **002**: The DSCL checks if the DA is authorized to create the resource for registration. Once these checks have been performed, the resource is created and the DSCL updates the relevant attributes not provided by the DA. The DSCL can also provide default values for optional attributes not provided by the DA – these are inferred from the SCL

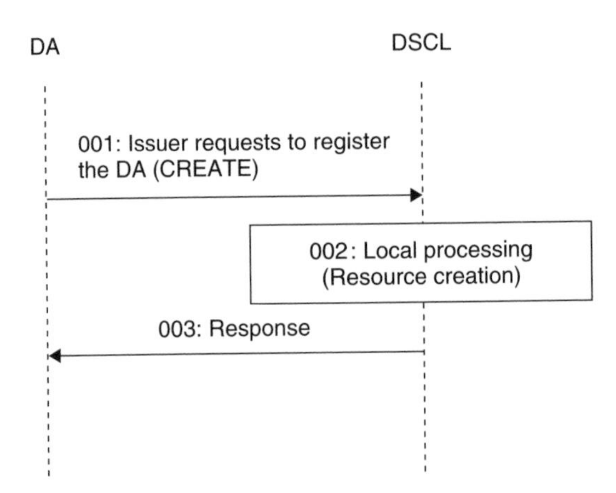

Figure 5.28 Application registration procedures example.

policies. For example:
  - the DSCL may reduce the expiration time suggested by the issuer.
  - if the applicationStatus has not been provided, the hosting SCL will set the applicationStatus attribute to "ONLINE".
- **003** The DA is informed about the successful registration.

5.7.4.6 Announcing a Registered DA to the NSCL

Assume now that the DA wants to announce its registration to the NSCL so that the NA can discover all the smart meter applications that have become operational. In this case, the DA will need to update a particular attribute, announceTo, of the DA resource, so as to request the announcement to the NSCL. The details of the corresponding message flow (see Figure 5.29) are:

- **001**: The DA requests to announce its DA resource; this is done through the use of the update primitive to set a specific value of the announceTo attribute.
- **002**: The DSCL confirms the validity of the request and then updates the announceTo attribute of the DA.
- **003**: A generic response is returned to the DA, confirming that the request to announce is acceptable to the DSCL, but it does not confirm that the actual announcement has taken place.
- **004**: The DSCL initiates the procedures to create an announcement resource on the NSCL.
- **005**: The DSCL requests the creation of a new sub-resource under the resource <protocol>://smartmetering.utility1.com/scls/meter12345/applications/. This sub-resource will be of type <applicationAnnc> that has the structure shown in Figure 5.30.
- **006**: The NSCL validates the received request and then creates an announced resource with the specified attributes.
- **007**: The NSCL returns a response indicating whether the creation was successful.

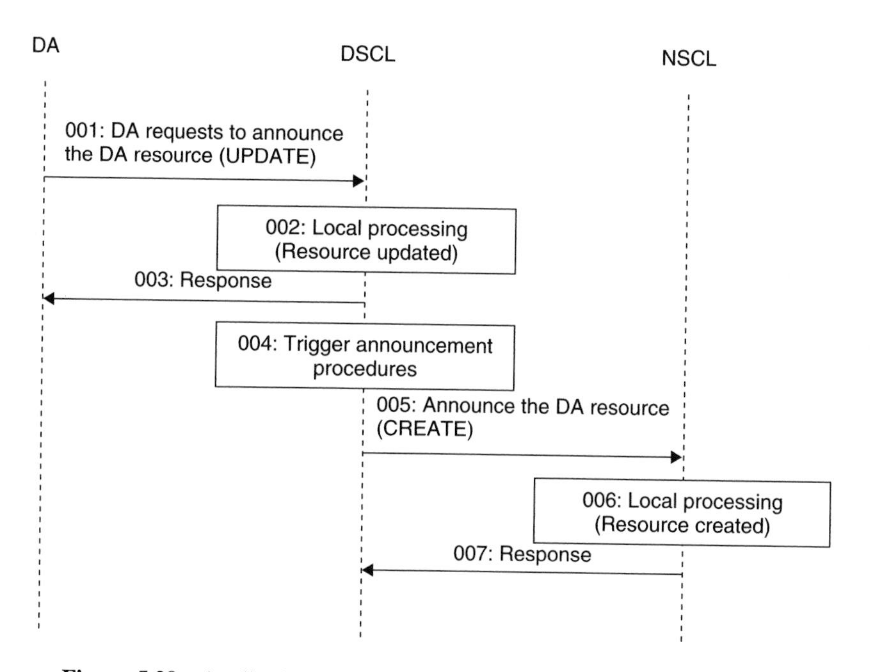

Figure 5.29 Application registration announcement procedures example.

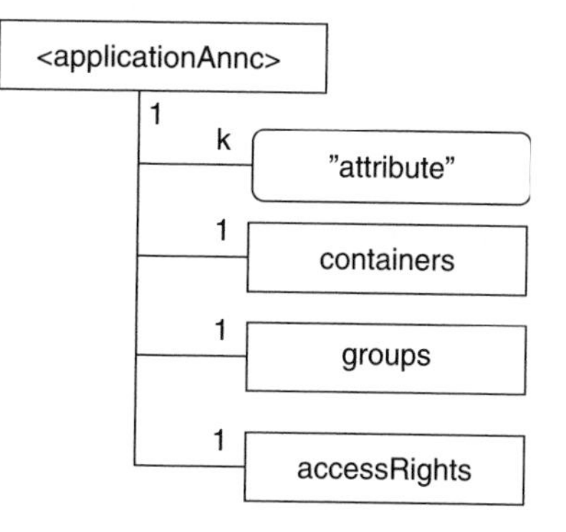

Figure 5.30 ApplicationAnnc resource structure.

Figure 5.31 provides a graphical representation of the NSCL and DSCL, once the DA has registered locally to the DSCL. This figure also takes into account the DSCL registration to the NSCL.

5.7.4.7 Reporting Meter Data through the Use of Container Resources

Assume now that the DA is programmed to report the application data. The ETSI 102 690 specification allows for multiple options to do this reporting. Exchanging application data is done through the use of the container collection resources. As shown in Figure 5.22,

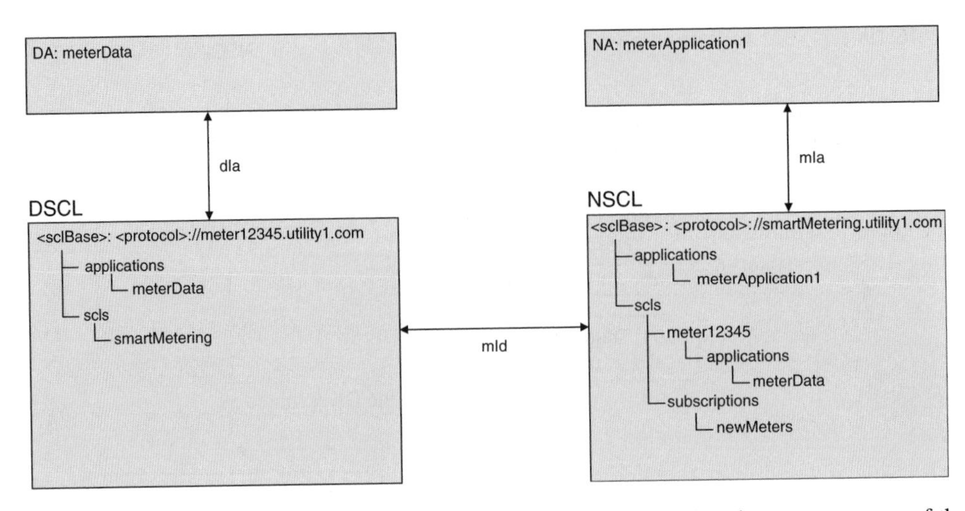

Figure 5.31 Graphical representation of the resource structure following the announcement of the local DA registration.

there are multiple resources where container sub-resources can be created, such as under <sclBase>, <scl>, or <application>. For our particular example, the logical choice is to create a container collection resource, either under the DSCL DA application resource <protocol>://meter12345.utility1.com/applications/meterData/containers/meterDataSamp les) or under the DSCL DA announced resource <protocol>://smartmetering.utility1.com/ scls/meter12345/applications/meterData/containers/meterDataSamples) where meter-DataSamples is the container collection created to report the meter data samples as sub-resources. The final choice needs to be made by the application developer based on its business needs, but also on several other parameters such as charging and network usage. For the sake of our example, it will be assumed that the actual meter samples are reported under the NSCL-announced DA resource. Figure 5.32 provides the message flow to initially create the containers collection.

- **001**: The DA requests the creation of a container resource under the NSCL resource <protocol>://smartmetering.utility1.com/scls/meter12345/applications/meterData/con tainers/. The parameters of this request contain the name of the container collection: meterDataSamples.
- **002**: The DSCL recognizes that this request pertains to a resource creation under the NSCL based on the URI of the parent resource. The DSCL performs the necessary request validation, the checking optionally including the validation of the access rights.
- **003**: The request is forwarded to the NSCL.
- **004**: The NSCL checks the validity of the request and validates that the issuing entity has the necessary access rights to create the resource. Then the NSCL creates the container resource and the DA announced resource. The resulting collection resource becomes addressable via <protocol>://smartmetering.utility1.com/scls/meter12345/ applications/meterData/containers/meterDataSamples.
- **005**: A response is sent back to the DSCL.

- **006**: The DSCL forwards the answer in 005 to the DA to confirm the creation of the resource.

The procedure used to create a container instance will be exactly the same as the one just described. Since the creation of the container instance will be performed under the collection, the Create request in Step 001 will need to use the resource location <protocol>://smartmetering.utility1.com/scls/meter12345/applications/meterData/containers/meterDataSamples/contentInstances.

5.8 Chapter Conclusion

This chapter outlines the ETSI M2M resource-based architecture. A description of the three most important capabilities for the ETSI TC M2M Release 1 was provided along with the resource structure and an overview of the most important procedures through an example.

ETSI TC M2M specifications represent an important step forward for providing the foundation standards for a horizontal M2M service platform. While the initial release mostly tackled data mediation, security, and device management, it is expected that future releases will go the extra mile in standardizing the other service capabilities.

At the time of writing, ETSI M2M specifications are still evolving, so some details may have slightly changed. The readers of this chapter should read it with a system/architecture approach in mind and refer to the final version of the specification for more details. This chapter focused on stage 2 aspects (architecture and message flows) and intentionally did

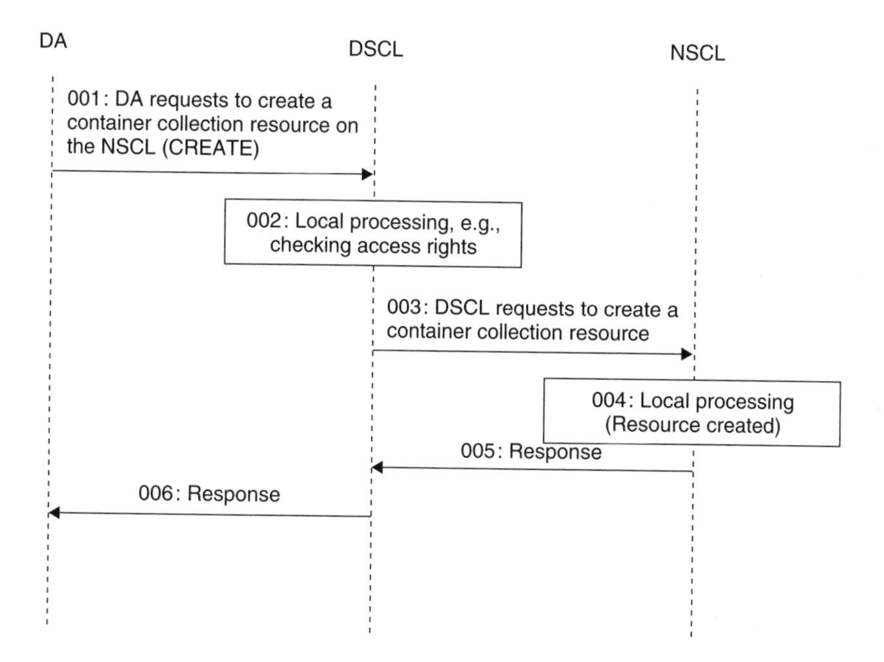

Figure 5.32 Container resource creation procedures example.

not handle protocol examples. Some protocol examples are provided in the companion book: “The Internet of Things, Key Applications and protocols”.

References

[3GPP TS 23.060] 3GPP TS 23.060. General Packet Radio Service (GPRS). Service description. Stage 2 (2011).

[3GPP TS 23.401] 3GPP TS 23.401. General Packet Radio Service (GPRS) enhancements for Evolved Universal Terrestrial Radio Access Network (E-UTRAN) Access (2011).

[BBF TR069] Broadband Forum TR-069. Amendment 3 CPE WAN Management Protocol (2010).

[ETSI TS 102 690] ETSI TS 102 690: Machine-to-Machine communications (M2M); Functional architecture; v1.1.1, 25 Oct 2011.

[G.hn] ITU-T Recommendation G.9960. Next generation Home Networking Transceivers Foundation (Copper-pair, Powerline and coax PHY Layer) (2010).

[OMA DM] OMA-AD-DM-V1_3. Device Management Architecture, Version 1.3 (2010).

[Yankee Group] Yankee Group (2010) A Closer Look at M2M Carrier Strategy, Market Research Report.

6

M2M Optimizations in Public Mobile Networks

Toon Norp[1] and Bruno Landais[2]
[1]*TNO, Delft, Netherlands*
[2]*Alcatel-Lucent, Lannion, France*

6.1 Chapter Overview

Many M2M applications use public telecommunications networks to transfer data from M2M devices to an M2M server. These telecommunications networks will have to be adapted to cope with the traffic generated by the projected growth of M2M applications. In the near future, many more devices will be connected to the existing networks, as more and more M2M applications are introduced. These M2M devices will have a very different effect on the telecommunications network compared to human-oriented devices. This chapter introduces the effects that M2M has on public telecommunications infrastructures and describes some of the methods that network operators will use to optimize their networks for M2M communications.

The telecommunications networks will provide the data communications and associated value-added services to the M2M applications. Note that, with M2M service-level offerings, telecommunications operators seek to provide more than simple M2M data communications. Some operators even provide complete home-security services on the basis of M2M. However, in this chapter we will limit ourselves to the data communications part of M2M communications.

M2M Communications: A Systems Approach, First Edition.
Edited by David Boswarthick, Omar Elloumi and Olivier Hersent.

The focus in this chapter is on M2M in mobile networks. For M2M application owners, mobile networks are often more suitable as they offer better possibilities for controlling the subscription and have end-to-end visibility of the data connection. From an operator point of view, mobile networks create more challenges whereby optimizations are needed to cater for large volumes of M2M traffic. Nevertheless, a lot of the issues, problems, and solutions discussed are generic and will also apply to fixed networks.

This chapter is intended to show what network improvements are needed to telecommunications networks to cater for M2M. Nevertheless, the chapter may also be useful in providing a better understanding of M2M network issues for M2M application owners, M2M application providers, and others beyond the mobile industry.

In Section 6.2, an overview is given of M2M over public telecommunication networks. In Section 6.3, the focus is on optimization of the public telecommunications network to make it better suited for M2M.

6.2 M2M Over a Telecommunications Network

6.2.1 Introduction

Many M2M applications use dedicated infrastructures for their data transport. For example, a private local wireless infrastructure is probably the simplest solution for monitoring the water levels of pot plants in a greenhouse. However, there are many M2M applications where the use of a public telecommunications network is more appropriate. For example, an eHealth application where out-patients are remotely monitored at home would have to rely on publicly available telecommunications networks.

In this chapter, we look at how M2M applications use public telecommunications networks and the requirements that M2M applications place on these networks. The impact that M2M applications have on the public telecommunications networks can be quite different from the impact of traditional telecommunications services. M2M communications predominantly consists of data communications. However, even when comparing M2M data communications with the current-day (mobile) broadband Internet services, we see many differences.

For telecommunications operators, it is important to consider how they have to prepare their networks for M2M communications. The assumption is that M2M communications will continue to grow for quite some time, possibly to a level where it would exceed traditional types of traffic. If the public telecommunications networks were optimized only for human-to-human communication and Internet access, then connecting large numbers of M2M applications would have a negative effect on the efficiency of these networks and the services delivered over them.

For M2M applications owners and M2M applications developers, it is of interest to understand the underlying public telecommunications networks over which their applications will run. A slight difference in how M2M applications organize their data communications can make a large difference in the impact that these applications have on the public telecommunications network. Designing M2M applications that are friendly to the underlying network is going to be cost-efficient for the M2M application owners, as operators are likely to take into account the impact that M2M applications have on their networks within their pricing structures.

Section 6.2.2 below will first introduce the M2M communication scenarios and where the public telecommunications networks play a role in these scenarios. Section 6.2.3 will then discuss whether to use fixed or mobile telecommunications networks for M2M communications. Finally, Section 6.2.4 will explain the kind of requirements that M2M applications have on data connections over a public telecommunications network.

6.2.2 M2M Communication Scenarios

In most M2M communication scenarios, a large number of M2M devices communicate with a central server. An example of such a device-to-server communication scenario would be an energy company that remotely collects the meter readings of all its customers. The number of devices can range from several dozen for a small trucking company that uses an M2M fleet management application to several million for a large energy company.

The device-to-server communication scenario, as shown in Figure 6.1, may not truly have a N:1 ratio of devices to servers. There may be multiple servers, for example, for load balancing or redundancy. However, the number of M2M devices is typically much larger than the number of M2M servers. Furthermore, most M2M devices do not have to be concerned about which particular M2M server they are communicating with. The M2M devices are simply configured to communicate with a server and do not have to make any form of server selection.

In a true device-to-server scenario, the public network operator provides the connectivity between the M2M devices and the M2M server. Typically, the network operator does not provide this connectivity to individual owners of M2M devices, but rather to the M2M application owners. In the basic form of a device-to-server scenario, the M2M application owner owns all M2M devices and the M2M server. Alternatively, the M2M application owner may not actually own the M2M devices, but owns the data communication subscriptions. For example, a manufacturer of vehicular navigation devices may sell navigation devices with a subscription for real-time traffic information and points of interest. Though the navigation device is sold to the end-customer, the manufacturer of the navigation device owns the mobile data subscription that is needed to transport the information to the navigation device.

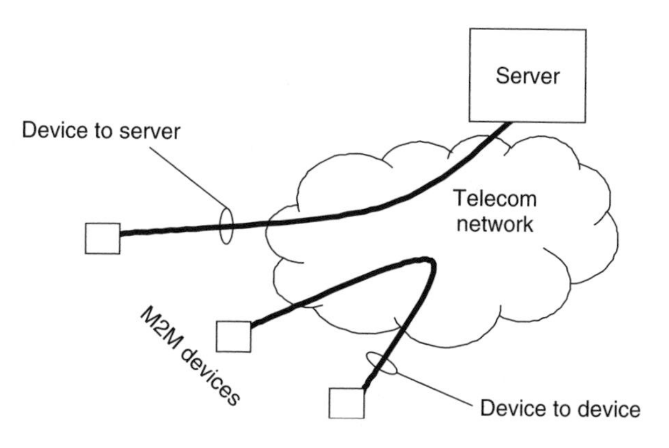

Figure 6.1 Device-to-server and device-to-device communication scenarios.

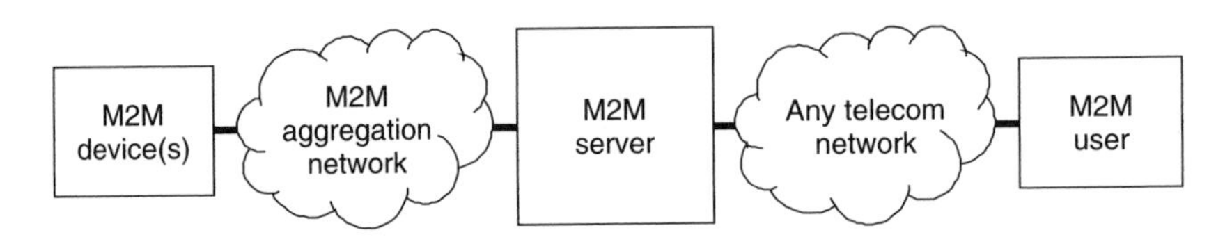

Figure 6.2 M2M communication path via M2M server.

Note that the network operator may offer more than the communication between the M2M devices and the M2M server. They may also provide service capabilities based on an actual M2M server. For the purpose of this chapter, however, we focus on the public network operator in its role of data communications provider. In that sense, the network operator may be seen as providing data communications services to its own internal M2M service provider.

Especially when the network operator provides M2M service capabilities with a network-based server, the end-to-end communication path (see Figure 6.2) will also involve an M2M user. This M2M user may, for example, be accessing metering data collected on the M2M server. The communication between the M2M user and the M2M server does not have specific M2M characteristics. Typically, an Internet-based interface is used that can run over any type of telecommunication network.

As shown in Figure 6.1, there is also a communication scenario where M2M devices communicate with each other directly – without the involvement of an M2M server. An example of such a scenario is an application where a photo camera remotely synchronizes with a media server in the customer's house. Another example could be a house alarm system that directly contacts the house owner by, for example, sending a message to his or her phone. Note that direct device-to-device communication is not the same as communication between two M2M devices via an M2M server. For example, in a multiuser game, two game consoles involved in the same game session could either both be connected to the same game server or could connect directly to each other without a server being involved. The device-to-device communication scenario is still far less common than the device-to-server scenario which constitutes the bulk of M2M communications. Nevertheless, when in the future more and more different kinds of devices become connected, the possibilities for direct device-to-device communication scenarios will increase.

With device-to-device communications, the M2M devices need to be able to select which other M2M device they want to communicate with. As such, there is an M:N connectivity. The business scenarios will also in all probability be different. There is not necessarily a single M2M application owner that owns all the M2M devices involved in the communication. In that sense, the device-to-device communication scenario is very much like the communication scenario that we all know from services such as telephony.

When there are large numbers of M2M devices in a single area, it may be beneficial to adopt an M2M gateway (GW) scenario (see Figure 6.3). In the M2M GW scenario, many M2M devices can share a single connection through the public telecommunication network. For the communication between the M2M devices and the M2M GW, a local networking technology is used (e.g., LAN, WLAN, or ZigBee).

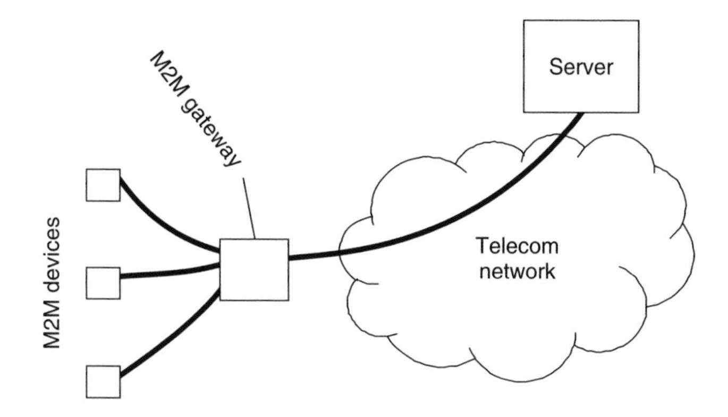

Figure 6.3 M2M gateway scenario.

6.2.3 Mobile or Fixed Networks

Broadly speaking, public telecommunications networks are either fixed or mobile telecommunications networks. Both types of networks can be used for M2M applications. Nevertheless, the focus of M2M nowadays is on mobile telecommunications networks.

There are many M2M applications for which mobile telecommunications are clearly the only option. A track-and-trace application for a trucking company will clearly not work with a fixed communications solution. However, also for those cases where the M2M device is not mobile, mobile telecommunications are often preferred. Especially for M2M devices in difficult locations, connecting to a fixed network may be costly. Examples are irrigation pumps, sluice gates, and other equipment for water management. This equipment is often located in a rural environment, far from any other buildings or infrastructure. It is therefore very expensive to get cables to these locations.

Also, in cases where M2M devices are installed in private residences or offices, mobile telecommunications are often the preferred solution. This is because the M2M application owner is often different from the owners of the fixed network subscriptions. In these cases, the M2M application owner does not want to become dependent on the owner of the fixed-network subscription.

In Figure 6.4, we consider the example of an energy company that wants to connect electricity meters. In this case, the energy company is the M2M application owner. Most of the customers (A to D) of the energy company will have some form of fixed-network connection that could be used to connect the energy meter. However, some customers use fixed network operator A and others are customers of fixed network operator B. If the energy company wants to use the existing fixed network connections, it will have to adapt its technical solutions to whatever fixed network operator the customer chooses. Cases where the customer does not have a fixed-network connection (or has forgotten to pay the bill) or where an electricity outage also disables the fixed network residential gateway further complicate a solution based on a fixed network. Alternatively, the energy company deals with one mobile operator A to get mobile subscriptions, and inserts a SIM card in all electricity meters.

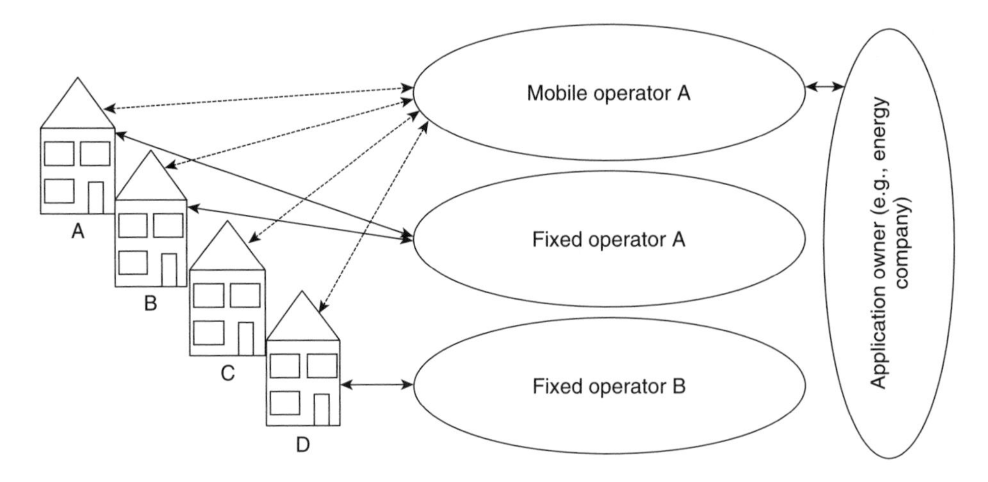

Figure 6.4 M2M via mobile networks creates independence from operators.

Using mobile networks for M2M devices that are not mobile also has its disadvantages. M2M devices may be located indoors in areas where the coverage of the mobile network is not particularly good. When using mobile networks for mobile M2M devices, bad mobile coverage in one location will even out with good coverage in another location. However, a stationary M2M device that is located in a bad coverage spot will never be able to send data.

A solution to the coverage problem is to use a roaming SIM card. When using a SIM card from an operator in a particular country, it will typically only work with its home network operator, since national roaming between operators in one country is not often supported. So in the event that the home network operator does not provide sufficient coverage in a certain location, there is no fall-back solution. However, if a SIM card from a foreign mobile network operator is used, the M2M device will first try the preferred network operator but, assuming the home network operator has the relevant roaming agreements, it can also use all the other available mobile networks. A disadvantage of using roaming SIM cards is the higher cost of data communications because of roaming charges. However, for M2M applications that send very little data, this is a minor issue. Roaming SIM cards are also used for logistical reasons; for example, a manufacturer of digital photo frames may want to avoid the logistical problems of using a dedicated SIM card for every country in which the photo frames are sold. It is generally expected that a very large percentage of mobile M2M devices will be roaming semipermanently.

6.2.4 Data Connections for M2M Applications

Different M2M applications have very different communication needs. It is clear that a surveillance camera application with a 24/7 video stream places higher demands on data communication than a vending machine application where it only needs to send small amounts of data when, for example, a particular product is out of stock. Even with similar M2M applications, the requirements of the individual applications can still be very different. Two energy companies that both wish to do introduce a smart metering application can have very different requirements on specific parameters, such as how often to record meter data.

It is not possible to list the characteristics of particular M2M applications, as there are simply too many different M2M applications with different characteristics. However, it is possible to distinguish particular aspects that can be used to characterize different M2M applications, such as:

- **data volume** – how much data does the M2M application send?
- **QoS requirements** – what kind of QoS requirements (if any) does the M2M application have for its data transport?
- **time sensitivity** – does the application need immediate data communications or can the application defer data communications, say, for a few hours?
- **communication direction** – with certain M2M applications, the M2M device always initiates the data communication, other M2M applications require the M2M server to also be able to initiate data communication.

Table 6.1 shows some examples of M2M applications with their communications needs. Note that these are only examples, as there are many more M2M applications, and similar M2M applications may have very different characteristics.

The communication direction requirements of M2M applications are important for the communication set-up in most networks. In networks that are connection-oriented, a connection first needs to be set up before data communication can take place. Many networks, such as the GSM/UMTS packet-switched network, do not support network-initiated connection set-ups.[1] In connection-less networks, there is no need for connection establishment. But in connection-less networks, firewalls and network address translations can be a bottleneck for network-initiated communications. Usually, the pinholes and translations in firewalls and network address translations are configured by device-originated communications. This becomes an issue when addressing M2M applications that need network-originated communications.

To establish a device-initiated connection at the request of the server, "triggering" is used. With triggering, the M2M server can send an indication to the device that it should establish communication with the M2M server. In current M2M applications over mobile networks, triggering is used due to the lack of a network-requested packet data protocol (PDP) context-establishment procedure. The M2M server sends a trigger to the M2M device in the form of an SMS or a circuit-switched (CS) telephony call set-up. Upon receipt of a specially formatted SMS, the device sets up a data connection (i.e., PDP context) to the M2M server. Also in the case of a CS call set-up, the device does not answer the call but sets up a data connection to the M2M server. Generally, the device is already configured with the address of the M2M server, so only trigger indication is needed.

In connection-oriented networks, it is important to consider when to set up and tear down the data connections. The signaling associated with a data connection establishment can be a significant overhead compared to the amount of data that the M2M application sends. For a device that sends data frequently, it will be more efficient to keep the data connection alive between consecutive data transmissions. The signaling associated with

[1] 3GPP standards specify a network-requested PDP context activation procedure, but this is rarely implemented.

Table 6.1 Different M2M applications have different communication needs. Reproduced by permission of TNO

	Data volume	Required quality of service	Time sensitivity	Communication direction
Surveillance cameras	High	High	High	Network-originated
Energy meters	Low	Low	Low	Network-originated
Fleet management	Low	Low	Medium	Device- and network-originated
eBook readers	High	Low	Medium	Device-originated
Media synchronization	High	Low	Medium	Device- and network-originated
Point of sales terminals	Low	Medium	High	Device-originated

the set-up and tear-down of the connection will consume more resources than keeping the connection open. But for applications that rarely send data it is more efficient to tear down the data connection in between data bursts.

For M2M in mobile networks, it is also important to consider whether to keep devices attached when they are not sending data. In many networks, the device also has to register and provide authentication to the network before communication is possible. In GSM/UMTS networks, this registration takes the form of an attach procedure. When a device is attached to the network, mobility management is performed. This implies that the network keeps track of the device location, with the granularity of location areas. The network can then set up a connection to the device by paging the device within the known location area. For M2M applications that need to be able to initiate data communications, the network needs to know where to page the device. Keeping the devices attached ensures that the network has this information. On the other hand, for a device that seldom sends data, the mobility management signaling will be a significant overhead

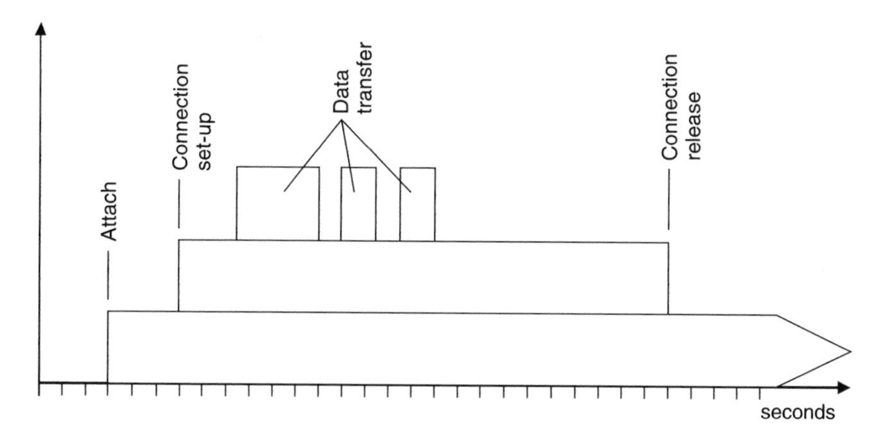

Figure 6.5 Data connection establishment in a cellular mobile network.

compared to the amount of user data sent. If there is also no need to trigger the device, then it is more efficient to detach the device when it is not communicating.

The basics of connectivity described above are important if we take a look at how different applications have different activity patterns. The activity pattern very much determines the efficiency of the data communications.

Figure 6.5 shows a typical example of the activity model of an M2M device. It could represent a vending machine that performs a status update. Firstly, the device attaches to the network before setting up the data connection. After a few bursts of data transfer, there is no more data to send and the connection is released. After the release of the connection, a detach procedure may follow (not shown).

The optimization of the data connection establishment for different M2M applications depends on their data communication pattern. Devices that rarely send data require different optimizations from devices that frequently send data. Figure 6.6 shows some examples of data communication patterns. The characteristics of the data communication patterns can differ widely between different M2M applications. Some applications send data constantly, whereas others might wait more than 15 years before sending a data burst.

The video surveillance application in Figure 6.6 will continuously generate data. In this case, it is more efficient to attach, set up, and maintain a connection. Also for remote control, it is best to maintain a continuous connection. In this case, data is sent in bursts at intervals of several seconds. It would be possible to tear down the connection in between the burst of data, but setting up and tearing down the connections consumes more resources than keeping the connection open. With a tracking and tracing application, the time between bursts of data becomes longer (e.g., once every 15 minutes). It now starts to become debatable whether it is best to keep the connection open between two bursts of data or whether to release the connection. For an e-Book reader application, the time between data bursts becomes even longer. There can be weeks between purchasing books online. There is no need to connect when not buying a book. With an eCall application, a car automatically contacts an emergency center when it is involved in a crash. In this case, it will hopefully be many years before the eCall application sends any data. It is wasteful to keep the M2M device attached for all those years, and especially to keep the connection open.

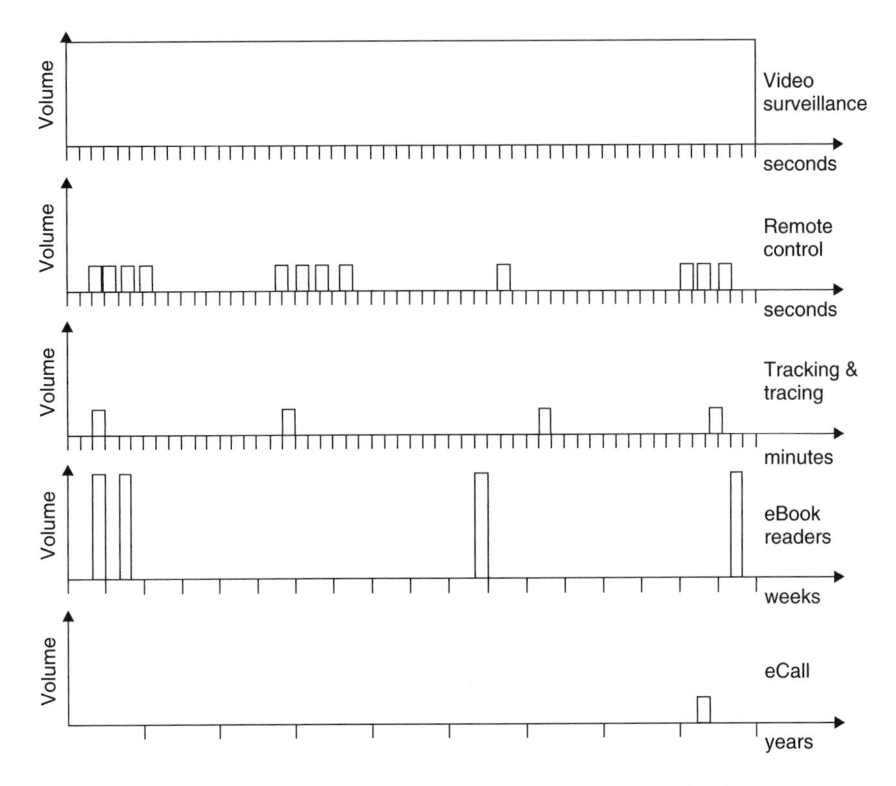

Figure 6.6 Different M2M applications with different communication patterns.

6.3 Network Optimizations for M2M

6.3.1 Introduction

Current-day public telecommunications networks are not designed for M2M applications. These networks were originally designed for telephony traffic. Even after the introduction of packet-switched communications and next-generation networks, the networks are still not optimized for M2M communications. Current-day packet-switched communication networks are largely optimized to provide broadband Internet access to end-users. However, providing M2M data communications to large numbers of M2M devices and providing services to M2M application owners rather than end-users implies a different optimization of the network. The architecture of current-day packet-switched telecommunications networks will not need a complete overhaul, but a significant number of changes will be needed to support an M2M market that really lives up to its promise.

Network optimizations for M2M are classified into five different categories:

- **cost reductions** – reducing the cost for network operators to provide M2M data communication;
- **value-added services** – enabling the network operators to provide M2M-specific value-added services;

- **trigger optimizations** – providing improved support for network-originated M2M applications;
- **overload and congestion control** – ensuring that high loads of M2M applications do not bring the network down;
- **naming and addressing** – ensuring that there will still be enough identifiers and addresses in an era where billions of M2M devices are connected.

The next section first introduces the 3GPP activities related to the standardization of network optimizations for mobile networks. The remainder of this section elaborates on the different categories of network optimizations for M2M.

6.3.2 3GPP Standardization of Network Improvements for Machine Type Communications

There are several standards development organizations working on M2M standardization. However, the mobile networks standardization organization 3GPP (Third Generation Partnership Project – www.3gpp.org) is clearly the most advanced in standardizing network optimizations for M2M.

In 3GPP, the service and requirements group 3GPP WG SA1 established a work item on network improvements for machine-type communications in 2008. The aim of this work was to define the service requirements for M2M. 3GPP refers to M2M as machine-type communications (MTC) to also take into account the possibility of a machine communicating with a human being (M2H or H2M communications).

The work in 3GPP WG SA1 has resulted in a requirements-specification document 3GPP TS 22.368 "Service Requirements for Machine-Type Communications (MTC); Stage 1" [22.368]. A distinction is made in this specification between common-service requirements that apply to all MTC devices and subscriptions and MTC features, which will only apply to specific subscriptions. The idea behind the MTC features is that operators can differentiate their offerings for particular M2M applications on a per-subscription basis. A first version of [22.368] was completed for 3GPP Release 10.

The MTC service requirements in [22.368] are the basis for architecture and protocol specification in other 3GPP groups. However, it has become clear that [22.368] contained far more requirements than the architecture and protocol specification groups in 3GPP could handle in a single release. For Release 10, only overload and congestion-control-related functionality has been fully specified. Even for Release 11, the architecture and protocol specification work has only focused on a small subset of the functionality defined in [22.368]. An indicative list of functionality for Release 10 and Release 11 is provided in Table 6.2.

Key issues and architecture solutions for network improvements for MTC are first studied in an MTC-specific technical report 3GPP TR 23.888 "System improvements for Machine-Type Communications (MTC) " [23.888] before being specified in various architecture and protocol specifications. 3GPP normative architecture specification for MTC can be found in a number of specifications (e.g., interactions with external M2M servers are specified in [23.682], MTC specifics of the packet-switched (PS) core network are specified in [23.060] and [23.401], extended access barring (EAB) is specified together with other access control mechanisms in [22.011]).

Table 6.2 Network optimization for machine-type communication features in 3GPP Release 10 and Release 11

3GPP Release 10	Extended access barring Low-access priority indications in radio-resource control Extended wait timers in radio resource control Throttling of downlink data notification requests APN-based congestion control Generic core network mobility management control Optimizations to prevent overload from PLMN reselection
3GPP Release 11	Triggering optimizations and triggering architecture Addressing optimizations and removal of dependency on telephone numbers

The 3GPP work on network improvements for MTC is likely to continue for quite some time. There are various requirements in [22.368] that will have to be implemented in releases beyond Release 11. Furthermore, 3GPP WG SA1 is still defining new requirements. In [22.888], new requirements such as device-to-device communications and gateway scenarios are being studied.

6.3.3 Cost Reduction

Many M2M applications have to be low-cost. For example, the cost of data communications for a smart meter should not significantly add to the cost of each household electricity bill. This also implies lower revenues for the network operators. Compared to the average revenues for a broadband Internet access subscription, the average M2M subscription will generate significantly less income. To make the provision of M2M communication services financially viable, the cost of providing M2M data communications will have to be reduced to a minimum.

For an operator it is always attractive to bring down the average cost per bit for data communications. But with M2M data communications, further cost reductions are possible by exploiting the specific nature of M2M applications. For example, a network operator may be able to negotiate with an M2M application owner that data communication for the M2M application is restricted to outside the busy hour of the network. For the network operator, this implies that no additional investment in network capacity is needed. For the M2M application owner (depending on the M2M application), it may be quite acceptable to be restricted to sending data once per day during off-peak hours.

For the network operator, it is important to distinguish the M2M data communications from other data communications, such as broadband Internet access. If specific restrictions for M2M data communication can lead to lower costs, then the network operator can negotiate a lower price for M2M applications that can support these restrictions. However, lowering the price for all data communications in order to reach an acceptable price level for the average M2M application owner would cannibalize the income from those other data communication users that can accept a higher price. The goal is to differentiate the M2M portfolio, with a right mix of restrictions – and value-added services – to appeal to the many different M2M applications.

In order to find opportunities for cost reductions in M2M applications, it is useful to first investigate the cost drivers and network cost components. Table 6.2 shows cost drivers for M2M in a mobile communications network. Defining the most important cost driver depends on the type of M2M application. The number of simultaneously attached devices is going to be an important cost driver in an M2M application for remote diagnostics of cars: every car will likely be monitored only a few times per year, but the M2M devices in the cars need to be attached all the time in order to be reachable. On the other hand, for a surveillance camera application in public transport buses, the number of simultaneously attached devices is not going to be an important issue. The amount of data transported with continuous video footage will be a much bigger cost driver (see Table 6.3).

Many M2M application owners have large amounts of subscriptions. For an energy company with a smart metering application, the number of subscriptions will extend into the millions. Most of these subscriptions are basically the same. For example, the service-profile options for the QoS profile will not differ between the subscriptions. It would therefore be possible to save on subscriber database storage by not replicating the service-profile data for every individual subscription, but to have all subscriptions refer to a common-service profile. This creates a group-based subscription.

For the M2M application owner, group-based subscriptions can also be easier to manage. If a change to the service profile is needed for the application, this change probably needs to apply to all the individual subscriptions. Performing this update one by one for every individual subscription represents a huge task, with a risk of errors and inconsistent data. It is easier to carry out the service-profile change for all the subscriptions by changing the common-service profile. One disadvantage of group-based subscriptions is

Table 6.3 Cost drivers for an M2M application in a mobile telecommunications network. Reproduced by permission of TNO

Cost drivers	Network cost components	Example applications
Number of group-based subscriptions	SIM cards; E.164 numbers; HLR capacity	Applications that rarely send data and are device-originated only (e.g., alarms)
Number of simultaneously attached devices	Mobility context data in network nodes; Mobility management signaling	Applications that rarely send data but have to be reachable (e.g., remote management)
Number of simultaneous always on data connections	Session context data in network nodes; Firewall capacity; IP addresses	Applications that continuously need to send and receive data (e.g., remote control)
Number of data sessions	Connection set-up signaling; RADIUS/Diameter capacity; CDR processing	Applications that regularly set up a connection to send small amounts of data (e.g., metering)
Volume of data throughput	Radio capacity; Transport network capacity; Capacity in network nodes	High-bandwidth applications such as video surveillance

that a management process is needed to indicate which subscriptions belong to which group.

Group-based subscriptions will also have an impact on the procedures for registration. Often in a registration process, service-profile data is downloaded from the subscriber database to the service control node (e.g., an SGSN, MME, or S-CSCF). When there is a change in the subscription data in the database, an update procedure is needed to change the data in the service control node. If this update procedure is not improved for M2M communication, a change in the common-service profile data would potentially result in millions of individual service-profile updates being simultaneously pushed to the service control nodes. This would generate a dangerous load on the network. One possible solution is to also refer to a common-service profile in the service control nodes. Whenever there is at least one subscription from the group of subscriptions registered on the service control node, the common-service profile data is downloaded. An upgrade of the common-service profile data now only requires an upgrade of the downloaded common-service profile data in the service control nodes.

Group-based subscriptions can also help to reduce costs related to charging by combining charging data for all subscriptions in a group. Without optimizations for group-based subscriptions, charging data records (CDRs) are generated for every individual subscription. This implies that for an M2M application with a million active devices, a million CDRs are being generated. Handling all these individual CDRs is expensive. CDRs need to be generated, collected, mediated, stored, and processed into a bill. Note that the amount of data in a single CDR may actually be bigger than the amount of M2M application-user data which it reports. The operator and the M2M application owner have to agree whether group-based charging applies. The network operator may provide a lower price for M2M application owners that agree with group-based charging. One disadvantage of group-based charging is that the M2M application owner cannot distinguish (unusual) usage patterns for individual devices, because such usage patterns may be used, for example, to identify cases of fraud.

Group-based charging data works best when a lot of subscriptions in the group use the same network node where CDRs are generated. It would imply that for each service node that generates CDRs, there is always only a single CDR for the group. A new option in the subscribed charging characteristics in the service-profile data would need to be added to indicate when group-based charging applies. The network nodes where CDRs are generated need to be informed about whether group-based charging applies. This can be done in the same way in which the other subscribed charging characteristics (e.g., post-paid/prepaid) are distributed.

Another group-based optimization is group-based policing. Network operators generally apply fair-use policies where, for example, a maximum volume of data transport is set for a particular subscription. A gateway node meters the transported data for that subscription and takes action if the limit is exceeded. For M2M applications that individually do not send a lot of data, such subscription-based fair-use policies do not make a lot of sense. But since a large group of M2M devices collectively may still generate large amounts of data, a group-based fair-use policy may make more sense.

With group-based policing, the gateway nodes measure usage data for all devices in a group collectively. The total usage for all devices in a group via all gateway nodes can be monitored on a central online charging services (OCS) platform. When it has been determined that the daily, weekly, or monthly limit has been reached, action is taken, such as to reduce the throughput for that application. This can, for example, be done by downgrading the maximum throughput of the individual data connections for each device. It would be easier to police the throughput on the gateway node, discarding packets when a particular throughput is exceeded. The problem with this approach is that this somehow necessitates "sharing" a throughput for devices that may be connected via different gateway nodes; there is no guarantee that a single gateway node is being used for all devices belonging to the same group.

A third group-based optimization is group-based triggering. When an M2M application owner wants to trigger a particular batch of M2M devices, a group-based trigger may be broadcast to all devices simultaneously. This assumes that the M2M application owner has some level of knowledge of the location of the M2M devices that need to be triggered. A network-wide broadcast will be inefficient for all but the very largest of M2M applications. Care should also be taken with what happens when very large groups of M2M devices are triggered simultaneously. If all these devices react by connecting to the M2M server simultaneously, this may cause a network overload.

M2M devices may potentially belong to different groups. For example, an M2M application owner may use one set of groups to indicate devices of a particular model or version, whereas another set of groups may be used to indicate particular types of charging arrangements with the operator. Any particular subscription could then belong to multiple groups. However, allowing subscriptions to belong to multiple groups complicates subscription management- and service-profile-related procedures significantly. 3GPP has therefore specified in 3GPP TS 22.368 that any subscription must always belong to one group only. The practical implications of this restriction are limited: an M2M application owner that wants to distinguish five different device models and two different charging arrangements can simply define 10 different groups.

A group identifier can be defined to identify particular groups in subscription procedures and service-profile-related procedures. In the event that roaming needs to be supported, the group identifier must be globally unique. This can be done by including an operator identifier in front of the group identifier. Figure 6.7 shows how a group identifier may be used in a data model for group-based subscriptions.

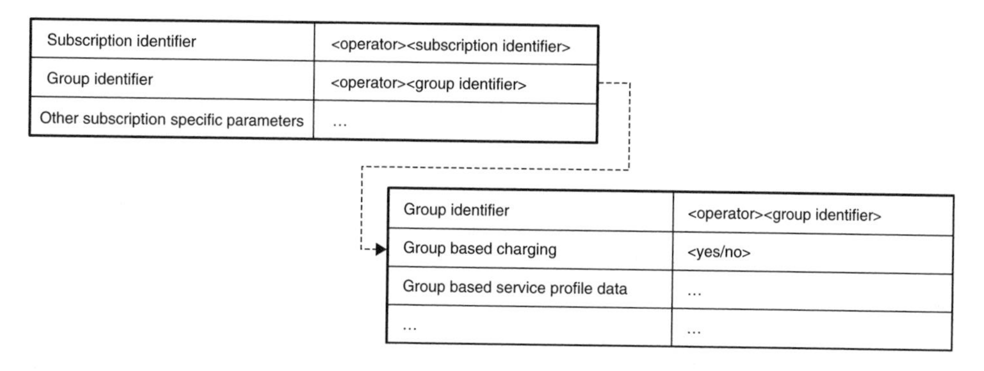

Figure 6.7 Example data model for group-based subscriptions.

6.3.3.1 Reduce Network Resources for Devices When They Are Not Sending Data

M2M devices consume network resources even when they are not sending data. An example is the session context data that is maintained in the network for connections or – in the case of mobile networks – the mobility management context for devices that are attached. Simultaneous data connections and simultaneously attached users are widely used dimensioning parameters for packet-switched mobile core network equipment. This implies that a network operator will invest more in network equipment for M2M devices that are kept connected or attached. Additionally, firewall capacity and the required IP addressing range depend on the number of simultaneous connections.

Depending on the communication pattern of the M2M application (see Section 6.2.4), it may be more efficient to disconnect or detach the M2M device when there is no data to be sent. There is a trade-off between saving network resources by removing context data and the signaling needed to attach and reconnect the next time that data needs to be sent. The ideal balance depends on the network technology used,[2] how easy it is to attach and/or connect, and on the costs associated with keeping the context data. There are no clear-cut rules, but an application that sends data every 3 minutes would be best kept connected, whereas an application that only sends data once a day would be best kept detached.

Reducing network resources for devices when they are not sending data is a cost optimization that is only applicable to the network operator. The M2M application owner will probably prefer to keep the devices connected and attached. For example, detaching the M2M devices implies that they are not reachable by the M2M server. Keeping the M2M devices disconnected implies that they have to be triggered to establish a connection. Some M2M network operators, however, provide incentives to the M2M application owners by providing price reductions for subscriptions with restrictions on how long devices can be attached or connected.

Network operators that wish to reduce network resources will want some mechanisms to influence the network state (connected/attached) of M2M devices. One option is, for example, to send policies to the M2M device that indicate what the device should do after sending data (e.g., detach after 15 minutes of no data or disconnect after 3 minutes of no data). This would be similar to policies that are sent to mobile devices for network selection. One alternative is to police the network state in the network. For example, the network may disconnect M2M devices after a pre-agreed daily maximum duration of access.

It should be noted that there is also another reason for limiting the context data stored in the network: end-users may have privacy concerns related to the network-storing context for their M2M devices. For example, if an M2M device for a vehicle management application stays attached to enable network-based connections, the network will also keep context data that enables location tracking of that vehicle. As this will cause privacy concerns for some car owners, there will have to be a way of opting out of such a vehicle management application. For government-mandated applications where there is no possibility to opt out, such as road pricing or eCall, maintaining a permanent mobility management context in the network will not be possible.

---

[2] Note that in the evolved packet system (EPS), a default connection is always set up when a mobile device attaches via E-UTRAN. For M2M applications where the device needs to be kept attached to be reachable but have little data to send, this may not be very efficient.

6.3.3.2 Avoid Network Signaling

With M2M communications, the ratio between the amount of network signaling and the amount of user data transmitted is relatively high. For M2M communications, it pays off to investigate how the amount of network signaling can be reduced.

One way of reducing the amount of network signaling is to charge for device-originated signaling (e.g., attach requests, location updates, and connection requests). M2M application owners will automatically receive an incentive to be as efficient as possible. However, for human-to-human networks, signaling is not charged. Creating CDRs for signaling may actually generate more overhead in CDR-handling than it saves in a reduction of network signaling.

In mobile networks, mobility management signaling can be reduced for low-mobility devices. Generally, mobility management signaling is repeated regularly to ensure that the M2M device is still attached to the network, although the timer values for this recurring mobility management signaling can be extended or even set to infinity. In that case, mobility management signaling is only performed when the M2M device detects a change of location. For many low- or reduced-mobility[3] M2M devices, this implies that they will very rarely generate mobility management signaling.

Another simple way to keep the amount of signaling down is to keep devices detached as much as possible when there is no data to be sent. M2M devices that are not attached will also not create any network signaling.

For devices that send data frequently, say, every few minutes, it may be more efficient to keep the connections open. The frequent setting-up and tearing-down of connections and/or attach requests may require more signaling than keeping the connection open would actually require.

For applications that send small amounts of data infrequently, for example, alarms, the signaling required to attach and set up the connection and then possibly to disconnect and re-attach represents a great deal of overhead compared to the small amount of data that is sent. There will easily be more data associated with the signaling than actual user data. For these kinds of applications, the best option is to append the user data in the signaling. The user data may already be included in the attach requests sent from the M2M device to the network. In the event that the authentication that is part of the overall attach procedure succeeds, the user data is forwarded to its destination.

6.3.3.3 Reduce Peaks in User Data

Network operators need to dimension their networks for peak usage. Any additional traffic that needs to be handled during the busy hour implies additional investment in network capacity. On the other hand, it implies that if M2M traffic can be moved off-peak, it is almost handled for free. The marginal cost for running traffic over a network that has plenty of spare capacity at that time is very low.

The concept of time-controlled, as defined in 3GPP, exploits the daily usage pattern in telecommunications by pushing non-time-sensitive applications to non-peak hours. There are many applications where, for example, data needs to be sent once per day but where

[3] A vending machine or printer/copier can change location occasionally, but generally is not mobile. For these kinds of devices, optimizations for low- or reduced-mobility management apply.

it does not particularly matter when this data is sent. In that case, costs can be saved by ensuring that the data is sent during off-peak hours.

With the time-controlled feature, the operator and the M2M application owner agree that data can only be sent within a particular access grant time interval. Once again, the assumption is that the operator provides an incentive to the M2M application owner by lowering the price for data transport for applications that can support this time-controlled restriction. The operator, since a discount is being given, wants to monitor whether the M2M application owners are indeed following the restrictions that have been set.

When an access request, such as an attach or a connection set-up, is received for an M2M device outside its access grant time interval, the request can either be rejected or accepted but charged at a different rate. Charging at a higher rate may be preferential for the operator as it still generates income. However, the disadvantage is that the charging solution becomes very complex. Changing the charging rate implies that the existing CDR needs to be closed to capture the amount of data sent with the old rate and a new one, for the data sent with the new rate, needs to be opened. The additional CDRs that need to be generated and processed may counteract the cost advantage of sending the data during off-peak hours.

For an operator, the off-peak period may change over time. Particularly if the operator is going to attract large M2M accounts with heavy volumes of traffic, the off-peak periods may actually start to fill up. It may also be that because of local differences in the daily pattern or different time zones, the local off-peak interval is different. Therefore, the time-controlled feature allows the operator to alter the access grant time interval to suit local needs. The altered access grant time interval needs to be communicated to both the M2M device and the M2M server. A cause value in the rejection (or acceptance) message for the access request can be used to inform the M2M device about any changes.

The time-controlled feature also includes a forbidden time interval. During the forbidden time interval, all access requests are rejected. This allows for service windows on the M2M server in which no data will be received, irrespective of the local network circumstances. The forbidden time interval cannot be altered by the network operator. It is clear that the forbidden time interval and the access grant time interval should not overlap.

It is important to randomize access from M2M devices within the access grant time interval. Otherwise, all M2M devices that had to stop their communications when trying at first outside the access grant time interval will all start communicating at precisely the beginning of the access grant time interval. The M2M device may, for instance, add a randomized time offset to the start of the access grant time interval.

Furthermore, it is interesting to determine what happens at the end of the access grant time interval. If some M2M applications do not manage to get their data sent within the access grant time interval and then get disconnected, they will never get their data across. However, it can be assumed that the access grant time interval is defined to be more than wide enough for the M2M application to send all necessary data. For example, in the case of an M2M application that needs approximately 5 minutes to send its data, a 1-hour access grant time interval would be appropriate. Different M2M applications are then spread out over the 1-hour access grant time interval. In that case, any connections that still exist after the end of the access grant time interval can simply be disconnected by the network since they violate the agreements that the M2M application owner made with the network operator (in the same way as a telephone call is disconnected when a prepaid account runs out of credit).

3GPP has defined the concept of low-priority communication as a mechanism to combat overload caused by M2M traffic in exceptional cases (see Section 6.3.7). But defining M2M devices and subscriptions as low-priority can also work as a way to reduce network investment by diverting M2M traffic away from more regular peaks in network usage. When a low-priority M2M device tries to access the network during a peak in traffic load, it can be deferred to a later moment by sending back a reject message with a back-off timer. Only after expiry of the back-off timer is the M2M device allowed back onto the network. This mechanism can therefore be used to prevent investment in additional peak capacity being needed to handle M2M traffic.

6.3.3.4 Separate Network for M2M

When the M2M traffic actually becomes very different from human-to-human traffic, it can be beneficial to implement a completely separate infrastructure for M2M traffic. The M2M-specific network can then be configured and scaled specifically for M2M traffic, while the other network can remain optimized for human-to-human communications. Core network entities in an M2M-optimized network could, for instance, be scaled to allow for a large number of subscriptions and related context data but with limited data throughput. Similarly, access network infrastructure for M2M traffic may be made cheaper by not enabling high throughput, but keeping a very low latency in establishing communications.

In the most extreme case, a completely different access and core network can be set up for M2M communications. The M2M device is then configured to work with the specific M2M network. A less extreme case is where the access network is still shared between the M2M and human-to-human traffic, but where there is a separate core network for M2M traffic. In that case, the access network has to be able to identify the M2M traffic in order to send that part of the traffic to the M2M-dedicated core network.

6.3.4 M2M Value-Added Services

In the previous section, we have seen that M2M applications can often support specific restrictions if this results in a lower price for the data communications. However, M2M applications often also have specific requirements for value-added services in addition to general data communications. With M2M value-added services, the operator can make basic connection services more appealing to M2M application owners and set themselves apart from operators that are only providing standard connectivity services.

Whereas cost reduction was geared toward enabling operators to offer a lower price to M2M application owners who do not wish to pay the price of normal data connections, M2M value-added services will be provided at additional cost. They enable operators to generate more revenue from M2M services.

It should be noted that M2M value-added services are not services at the M2M application level. The M2M value-added services relate to the data communication services themselves, not to the application in the M2M device and M2M server. Examples of M2M value-added services are:

- QoS and priority differentiation;
- charging and subscription management;
- device management;
- connection monitoring;
- fraud control;
- secure connections.

6.3.4.1 Higher QoS and Priority

Different M2M applications will have different QoS requirements. Although many M2M applications have no stringent QoS requirements and can deal perfectly well with best-effort QoS, some M2M applications have higher QoS or priority requirements than normal data services. Operators are increasingly providing QoS differentiation in their packet-based networks. This will represent an added value for many M2M applications.

Some QoS requirements for M2M applications depend on the type of media being transported. For example, video surveillance requires a streaming-class QoS with sufficient bandwidth in very much the same way as other video-streaming services have QoS requirements.

Other QoS requirements are more specific and are less frequent in current human-to-human data services. One of these is a more stringent requirement on transfer delay. Some remote-control applications require a lower latency than is currently the norm, at least in 2G and 3G mobile networks, for example, a feedback loop for controlling a generator would not be possible with round-trip delays of several hundred milliseconds. Fixed networks will be more suited to getting a sufficiently low transfer delay. However, the low latency of long term evolution (LTE) also provides an opportunity for M2M applications. Whereas latency in UMTS is generally in the order of 200 ms, the latency in LTE is typically as low as 15 ms. LTE was specifically designed to provide a lower latency and that provides possibilities for new M2M applications, such as multiuser gaming.

Another example is the allocation and retention priority (ARP). The ARP determines the priority that a device gets or maintains connectivity in the case of congestion in the network. Contrary to the different traffic classes, the ARP works on the connection instead of on individual IP packets. Many M2M applications can deal with a lower ARP than, for example, standard Internet connectivity. These M2M applications are not time-critical and can delay their data transfer until the congestion is over. But some M2M applications require a higher ARP. For example, seismic sensors need to be able to warn against earthquakes, even in the event that the earthquake results in congestion in the network. Applications can delay their data communication until congestion is over. Patients with a heart-monitoring device would also probably like their devices to have a somewhat higher priority than other data traffic in the mass of mobile communication on New Year's Eve.

The QoS of a data connection is defined in the QoS profile. Tables 6.4 and 6.5 show the various parameters in the QoS profile for UMTS, that is, different parameters apply depending on the traffic class. The QoS profile for a connection can be set upon the connection being established. The QoS profile in the subscription determines the maximum

Table 6.4 QoS traffic classes in UMTS. Reproduced by permission of 3GPP

Traffic class	Conversational class	Streaming class	Interactive class	Background
Fundamental characteristics	Preserve time relation (variation) between information entities of the stream Conversational pattern (stringent and low delay)	Preserve time relation (variation) between information entities of the stream	Request response pattern Preserve payload content	Destination does not expect the data within a certain time Preserve payload content

Table 6.5 QoS parameters applicable to the different UMTS traffic classes. Reproduced by permission of 3GPP

Traffic class	Conversational class	Streaming class	Interactive class	Background class
Maximum bitrate	✓	✓	✓	✓
Delivery order	✓	✓	✓	✓
Maximum SDU size	✓	✓	✓	✓
SDU format information	✓	✓	–	–
SDU error ratio	✓	✓	✓	✓
Residual bit error ratio	✓	✓	✓	✓
Delivery of erroneous SDUs	✓	✓	✓	✓
Transfer delay	✓	✓	–	–
Guaranteed bit rate	✓	✓	–	–
Traffic handling priority	–	–	✓	–
Allocation/retention priority	✓	✓	✓	✓
Source statistics descriptor	✓	✓	–	–
Signaling indication	–	–	✓	–
Evolved allocation/retention priority	✓	✓	✓	✓

QoS that can be requested for connections for a particular subscriber. The possibilities provided by QoS control in current fixed and mobile networks seem adequate for M2M communications.

A specific value-added service defined in 3GPP is the priority alarm message. 3GPP has defined various MTC features that enable the mobile operator to reduce cost, but also imply a restriction for the M2M application owner. For example, with the MTC feature time-controlled, the M2M device can only send data within a particular access grant time interval. The priority alarm message is intended to be able to override such restrictions, should this be needed. As the priority alarm message also overrides the cost reductions for the mobile operator, the priority alarm message will be priced accordingly.

6.3.4.2 Billing and Subscription Management

In fixed and mobile telecommunications, it is customary for specific billing- and subscription-management arrangements to be provided for corporate customers. A corporate customer that has mobile subscriptions for a few thousand employees will demand specific features for a telecommunication manager within the company who can, for example, more easily assign telephone numbers to new employees. Also for billing, specific added value has to be foreseen to provide, for example, a single bill covering all subscriptions, and various reporting capabilities.

One M2M application owner may have up to a few million, rather than a few thousand, subscriptions. This will also imply specific requirements concerning subscription and billing. For M2M application owners, the M2M application is often part of its primary process. This implies that integration is required for the subscription management with the IT processes related to that primary process. For an electricity company, providing electricity to a new address implies that an M2M subscription for the smart meter at that address needs to be provided. Without integration of IT systems, this would be an impossible task.

M2M application owners in the consumer electronics industry have other requirements related to subscription management. They often manufacture devices in one particular country, and then ship these devices all over the world. In consumer electronics devices, such as navigation devices or cameras, the SIM card required to access the mobile network is often embedded in the device. The device with its SIM card will have to be tested in the factory. But the M2M application owner may not want to pay roaming charges for these kinds of tests. Subsequently, the M2M device enters a supply chain until it ends up in stores anywhere around the world. The M2M application owner does not want to pay for subscription charges while the device is still in a warehouse. Only when the device is sold to an end-customer should the subscription be activated. All these M2M-specific logistics with SIMs that are embedded in the device at the factory will require specific M2M subscription provisioning.

Ideally, the M2M application owner would like the flexibility to choose the mobile operator further along in the supply chain. The M2M application owner would prefer to put a "white-label" SIM card in the device. When, at a later stage, it is determined in which country the device will be sold, an appropriate mobile operator for that country is selected. To allow this, the SIM card can be provisioned "over-the-air" with the subscription details (international mobile subscriber identities (IMSIs), security keys, and encryption algorithms) for that particular operator. Although operators are understandably reluctant to introduce such remote management of SIM cards, as it reduces their grip on subscriptions, standardization development is ongoing in the GSM Association (GSMA), the ETSI smart card platform and 3GPP to make remote management of SIM cards possible.

6.3.4.3 Device Management

For an M2M application owner that manages several hundred thousand M2M devices, device management is a very important value-added service. It is very costly to service all the devices, and very often the M2M application owner has little or no possibility to get physical access the device – for example, where a manufacturer of navigation devices sells the devices with a map update service. When the manufacturer subsequently wants to update the network addresses of the server, the navigation devices that are already sold to end-customers also have to be upgraded. Another example would be a remotely controlled highway information sign. It is often very expensive to service these kinds of M2M devices because getting physical access to them requires that a part of the highway be closed off. A solution where the M2M application owner can remotely manage settings in the M2M device is therefore most welcome.

Device management can be considered at M2M application level, for example, updating the map data on a navigation device, or at a communication level, for example, updating the access point name (APN) or the IP address for the M2M server. Application-level device management can also be provided to the M2M server by the M2M application owner itself, and is not seen as the responsibility of the network operator. The network operator would – in general and also for non-M2M devices – focus on communication-level device management. Nevertheless, the network operator may use its available facilities for device and network management to provide application-level device management services to the M2M application owner.

Special attention should be given to device management upon initial activation of the M2M device. In this situation, the M2M devices still need to obtain the parameters (e.g., APN, IP address) needed to get their connection to the M2M server. Particularly in the case of initial activation, the network operator can provide added value.

Network operators have existing mechanisms for device management at their disposal, which are also used for other data communication services. Mobile network operators can use the Open Mobile Alliance-device management (OMA-DM) framework [OMA DM]. Fixed network operators may use frameworks, such as the TR-069 framework from broadband forum [TR-069].

6.3.4.4 Connection Monitoring

With normal data communications, the end-user generally has to monitor only one device and one connection. However, an M2M application owner may have to monitor millions of devices and connections. Furthermore, for an M2M application owner it is often difficult or costly to get access to the device. For a company, say, that uses M2M connections for remote servicing of copier machines, a service call-out to the copier simply implies the inconvenience of a service call by an engineer to the customer, but for a company that uses an M2M application to monitor wind turbines at sea, a visit to the M2M device will really be a major cost. Some M2M application owners will never be able to get access to the M2M devices; for example, a company that sells navigation devices may get a call from a customer complaining that the remote map update feature is not working. Remote diagnosis of the problem is then the only option.

It is often important to monitor the connection state of the M2M device. For a security alarm application, it is important to know whether the fact that no alarms are being received is an indication that everything is OK or whether it is an indication that the security application cannot connect to the M2M server. The M2M device may, for example, be out of cellular coverage.

Monitoring can be done at application level. Sending regular diagnostic messages between device and server will enable the M2M application owner to detect when there is something amiss with the connection between the M2M device and the M2M server. However, for many applications, sending regular diagnostic messages is not very efficient. For alarm applications, diagnostic messages will generate much more data than the actual user data. Particularly when data connections are used as a backup, it is important to know whether the backup connection is still available without having constantly to send data.

A better option is to use network-based monitoring of the status of the device. One option is to monitor the reception of regular mobility management signaling. When the regular mobility management signaling is received, the M2M device has to have cellular coverage and be able to contact the network. The idea is to detect status information about the connection to the M2M device without generating a lot of additional data or signaling.

3GPP has defined requirements for an M2M monitoring feature in 3GPP TS 22.368. The M2M application owner can define what events need to be monitored, for example, loss of connectivity, removal of the universal identity chip card (UICC), or mobility outside a particular area. The M2M application owner can also define what needs to be done when a particular event is detected. In most cases, the M2M application owner will be notified. Alternatively, the network may automatically restrict the service to the M2M device if the event detected is an indication of fraud.

6.3.4.5 Fraud Control

Many M2M application owners have very large numbers of M2M devices. Furthermore, these M2M devices are often in public places and are vulnerable to fraud or theft. For instance, a local energy company might use M2M devices to switch streetlights on and off. The M2M device may be located somewhere high up in a streetlight structure, but are in the open and accessible to the public. Somebody might try to open up such an M2M device, take out the SIM card and use that SIM card for something else.

For an M2M application owner, it is important to detect fraud as soon as possible. The network operator can help by providing network-based information that is indicative of fraudulent behavior. For instance, if a particular subscription suddenly transmits 10 times more data than the average of the other subscriptions within the same application, it is possible that some form of fraud is taking place. The device or the SIM card may have been stolen and used for something else, for example, to send large amounts of personal data.

The first line of defense is to monitor the usage of all devices in (or near) real time. Whenever particular subscriptions show unexpected patterns of behavior, the M2M application owner can take action, for example, by blocking this particular subscription.

Other aspects can also be used to indicate that there is something wrong. For instance, in an M2M application with low mobility (e.g., smart metering), it is not expected that the M2M device will suddenly report from the other side of town. If the M2M device is more mobile than expected, this is another indication of fraud.

Another option for protecting the SIM card from misuse is to restrict the usage of SIM cards to particular devices. The relation can, for example, be based on the international mobile equipment identity and software version (IMEISV) range of the M2M device. USIMs will then only work with particular types of devices. It will be possible to replace the device by a similar device but not to insert the SIM card in a different type of device.

A third possibility for fraud control is to restrict the addressable destinations for data communications. For example, it is possible to restrict APNs and/or restrict the IP addresses which the M2M device is allowed to use.

6.3.4.6 Secure Connections

A network operator generally will ensure security protection – for example, integrity, confidentiality – of the data transported over its network. For example, mobile network operators apply encryption to the data that is transmitted over the radio interface. The interface between the gateway nodes in the operator's network and an M2M server at the customer is also generally protected with, for example, an IP VPN.

Some application owners, however, want additional end-to-end security protection; for example, for a mobile payment application, a secure end-to-end IPsec tunnel may be used between the payment device and the M2M server in the network. The network operator does not provide this kind of end-to-end security, which has to be ensured at application level.

Nevertheless, 3GPP defined in [22.368] an M2M value-added service related to security. The operator can provide assistance to the application owner with the key exchange needed for the end-to-end encryption. Similar to the use of the generic bootstrap architecture (GBA) [33.220], the association between the SIM application on the UICC and the network is used to generate encryption keys in the device and in the M2M server.

6.3.5 Numbering, Identifiers, and Addressing

The anticipated growth of M2M implies that there will be many more M2M devices and many more M2M-related subscriptions. All these devices need numbers, identifiers, and addresses. The issue is whether the existing numbers, identifiers, and addresses can cope with the anticipated growth of M2M. Are the structures of these numbers and identifiers long enough to cater for billions of M2M devices and subscriptions?

3GPP [22.368] assumes that for M2M two orders of magnitude more devices and subscriptions need to be identified and addressed than for human-to-human communications. If there are ~6.5 billion people on earth, then an identifier structure that can identify 10 billion subscriptions or devices should be sufficient for human-to-human communications. The 3GPP requirement implies that an identifier structure should be able to hold 100 × 10 billion = 1 trillion unique identifiers.

6.3.5.1 E.164 Numbers

E.164 numbers or telephone numbers is where there is an urgent shortage of numbers. The planning and allocation of telephone numbers is the responsibility of telecommunication

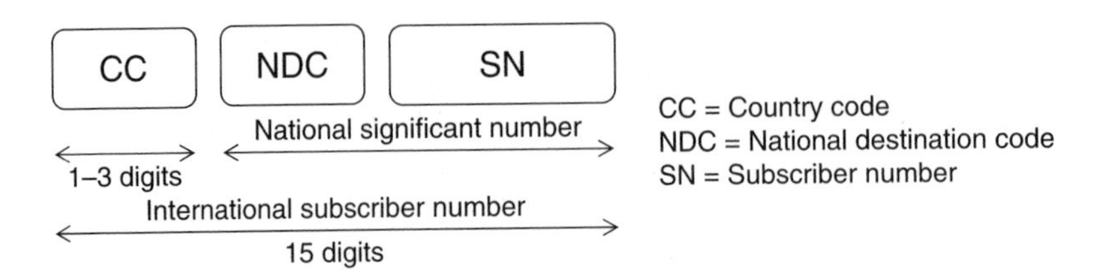

Figure 6.8 E.164 number format for telephone numbers.

regulators. However, regulators have indicated that in some countries E.164 numbers for M2M communications are running out [ECC 153].

Telephone numbers are intended for telephony applications, where the A-party "dials" the B-party's telephone number to indicate the intended destination of the call. In data communication, which covers the majority of M2M communications, a telephone number is not strictly needed.[4] To identify the destination in data communications, an IP address seems more suitable. Nevertheless, there are a number of places in telecommunication networks where telephone numbers are still relied upon, for example:

- **in billing** – the CDRs used to collect and transport charging data contain telephone numbers;
- **in provisioning** – a subscription profile record in a Home Subscriber Server without a telephone number is considered to be inactive. Adding the telephone number to the subscription profile implies that the subscription is activated;
- **in over-the-air device management and SIM management in mobile telecommunications**, which often imply sending an SMS, a telephone number is needed to indicate the destination of the SMS.

The structure of telephone numbers, as shown in Figure 6.8, is defined in the ITU-T Recommendation E.164 [E.164].

The maximum length of a telephone number is 15 digits. With 15 digits, it should be possible to identify all M2M devices. Even if we take away the maximum three digits for the country code, the 12 remaining digits can still identify 1 trillion unique telephone numbers in every country. The reason for the telephone number shortage is that in most countries, telephone numbers are shorter. For example, the North American numbering plan uses 11-digit telephone numbers (including the country code). Moreover, the national destination code is used to indicate particular geographic areas or types of non-geographic numbers. This makes the allocation of numbers less efficient. Most numbers for M2M communications were requested from the mobile number ranges. That is where the number shortage is most urgent.

In some countries, a short-term solution may be feasible to define new number ranges within the existing number plan for M2M communications. A slightly more future-proof solution is to define longer telephone numbers for M2M communications. The preference

[4] In the 3GPP packet-switched mobile network standards [23.060, 23.401], telephone numbers are not used in any of the connection set-up or mobility management procedures. The telephone number is only transported in signaling messages in order to be able to record the telephone number in CDRs.

for shorter, user-friendly telephone numbers that people use does not apply in M2M communications, and hence longer telephone numbers can be assigned.

To stay safely within the maximum number length defined by E.164 [E.164] a telephone number length of 12 to 14 digits is chosen for numbers for M2M communication. Such telephone number lengths should safely be handled by all international telecommunications networks. However, increasing the length of telephone networks will have a significant impact on most IT systems for billing and provisioning. Operators are generally not happy to use longer telephone numbers for M2M communications.

Another option is to reuse an E.164 number range across operators, thus allowing a particular number range to be used by all operators in the country. In this case, numbers are not exchanged between operators, nor is interconnection supported for these numbers. For many applications, there is no problem restricting the use of telephone numbers to just between the M2M application owner, the network operator, and the M2M devices. The M2M devices do not have to be reached by parties other than the application owner. Only where multiple parties are involved in an M2M application does the solution of shared number ranges not work.

Number portability cannot be provided with these shared number ranges, because a particular telephone number may already be assigned in the other mobile network. However, number portability is not really an issue with most M2M applications. M2M application owners may want to switch to a new network operator, but changing the telephone numbers is then only an internal issue for the M2M application owner. There are no other parties that have to be informed about a change of telephone number.

For the future evolution of M2M services, alternatives are being proposed for E.164 numbers. It should be possible to provide packet-switched telecommunications without the need for E.164 numbers. Different solutions to remove the dependency on E.164 numbers are studied in 3GPP [23.888] and [22.988]:

- One option is to simply not use a telephone number (MSISDN) and to rely on the subscription identifier (IMSI) and IP addresses only. For billing purposes and provisioning between operator and subscription owner, the IMSI is quite suitable. However, there are some security concerns arising from the use of the IMSI outside the relationship between the operator and subscription owner, which makes the use of IMSIs difficult in scenarios where third parties need to connect to the M2M devices. Nevertheless, in many M2M application scenarios, both the subscriptions in the M2M devices and the M2M server belong to the same M2M application owner, and therefore IMSIs could also be used. Sufficient security (e.g., an IP VPN) is needed to protect the IMSI on the interface between the mobile network and the M2M server. The benefit of using existing IMSIs and IP addresses only is that it does not require the definition of new identifiers; the drawback is that it cannot handle all M2M application scenarios.
- In case there are multiple parties involved in a single M2M application, a new identifier to replace the MSISDN has benefits; for example, when two people want to play a direct multiuser game, they need to be able to identify the other game console in order to set up communication between the two. Using IP addresses only does not work because the game consoles may change their point of attachment to the network and thus their IP address. The IMSI is not really suitable in this case because IMSIs were not designed to be used as an external identifier in the public domain. A new identifier

is needed to replace the role of the MSISDN. One option is to use a fully qualified domain name (FQDN). A DNS is used to store the IP address of the M2M device that belongs to that identifier. Another option is to use an SIP URI. In that case, an SIP/IMS registration is used to store the relationship between the IP address and the identifier. The disadvantage of replacing the MSISDN with a new identifier is that a new identifier structure needs to be defined. Moreover, a new identifier also implies a new body to assign these identifiers. Setting up such a framework, including the associated business model and regulatory issues, such as portability requirements, may take quite some time.

6.3.5.2 Other Identifiers

Other identifiers may also run out of space with the expected number of M2M devices. We will look at a three relevant identifiers: IMSI, IMEI, and the Integrated circuit card identifier (ICCID), to see whether they can cope with the predicted growth of machine type traffic.

The IMSI is used in mobile networks to identify a particular subscription. The structure of the IMSI is as shown in Figure 6.9 [23.002].

With a mobile subscriber identification number (MSIN) of a minimum of nine digits, the IMSI structure can hold at least 999 999 999 subscriptions for one mobile operator. But if we assume that M2M communications requires 100 times more identifiers than needed for human-to-human communications, then an operator with more than 10 million human subscribers would already be in difficulty. There are numerous mobile operators that currently have more than 10 million subscribers, so an IMSI with a nine-digit MSIN may not be long enough. With 10-digit MSINs there is more room, as few operators have more than 100 million customers. Nevertheless, it cannot be taken for granted that the IMSI structure is long enough to cope with the long-term growth in M2M communications. One escape route may be to assign multiple mobile network codes (MNCs) to a single operator, thus multiplying the maximum number of subscriptions for that operator with the number of MNCs.

The IMEI or international mobile equipment identity and software version numbers (IMEISV) are used to identify individual mobile devices. The structure of the IMEI is shown in Figure 6.10 [23.003].

The structure of the IMEISV is similar to that of the IMEI, a two-digit software version indicator replaces the check digit, as shown in Figure 6.11 [23.003].

The type allocation code (TAC) identifies a particular model of mobile device. The SNR is a serial number for all devices of that type.

The total number of devices that can be uniquely identified with an IMEI is 10^{14}, which seems adequate. However, most of the 14 digits are taken up by the TAC. With a

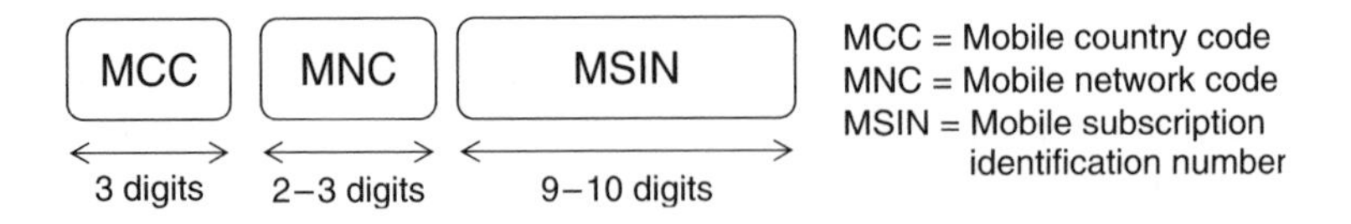

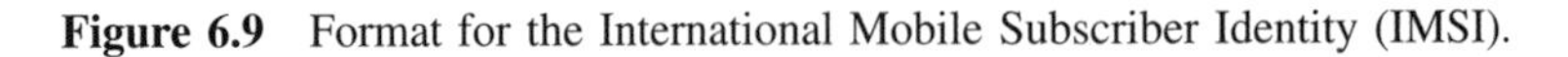

Figure 6.9 Format for the International Mobile Subscriber Identity (IMSI).

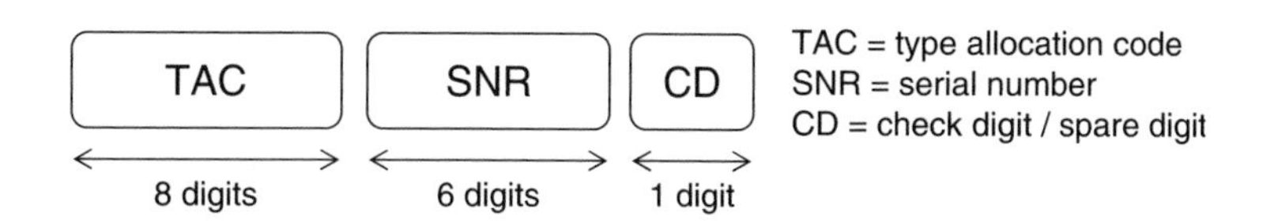

Figure 6.10 Format for the International Mobile Equipment Identity (IMEI).

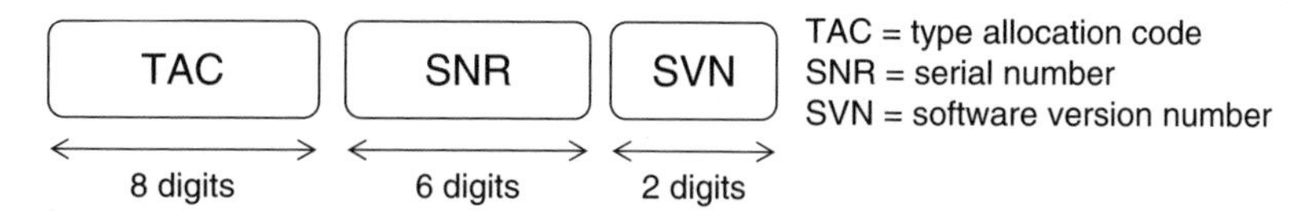

Figure 6.11 Format for the International Mobile Equipment Identity and Software Version Number (IMEISV).

six-digit SNR, the maximum number of devices of a single type is only 1 million. It is quite conceivable that more than 1 million units of a particular device model are manufactured. One possible escape route is through allocating multiple TACs to what is essentially the same type of device.

The ICCID identifies individual UICCs. The structure of the ICCID, ss defined by ITU-T in [E.118], is shown in Figure 6.12.

With at least 12 digits for every issuer identification number for individual UICCs, the structure of the ICCID seems adequate.

Where identifier structures appear to be restrictive, adaptations for M2M are necessary. It is up to the large mobile operators to decide whether they can live with the restriction of subscriptions that can identify with the IMSI or whether an extension or replacement of an IMSI is necessary. Replacing or extending the IMSI is going to have a large impact on operator telecommunications infrastructure and IT. The structures of the IMEI/IMEISV and ICCID seem to provide less of a problem.

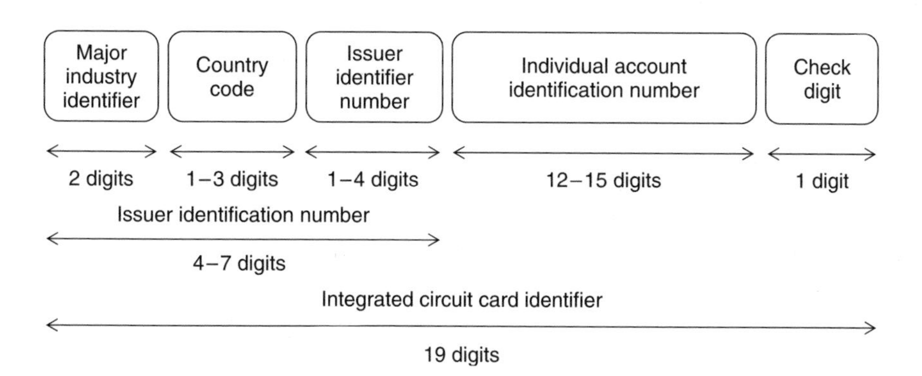

Figure 6.12 Format of the International Circuit Card Identifier (ICCID).

6.3.5.3 IP Addresses

Worldwide, there are major concerns that IPv4 addressing space is running out fast. The use of public IPv4 addresses for large-scale M2M applications is therefore not really an option. IPv6, on the other hand, provides such an enormous amount of addresses – 3.4×10^{38} to be precise – that it is unlikely that M2M applications will ever create a problem.

Ideally, M2M devices and M2M servers will all be issued with public IPv6 addresses, in which case they will all share the same IPv6 address space. This is shown in Figure 6.13.

Unfortunately, migration toward IPv6 is not quite there yet. There are still many Internet destinations that can only be reached with IPv4 addresses. Many operators and service providers are still not quite ready for IPv6. Therefore, it is expected that IPv4 will still be used for quite some time with M2M applications. The shortage of IPv4 addresses is circumvented by using private IPv4 addresses. A typical scenario is shown in Figure 6.14.

The M2M devices in Figure 6.14 are assigned private IPv4 addresses by the network operator. It would be a waste of IPv4 resources to provide public IPv4 addresses to all these M2M devices. Network address and port translation (NAPT) allows a large number of M2M devices to share a single public IP address. The M2M server is assigned a public IP address. This makes the M2M server easily accessible from different networks, such as from different fixed and/or mobile network operators. As the number of M2M servers is much lower, IPv4 address shortage is less of an issue.

One issue with NAPTs is that addressing from the M2M server to the M2M device is not always possible. For the M2M server to send an IP packet to the M2M device, the NAPT needs to maintain an association between the public IPv4 address and port number representing the M2M device at the NAPT and the private IPv4 address of the

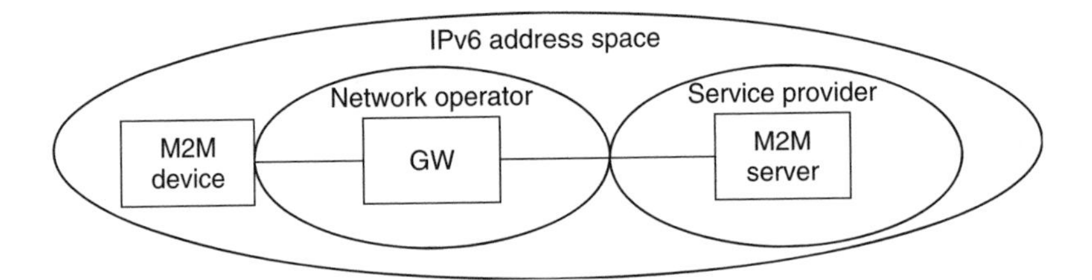

Figure 6.13 M2M communication scenario using IPv6 addressing.

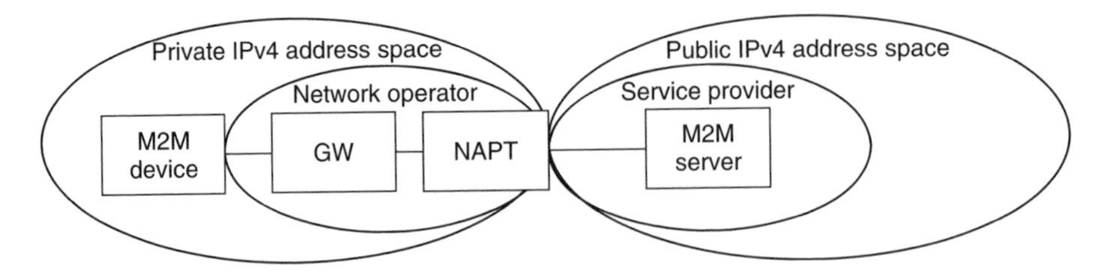

Figure 6.14 M2M communication scenario using private and public IPv4 addressing.

M2M device. This association is set up when the M2M device sends data to the M2M server and is only kept alive as long as the M2M device keeps sending data. Either the M2M device has to keep sending keep-alive messages or a new data session has to be initiated by the M2M device to enable communication. The M2M server will have to trigger (see Section 6.3.6) the M2M device to set up a data session in case it wants to initiate communication with the M2M device.

Another issue with NAPTs is the mapping of identifiers. For example, in case FQDNs are used to replace telephone numbers, a mapping between the FQDN and the IP address could usually be maintained in a DNS. If the M2M server needs to send something to an M2M device identified by the FQDN, it can look up the corresponding IP address in the DNS. However, with NAPT, the IP address and port number for a particular M2M device can change with every data session.

With IPv6 addressing, both the M2M device and the M2M server can be in the same IP addressing domain, so there is no need for NAPT. Nevertheless, with IPv6, there may also be in-between boxes, such as firewalls in the data path. The network operator will probably not allow everyone to send IP packets to the M2M devices. The network operator may also want to prevent the internal IP addresses and address structure from being visible from the outside. This implies that IPv6 addressing will have similar issues to IPv4 in relation to the NAPT; that is, addressing of M2M devices by the M2M server without prior data exchange and mapping of identifiers on IP addresses can also be widely practiced in IPv6.

When the M2M server only has to work with one or a few networks, one option is to use the same private IP addressing domain for both the M2M devices and the M2M server (see Figure 6.15). One benefit is that no NAPT is needed. If the M2M server is not physically located in the same network – for example, it is not owned by the network operator – a tunnel (e.g., an IP VPN) is set up between the gateway (GW) and the M2M server. The M2M devices select a specific gateway where this tunnel to the M2M server has an end-point. In a packet-switched mobile network, this is implemented by using a specific APN for the M2M server/application. The private IP addresses can either be assigned by the network operator or by the service provider that owns M2M server. When an IP address is assigned to an M2M device through RADIUS/Diameter, the RADIUS/Diameter server can inform the M2M server. This way, the M2M server can immediately send data to the M2M device when connection is established, even if the M2M device has not yet sent any data. There is also no problem mapping identifiers to IP addresses. The scenario with private IPv4 addressing only is not always suitable

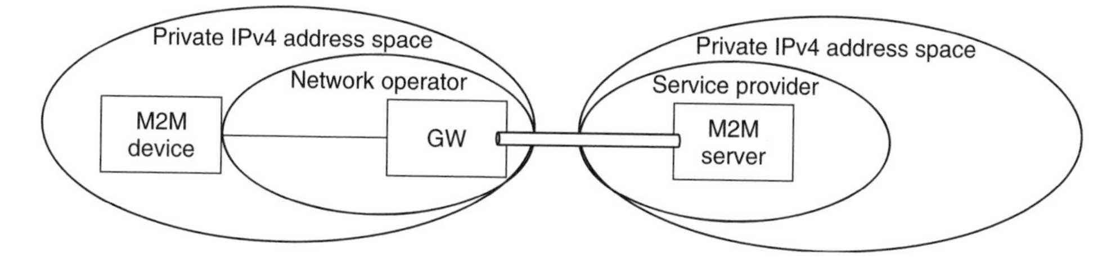

Figure 6.15 M2M communication scenario using private IPv4 addressing only.

because it places restrictions on where the M2M server can be accessed, but its simplicity makes it an attractive solution.

6.3.6 Triggering Optimizations

Many M2M applications need both device- and network-originated communications. In some cases, it is sufficient if the M2M device can initiate a data connection between the M2M device and the M2M server, but often the M2M server also needs to be able to initiate the data connection. The establishment of data connections in fixed and mobile networks, however, is generally device-originated. For example, the PDP (packet data protocol) context – the data connection in GSM and UMTS networks – is set up by the user equipment (UE). Furthermore, most IP middle boxes, such as firewalls and network address translators, work on the assumption that a data session was initiated by the device. In order to set up a connection from the network side, the network needs to inform the device that is should set up a connection. This process is called triggering.

Triggering is common in modern-day M2M communications, although there are some issues with triggering in its current form, requiring some network improvements in this respect, for example:

- One particular way of triggering, which is still common practice, is to set up a CS telephone call to the M2M device. The M2M device does not answer the call, but instead treats the incoming call request as a trigger and establishes a data connection to the M2M server. This way of triggering is particularly unattractive for the operator. The M2M device requires capacity in the CS network, but it will never generate revenue for those CS resources, since the call remains unanswered. Ideally, M2M subscriptions would be packet-switched only.
- Another way of triggering uses SMS messages. When the M2M device receives a (specially formatted) SMS, it will treat that as a trigger. It is possible to send SMS messages via the packet-switched domain. The issue here is that sending an SMS requires a telephone number to specify the destination. There is a shortage of telephone numbers for M2M (see Section 6.3.5).
- Another issue with SMS is that it is a text-messaging service, which is now "misused" as a means of transferring signaling. Mobile operators are introducing equipment for SMS value-added services, such as for spam blocking. This equipment will have to be scaled to handle large numbers of trigger messages that essentially are signaling messages and which do not benefit from text-messaging-related value-added services.

Because of the importance of triggering, it was one of the tasks that received the highest priority for 3GPP Release 11. Triggering should be made more efficient and it should be possible without relying on telephone numbers.

6.3.6.1 The Basics of Triggering

When the M2M server decides to trigger an M2M device, it generally includes the following information in the trigger message:

- the identity of the target M2M device;
- the identity of the application;
- a request counter associated to this request allowing for the detection of duplicated requests, for the correlation of requests with their acknowledgment, and to allow the application to cancel a request;
- (optionally) the IP address (or FQDN) and/or port numbers of the application server that the UE has to contact;
- (optionally) an urgency request indication;
- (optionally) a validity timer, allowing for the removal of triggers that are stored in the network when the device cannot be reached, for example, with SMS;
- (optionally) the area where triggering needs to be sent;
- (optionally) a limited amount of application-specific information, for example, to instruct the M2M device to do something before establishing communication with the M2M server.

Before the M2M server can send the trigger message, it needs to determine where to send the trigger request. 3GPP is defining a machine-type communication gateway (MTC GW) that will act as an entry point in the mobile network for control messages from the M2M servers. The MTC GW may be pre-provisioned in the M2M server in case the M2M server only communicates with one network operator. Otherwise, the M2M server first needs to determine to which operator a trigger request needs to be sent. Based on the device ID or IMSI, the M2M server should be able to find the correct mobile operator network.

Figure 6.16 shows how all device triggers are sent via the MTC GW. Only when there is already an active data session between the M2M device and the M2M server can the M2M server send data to the M2M device by simply sending an IP packet with the right IP address to the gateway GPRS service node (GGSN) or packet data network gateway (P-GW).

From the MTC GW onwards, there are a number of potential trigger solutions. A number of solutions are suggested in [23.888]. The MTC GW may be assigned the role of deciding which trigger method to choose, for example, based on status information

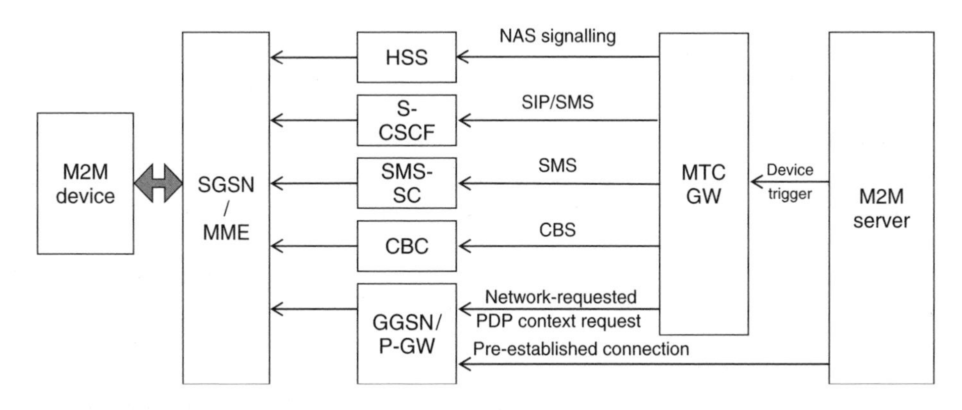

Figure 6.16 Overview architecture with different triggering solutions.

from the network. However, the M2M Server may also remain the decision point on which trigger method to use. The M2M server has the most knowledge about the M2M application, the M2M devices associated with that M2M application, and the kind of triggering methods these M2M devices would support.

Triggering is possible in different connection states. In [22.368], 3GPP specifies that triggering is needed in a connected state and a non-connected state, and while attached and while non-attached.

When the M2M device is in a connected state, the existing connection can be used to contact the M2M devices. It is assumed that the M2M server knows the IP address for the M2M device and there are no issues with middle boxes. The IP packet can then be delivered to the M2M device. In some cases, the M2M device is in a connected state in the sense that it has a logical data connection (e.g., PDP context), but without having a radio-resource connection. In that case, existing paging methods [23.060, 23.401] can be used to re-establish the radio connection and deliver the IP packet.

When the M2M device is attached, but does not have data connection, triggering can be used to set up communication from the M2M server. Additionally, when the M2M device is connected, but the M2M server does not have a usable IP address of the M2M device, the M2M server needs to trigger the M2M device.

Also when the M2M device is not attached, it is useful to be able to trigger that M2M device. Generally, being non-attached implies that the M2M device is not reachable. Without being attached to the network, no mobility management context is available in the network and it is not known where to page the M2M. But it may be possible to trigger devices when the location is known through other means. For example, if the M2M device is stationary, the location of the M2M device might be stored in the home subscription server (HSS). Alternatively, the M2M application may know the location of the M2M device. For example, a vending company will have information in its IT systems on where its vending machines are located. The M2M server can use that information in order to instruct the network to broadcast a trigger message in this location. The location information about the M2M device can also come from the M2M device itself in the event that the M2M device reports its location every time it senses its location has moved (e.g., a copier is being moved to a different building).

Triggering very-low-power M2M devices poses a specific problem. Very-low-power M2M devices do not want to listen to the network all the time because of the power this requires. These M2M devices only occasionally switch on to listen to the network. The problem is that to trigger very-low-power devices, it has to be known when they are listening to the network. If a trigger is sent when they are not listening, the trigger will not be received. An alternative is to use a store-and-forward mechanism that will send the trigger message when the M2M device registers to the network again. For example, an SMS will be stored in the network and be delivered when the M2M device attaches to the network again after been detached and switched off.

For security reasons, triggering is often restricted to authorized sources only. A trigger causes the M2M device to set up a connection to the M2M server. The M2M application owner will have to pay for the data that is sent over that connection, so if an M2M device is mistakenly triggered by an unauthorized source, it would still cost the M2M application owner money. Even worse is if the triggering of the M2M device is done for malicious reasons. For example, repeatedly triggering M2M devices may cause a denial of service attack. The MTC GW can authenticate the source of the trigger requests to ensure that the source is authorized to send triggers to that particular M2M device.

In [23.888], a number of different mechanisms are described to trigger an M2M device. 3GPP has not yet selected which of the trigger mechanisms will be formally standardized. In all likelihood, a number of solutions will be standardized, because the different triggering mechanisms each have strong and weak points. Here are five different triggering mechanisms:

- triggering using mobile-terminated SMS;
- triggering using IMS messages;
- triggering using cell broadcast;
- triggering via HSS and non-access stratum (NAS) signaling;
- triggering via network-requested PDP context establishment.

6.3.6.2 Triggering Using Mobile Terminated SMS

Figure 6.17 shows how the trigger sent by the M2M server is forwarded by the MTC GW to the SMS service center (SMS SC). From there on, the procedure is almost the standard mobile-terminated SMS procedure [23.040]. The main difference is that in normal SMS procedures, an MSISDN is used as the device identifier. With M2M communications, either the IMSI or an MSISDN replacement will be used. This implies a change to the SendRoutingInfoforSMS procedure.

The advantage of triggering using mobile-terminated SMS is that it does not require many changes to the existing networks and standards. The disadvantage is that it does not address all the requirements for an improved M2M trigger solution. Even if the reliance on MSISDNs can be circumvented, there is still an issue that it is very difficult to prevent unauthorized SMS messages. A malicious sender could use a standard interoperator interface to send unauthorized ForwardShortMessage messages to the SGSN where they expect to the destination of the M2M devices to be. As many M2M devices are not mobile, there is quite a good chance of finding the M2M device without requesting routing information from the HSS. It is very difficult to differentiate between such unauthorized ForwardShortMessage messages and genuine incoming SMS messages.

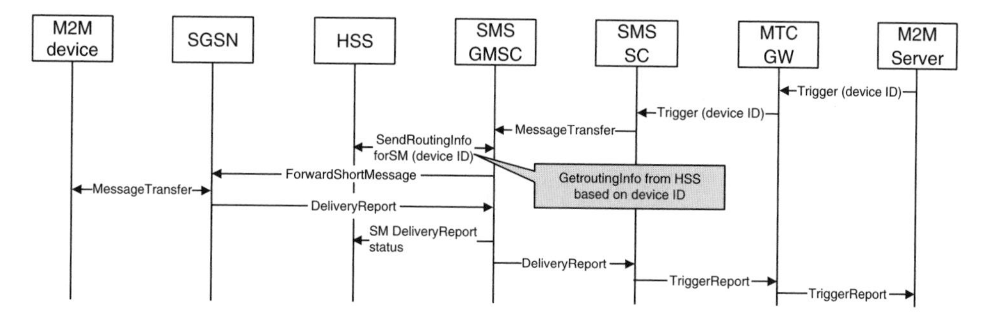

Figure 6.17 Triggering using mobile terminated SMS procedures.

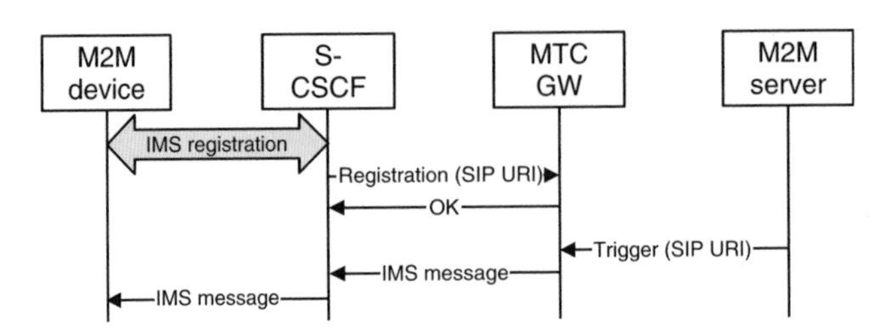

Figure 6.18 Triggering using IMS message.

6.3.6.3 Triggering Using IMS Message

In LTE, there is no native SMS support, as text messaging is implemented using IMS instant messaging. So whereas in LTE SMS-based triggering may be difficult, an alternative may be to use SIP Messages. Figure 6.18 shows how IMS message can be used to trigger an M2M device. The M2M device first needs to register in IMS in order to be able to receive IMS Messages. The MTC GW is seen as an IMS Application Server and the registration of M2M devices in the serving call session control function (S-CSCF) is forwarded to the MTC GW. When a trigger arrives at the MTC GW, it starts off an SIP Message toward the M2M device via the IMS service control interface with the S-CSCF.

The drawback of this solution is that the M2M device needs to maintain an IMS registration. This requires IMS credentials and authentication, recurring re-registration, S-CSCF capacity, etc. – a heavy load for M2M applications that may only send and receive small amounts of data. The benefit is that it requires very few changes to existing standards.

6.3.6.4 Triggering Using Cell Broadcast

Figure 6.19 shows triggering using cell broadcast. The assumption behind this solution is that the M2M server knows the area where the M2M device is located. A trigger request with the area is sent to the MTC GW, which forwards it to the cell broadcast center (CBC). From there, standard procedures of cell broadcast [23.041] are used to broadcast the trigger message within the indicated area. The M2M device listens to the broadcast messages. If there is a match between the identification sent in the trigger message and its own identification, the M2M device will set up a connection to the M2M server.

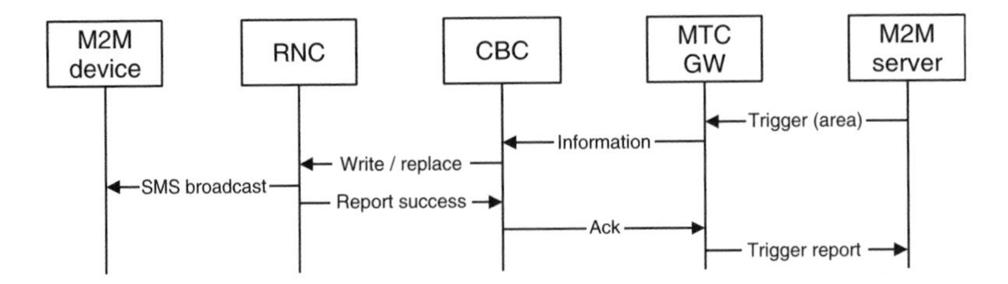

Figure 6.19 Triggering using cell broadcast.

The identification used in the trigger message can be any application-level identifier; it is completely transparent to the mobile network. For example, a smart metering application could use the serial numbers of the electricity meters. The identifiers may also identify a group of M2M devices. This allows for an efficient way to trigger a batch of M2M devices to carry out a software upgrade, for example. Care should be taken, however, to ensure that there is not too large a group of M2M devices that will connect to the M2M server simultaneously. The response to the trigger message should be spread out over time.

The M2M server determines in which area the trigger message is broadcast, in the same way as is done with cell broadcast. To minimize capacity requirements on broadcast channels, trigger messages should ideally be broadcast in a limited area. For a non-mobile M2M device, broadcasting in a few cells should be enough. Nevertheless, it is possible to broadcast trigger messages in a larger region; for example, to trigger a tracking device in a vehicle that is reported stolen, it will just cost more than a broadcast in a smaller region. For the M2M application owner, cell broadcast may be an efficient way of triggering M2M devices, whereas for the mobile operator it provides an extra source of income and return on investment on the cell broadcast infrastructure.

Triggering via cell broadcast can also work for M2M devices that are not attached. This will require that the M2M device continues to listen to the broadcast channels when it is detached. Although the notion that a device that detaches is powered-off is clearly old-fashioned, there will be changes needed to the mobile device standards to achieve this.

6.3.6.5 Triggering via HSS and NAS Signaling

Figure 6.20 shows a message sequence for triggering using existing NAS signaling. The idea behind this concept is that the trigger message can be piggy-backed onto existing signaling between the M2M device and the network.

The trigger request is sent via the MTC GW to the HSS. The HSS stores the request and, when the M2M device is attached, sends it in an insert subscriber message to the SGSN or MME where the M2M is registered. When the M2M device next signals to the SGSN/MME; for example, for a periodic routing area update, the SGSN/MME piggy-backs the trigger message in the response to the M2M device. When the trigger message is successfully received, the trigger request is deleted from the SGSN/MME and the HSS. In case the M2M device is not attached, the HSS will store the trigger request until the next time the M2M device re-attaches to the network.

The benefit of this mechanism is that no additional signaling is needed across the radio interface to transfer the trigger message. The store-and-forward nature of the mechanism has both advantages and disadvantages: there is no immediate feedback as to whether the trigger message has reached the M2M device, but the “fire-and-forget” mechanism implies that the M2M server does not have to keep track of which trigger messages were received and which need resending.

6.3.6.6 Triggering via Network-Requested PDP Context Establishment

The last of the five trigger mechanisms is depicted in Figure 6.21. The network-requested PDP context activation procedure was specified a long time ago in [23.060]. The only new

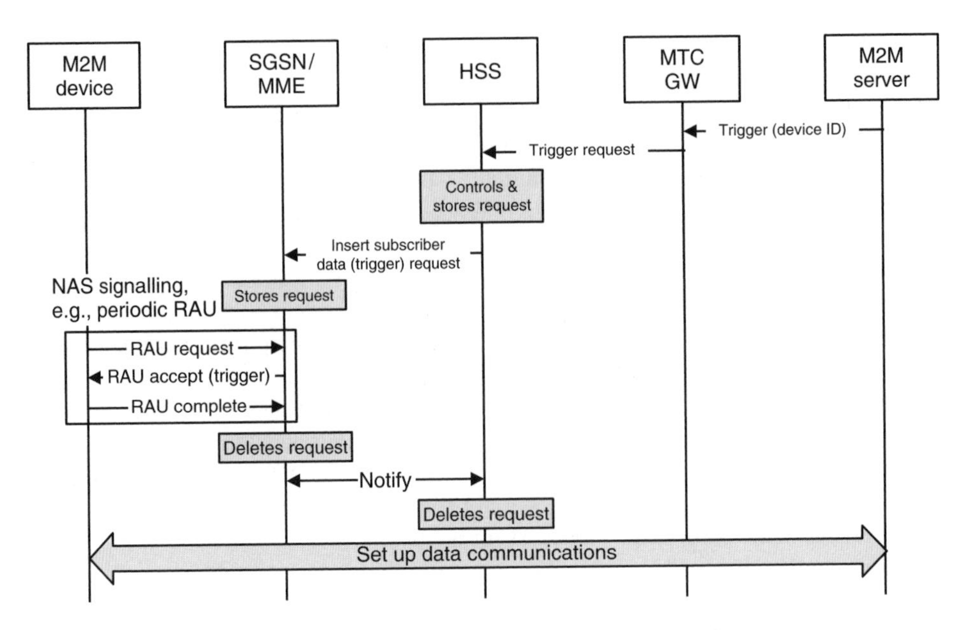

Figure 6.20 Triggering via HSS and NAS signaling.

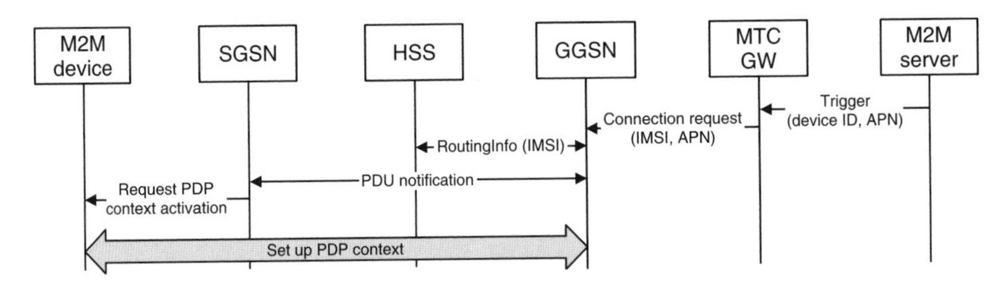

Figure 6.21 Triggering via network requested PDP context activation.

aspect is the connection request from the MTC GW to the GGSN. The existing network-requested PDP context activation procedure is not initiated by a signaling message, but by the receipt of an IP packet for the device.

The routing info to determine to which SGSN to send the trigger message is found in the HSS on the basis of the IMSI. This is how the network-requested PDP context activation is standardized, but this IMSI-based query of the HSS is not often implemented. It may also be an option to use a new device ID replacement of the MSISDN to query the HSS.

6.3.6.7 Conclusions on Trigger Mechanisms

There are a number of different proposals for trigger mechanisms in 3GPP. It is not yet clear which of these mechanisms will be selected for normative standardization specifications. Most probably a mixture of the solutions will be standardized. The M2M server can choose which trigger mechanism suits the application and the M2M devices best.

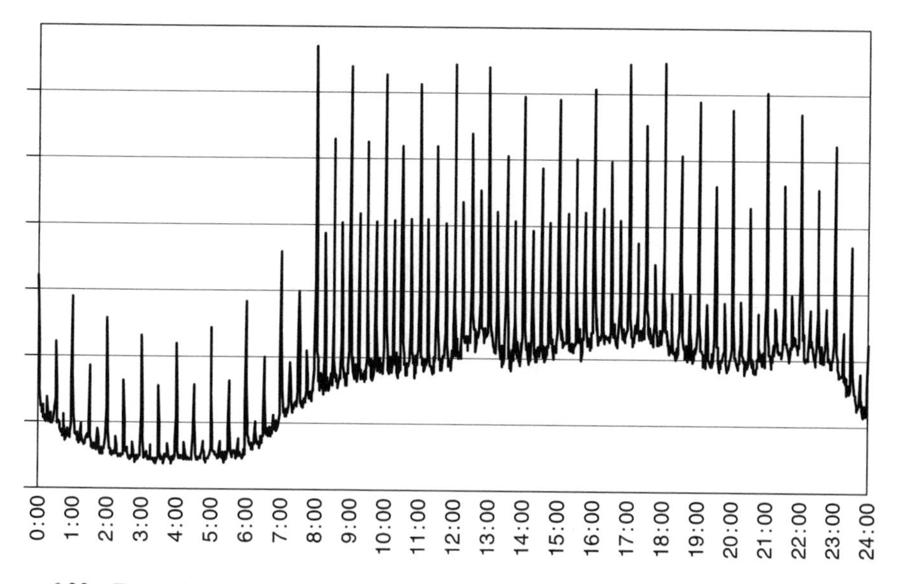

Figure 6.22 Example of synchronized data access from M2M applications (load on a RADIUS server). Reproduced by permission of TNO.

6.3.7 *Overload and Congestion Control*

In the long run, it is expected that there will be two orders of magnitude more M2M devices than personal communication devices. The number of M2M devices directly connected to mobile network operators is already significant and there are various cases where large numbers of M2M devices have incurred significant congestion in the mobile network. Overload and congestion caused by M2M devices is *mainly* the result of synchronized behavior of M2M devices that all simultaneously access the network. Overload and congestion is equally caused by both the mobile operator's own M2M subscriptions and by M2M devices roaming on an operator's network. In this latter case, the traffic patterns may be much less predictable.

M2M-related signaling congestion and overload may, for example, be caused by:

- M2M applications generating recurring data transmissions at precisely synchronous time intervals (e.g., precisely every hour or half hour – see Figure 6.22).
- an external event triggering massive numbers of M2M devices to attach/connect all at once, for example, high numbers of metering devices becoming active almost simultaneously after a period of power outage.
- large numbers of sensors all triggering at once. A particular example is an application for monitoring a bridge. When a train passes over the bridge, all the sensors transmit the monitoring data almost simultaneously. The same thing happens in hydrology monitoring during periods of heavy rain and in building monitoring when intruders break in.
- a malfunctioning in the M2M application and/or M2M server, for example, when the M2M server does not acknowledge receipt of data sent by the M2M devices, and all M2M devices keep resending their data.

- many M2M roaming devices lose network coverage following a base station outage. These M2M devices will all simultaneously roam onto another local competing mobile network.

Some overload and congestion cases may occur regularly, whereas others will be very rare events. The synchronous peaks of data access from M2M applications will have to be treated as a normal situation, since other applications should not suffer from the behavior of M2M applications. Similarly, when a single M2M application causes congestion, other M2M and non-M2M applications should not be affected. However, there are also more unusual events, such as network outages, earthquakes or other disaster situations. In such events, there is a possibility of a combination of massive numbers of M2M devices that try to access the network – for example, fire/burglar alarms triggered by an earthquake – and the network that has its capacity reduced by a network outage or disaster. These are exceptional cases, where the aim is to prevent a complete network collapse, with only priority and emergency services being guaranteed to function.

M2M congestion and overload cases mainly relate to control-plane congestion and overload. Although it is not unthinkable that M2M applications may generate very high user-plane data loads, most M2M devices only send limited amounts of user data. M2M-related congestion and overload may affect both the mobile radio network part – say, when many M2M devices need to transmit data almost simultaneously in a particular area – and/or the mobile core network, for example, when a large number of metering devices need to transmit data almost simultaneously toward the same M2M server as illustrated in Figure 6.23.

The mobile core network nodes that may suffer from M2M-related signaling congestion include:

- all PS domain control-plane nodes and gateways. With large-scale attach requests, the serving control node (SGSN/MME) is principally vulnerable. With data connection requests, the SGSN/MME is also vulnerable because this node has a relatively large load per connection request. The gateway nodes (GGSNs/PGWs) are especially vulnerable because M2M applications often use a dedicated APN that may be terminated at one GGSN/PGW. All connection requests for that particular application will then have to be handled by a single GGSN/PGW. M2M devices may concurrently attempt signaling interaction only in a limited area. That means that the signaling congestion could occur at just one or several particular signaling links and no overall congestion appears on network nodes.
- MSC/VLRs in the CS domain. Due to the need for mobile operators to configure devices via SMS, and the limited availability of PS-domain SMS, most existing M2M devices need to attempt to access the CS domain. A large number of M2M devices may attempt to simultaneously register on the CS network in the case where local competing networks fail.

It is essential for mobile network operators to be in a position to protect their networks from potential M2M-related overload so as not to degrade the quality of the CS or PS services that they currently provide to their subscribers, for example, voice calls or SMS.

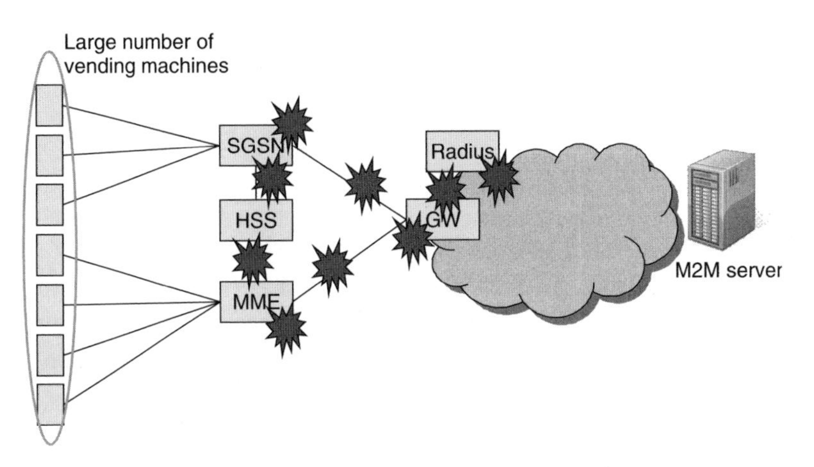

Figure 6.23 M2M signaling congestion in the mobile core network. Reproduced by permission of 3GPP.

6.3.7.1 Overload- and Congestion-Control Mechanisms

Generic overload- and congestion-control mechanisms are necessary to protect mobile operator networks from a complete collapse when the overload situation relates to an abnormal usage from a multitude of M2M applications and devices. The protection mechanism aims to manage the network load from all M2M devices/applications independently from other "traditional" devices, and will affect all or a significant number of M2M applications.

Application-specific congestion-control mechanisms are necessary to manage the network load from a particular M2M application or M2M devices of a particular M2M user, for example, to protect the mobile operator network from a badly behaving application. Since the congestion caused by one application should not result in adverse effects for other (M2M) applications, the protection mechanism is specifically targeted at that application without restricting non-M2M traffic or traffic from other M2M applications that are not causing a problem. Dedicated APNs or M2M group identifiers are possible means for identifying particular, large-scale M2M applications.

Rejecting connection or attach requests should not result in an M2M device immediately re-initiating the same request with the same or a different network. The network should be able to instruct M2M devices not to initiate a similar request until after a back-off time. This back-off time may also be used to instruct M2M devices with recurring applications to change their timing of attach/connection requests.

In many cases, M2M devices will be used as part of a contract between one network operator and a large multinational company. Typically, a multinational M2M customer, such as a car manufacturer, will want to have a contract with just one or a few mobile network operators, regardless of the various countries where the M2M devices are used. This inevitably leads to a great many M2M devices roaming most of the time. When many M2M devices are roamers and their serving network fails, then all these M2M devices will simultaneously move on to local competing networks, potentially overloading the networks that have not (yet) failed. Mechanisms are needed to provide protection from this kind of domino effect of failing networks.

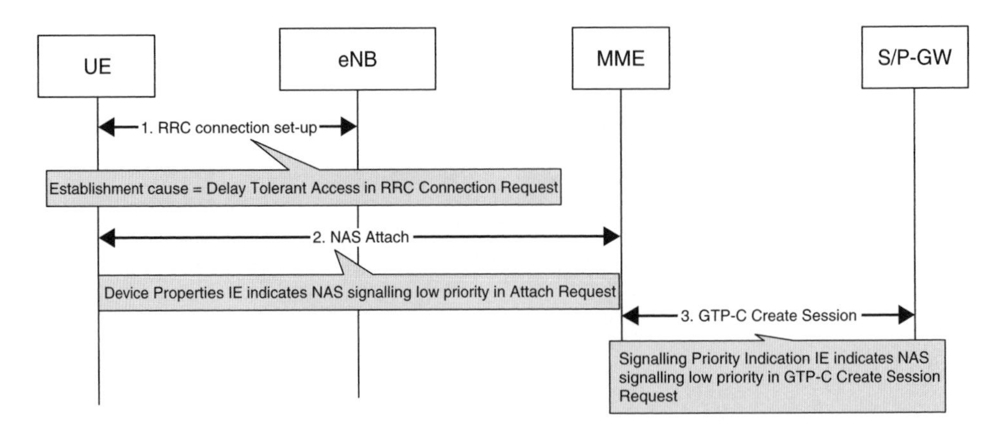

Figure 6.24 Attach procedure by a UE configured for low-access priority.

The following features have been defined in 3GPP Release 10 to protect mobile networks from M2M-related signaling congestion and overload. Although primarily motivated by M2M scenarios and use cases, these features are generally applicable: they have been specified in a generic way, allowing them to be used by any mobile device. The features can be activated or deactivated by the operator's configuration of the devices or by setting subscription data.

6.3.7.2 Network Overload Control for Mobile Devices Configured with "Low-Access Priority"

M2M devices using delay-tolerant M2M applications may be configured by post-manufacturing configuration "for low-access priority" pending agreement between the mobile operator and the M2M subscriber.

For mobile-originated services, an M2M device configured for low-access priority signals its "low-access priority" to the radio access network during the radio resource control (RRC)-connection establishment procedure and to the CS and PS core network (MSC, SGSN, and MME) during NAS[5] signaling procedures, for example, when registering to the CS or PS core network or when requesting the establishment of a data connection. During a data connection establishment, the PS core network nodes (MME or SGSN) forward the low-access priority indication in the connection-establishment request message it sends toward the gateway nodes (serving GW and packet data network GW).

Figure 6.24 depicts a high-level view of an attach procedure initiated by a mobile device (UE) configured with low-access priority, representing only the set-up of the RRC connection (i.e., the radio signaling connection), the initiation of the attach procedure toward the MME, and the establishment of the corresponding data connection toward the P-GW.

The low-access priority indication enables the radio access network nodes (RNC/eNodeB) and the core network nodes (MSC, MME, SGSN, SGW, and PGW) to decide whether or not to accept the RRC connection set-up or the NAS request based

[5] NAS signaling is signaling directly between the mobile terminal and a core network node, which transparently traverses the radio access network.

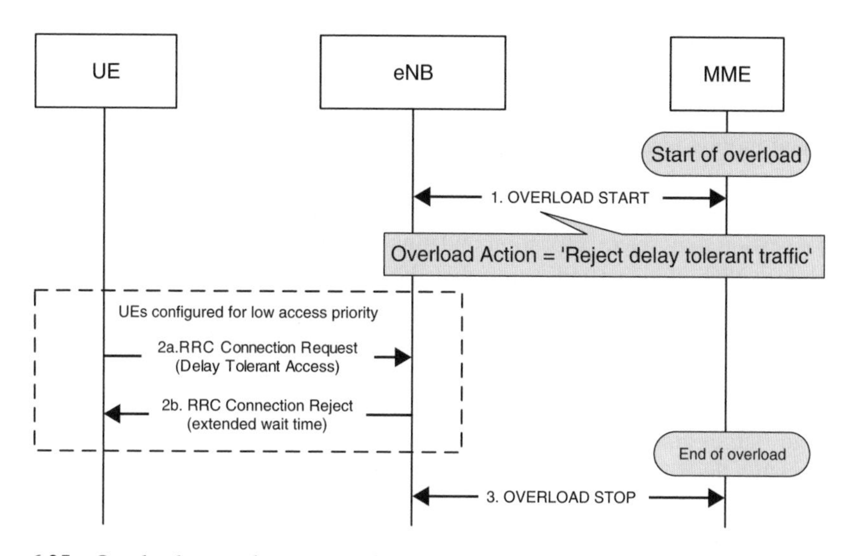

Figure 6.25 Overload procedure to restrict network access to UEs configured for low-access priority.

on the current network load. Radio access network and core network nodes experiencing general overload may reject requests from mobile devices configured with low-access priority before rejecting request messages without the low-access priority indicator. An MME experiencing general overload may also request the radio access network to restrict access to the network from devices configured for low-access priority by sending an "OVERLOAD START" message to a selection of eNodeBs. If necessary, the MME may also further request the radio access network to reject RRC-connection requests initiated by devices not configured for low-access priority, that is, to reject all RRC-connection requests that are for non-emergency and non-high-priority mobile-originated services.

When requested to do so, the eNodeB rejects RRC-connection establishment requests initiated by devices configured with low-access priority with an extended wait-timer during which further requests to the network are rejected.

When the MME has recovered and wishes to increase its load for mobile devices configured for low-access priority, the MME sends an "Overload Stop" message to the eNodeB(s).

The overall procedure is depicted in Figure 6.25.

Similar mechanisms have also been defined for an SGSN experiencing general overload. In an overload situation, an SGSN can request the radio network controllers to restrict the load of mobile devices configured for low-access priority. In such a case, the radio access network rejects the connection request.

RRC-connection establishments for the corresponding subcategory of devices with an appropriate timer value that limits further RRC-connection requests.

6.3.7.3 Generic Core Network Mobility Management Congestion Control

Under general overload conditions, an MME or SGSN may reject mobility management signaling requests from mobile devices (i.e., attach, tracking area update, or routing area update requests). When rejecting an NAS request, the MME or SGSN may provide the device with a mobility management back-off timer, during which time the mobile device will not initiate any further NAS request except for high-priority services, emergency services, and mobile-terminated services.

Similar procedures have also been defined for the CS domain. An MSC/VLR may also perform congestion control during mobility management, that is, an MSC/VLR experiencing congestion may reject a location updating request or IMSI attach message initiated by a mobile device, with an indication that the rejection is due to congestion and with a specific mobility management back-off timer for the CS domain. While the mobility management back-off timer is running, the device is not allowed to initiate any mobility management procedures, except for priority/emergency services and mobile-terminated services.

To avoid large amounts of mobile devices initiating deferred requests (almost) simultaneously, the MSC/VLR, MME, and SGSN should respectively select mobility management back-off timer values so that deferred requests are randomized.

The decision to apply mobility management congestion control may take into account the low-access priority indication if signaled by the mobile device.

6.3.7.4 Selective Throttling of Downlink Low-Priority Traffic Received for M2M Devices in Idle Mode

The above-presented mechanisms enable uplink signaling originated by mobile devices to be controlled or throttled. There is also a mechanism to protect the MME from massive simultaneous downlink traffic from M2M servers toward devices with established low-priority data connections. When a downlink user plane packet arrives at the serving GW for a mobile device in idle mode, the mobile device will need to be paged even though there is a logical data connection. This paging procedure is handled by the MME. An MME experiencing overload may restrict the signaling load that serving GWs are generating because of downlink user plane packets. This mechanism is depicted in Figure 6.26.

When the serving GW receives a downlink user plane packet for a mobile device in idle mode, it will send a downlink data notification to the MME. The MME may reject such downlink data notification requests for low-priority traffic or, to further offload the MME, it may request the serving GWs to selectively reduce the number of downlink data notification requests according to a throttling factor and for a specified duration.

The serving GW and the MME determine whether a bearer relates to low-priority traffic or not, on the basis of the bearer's allocation and retention priority (ARP) level and operator policy (i.e., the operator's configuration of the ARP levels to be considered as priority or non-priority traffic). The MME receives the ARP priority level from the serving GW in the downlink data notification message.

When throttling, the serving GW drops downlink packets received on all its low-priority bearers for mobile devices known as non-user-plane-connected (i.e., the serving GW context data does not indicate any downlink user plane tunnels) in proportion to the throttling factor, and sends a downlink data notification message to the MME only for the non-throttled bearers.

The serving GW resumes normal operations upon expiry of the throttling duration.

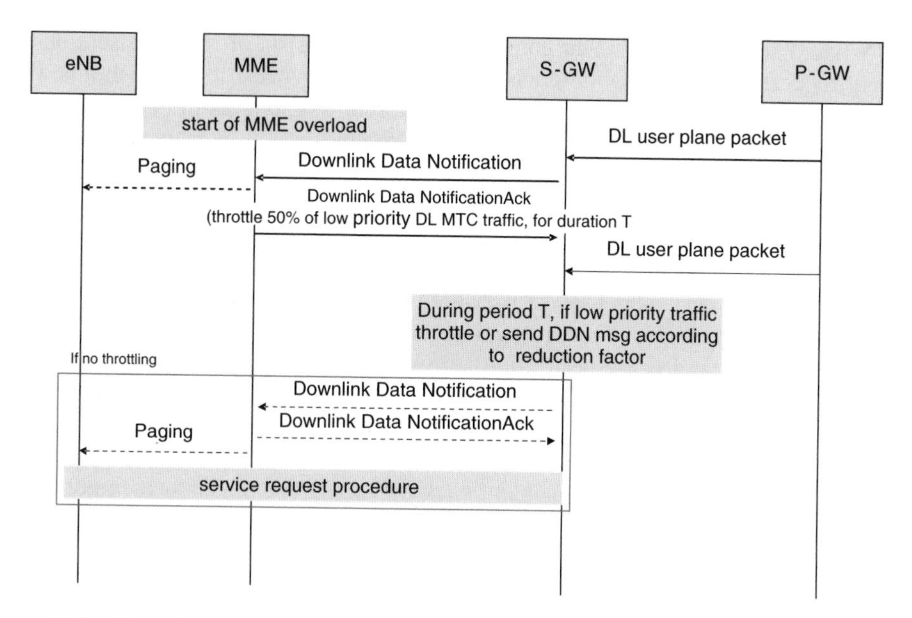

Figure 6.26 Throttling of downlink low-priority traffic received for mobile devices in idle mode.

6.3.7.5 Application-Specific Congestion Control

Application-specific congestion control is a protection mechanism for avoiding and handling signaling congestion associated with a particular APN, that is, which only targets the M2M applications using specific APNs.

A packet data network GW may detect APN-related congestion and start and cease performing overload control based on criteria such as:

- maximum number of active bearers per APN;
- maximum rate of bearer activations per APN.

When performing overload control, the packet data network GW rejects data connection establishment requests. In the response, it may include a session management back-off time indicating for how long the MME should refrain from sending subsequent data connection establishment requests. Upon receipt of the rejection from the packet data network GW, the MME may select an alternative packet data network GW serving that APN – if any – or may reject the data connection request from the mobile device.

An MME/SGSN may detect session management congestion associated with a particular APN, and start and cease performing APN-based session management congestion control by monitoring criteria such as:

- the maximum number of active bearers per APN;
- the maximum rate of bearer activations per APN;
- one or multiple packet data network GWs of an APN are not reachable or have indicated congestion to the MME.

APN-based session management congestion control consists of rejecting session management requests from the mobile device, for example, a request to establish a new data connection, with a session management back-off timer. As long as the back-off timer is running, the mobile device is not allowed to initiate any session management procedures for the congested APN except to release the connection. The mobile device may still initiate session management procedures for other APNs. The mobile device is also allowed to initiate session management procedures for priority or emergency services or to respond to paging. The mobile device will maintain a separate session management back-off timer for every APN that the mobile device may activate.

An MME/SGSN may detect mobility management congestion associated with a particular APN, and start and cease performing APN-based mobility management congestion control by monitoring the maximum rate of mobility management signaling requests associated with a particular APN. APN-based mobility management congestion control consists of rejecting attach requests initiated by mobile devices with a particular subscribed APN with a mobility management back-off timer. The device is then not allowed to initiate any mobility management procedures until the timer expires, except for priority or emergency services.

6.3.7.6 Optimizations to Prevent Overload from Network Reselection

The following mechanisms have been defined to protect a visited network from any overload caused by the failure of one (or more) other networks in that country.

Mobile devices may be configured by post-manufacturing configuration:

- with a long minimum periodic network search time limit to have an increased minimum time in between their searches for more preferred networks.
  Following the failure of a more preferred network, mobile devices may change to other local competing networks. Expiry of the search timer will lead to the mobile device re-attempting to access the preferred network, and then, if that network has not yet recovered, re-accessing one of the local competing networks. Using a too-short timer for the periodic network search can both prevent the failed network from recovering and impose more load on the competing networks.
- to perform an attach procedure with IMSI upon change of network, rather than doing a tracking area update procedure.
  A tracking area update procedure in general is more efficient than an attach procedure. But in the event of a network change, a tracking area update is generally not accepted, resulting in a subsequent attach procedure using the IMSI. Avoiding the tracking area update procedure reduces the message processing load on the local competing networks and hence makes it more likely that this network survives the failure of the other network.

It is also possible for an MME, SGSN, and MSC/VLR to signal longer periodic tracking area update, routing area update, and location update timers to mobile devices, as well as the option for mobile operators to foresee those periodic timers in the subscription data. Using these subscription-specific timers, as well as using the low-access priority indication signaled by mobile devices, the MME, SGSN, and MSC/VLR may allocate longer periodic update timers to M2M devices. This is likely to slow down the rate at which a UE detects network failure and thus slows down the rate of movement of devices from a failed network to other local competing networks. Using longer periodic timers also enables the reduction of the signaling load generated by M2M devices.

6.3.7.7 Extended Access Barring

EAB is a mechanism to allow the operator to control mobile-originating access attempts from particular M2M devices in order to prevent overload of the access network and/or the core network. Mobile devices that should be subject to EAB are configured to adhere to the EAB indications. In congestion situations, the operator can restrict access from mobile devices configured for EAB, while permitting access from other devices. Mobile devices configured for EAB are considered more tolerant to access restrictions than other devices. When an operator determines that it is appropriate to apply EAB, the network broadcasts the necessary information in a specific area.

EAB may also help to prevent overload from network reselection. EAB information may indicate that barring is applicable only to roaming mobile devices or only to mobile devices that are not on their preferred networks.

6.3.7.8 Evaluation of Congestion- and Overload-Control Mechanisms

There is not a one-size-fits-all overload mechanism that is applicable to all kinds of overload scenarios. Mobile operators will have to employ a range of overload control mechanisms to protect their networks. Table 6.6 illustrates the overload mechanisms that may be activated in some examples of overload scenarios.

Table 6.6 Overload mechanisms activated in different overload scenarios

Overload scenarios	Overload control mechanism(s)
External event triggering massive numbers of M2M devices to attach/connect all at once (e.g., M2M devices becoming active at the same time after power outage)	Network overload control for mobile devices configured for low-access priority, triggered by radio access or core network if a node experiences general overload
Malfunctioning in the M2M application and/or M2M server	APN-based congestion control targeting the malfunctioning application
A network serving many roaming M2M devices fails	Extended access barring Long minimum periodic network search time limit Attach with IMSI at network change Long periodic TAU/RAU/LAU update timers Generic core network mobility management congestion control
One-hour peaks, New Year's Day	M2M low-priority traffic can be rejected during those periods by network overload control for mobile devices configured for low-access priority triggered by radio access or core network

References

[22.011] 3GPP TS 22.011. Service Accessibility.
[22.888] 3GPP TR 22.888. Study on Enhancements for Machine-Type Communications.
[22.988] 3GPP TR 22.889. Study on Alternatives to E.164 for Machine-Type Communications.
[22.368] 3GPP TS 22.368. Service Requirements for Machine-Type Communications (MTC); Stage 1 (2011).
[23.888] 3GPP TR 23.888. System Improvements for Machine-Type Communications (MTC).
[23.682] 3GPP TS 23.682. Architecture Enhancements to facilitate communications with Packet Data Networks and Applications (2011).
[23.060] 3GPP TS 23.060. GPRS Service Description; Stage 2 (first published 2000).
[23.401] 3GGP TS 23.401. GPRS Enhancements for E-UTRAN Access (2008).
[OMA DM] Open Mobile Alliance Device Management Architecture (2006).
[TR-069] Broadband Forum TR-069 (2004). CPE WAN Management Protocol.
[33.220] 3GPP TS 33.220. Generic Authentication Architecture; Generic Bootstrapping Architecture (GBA) (first published 2004).
[ECC 153] Electronic Communications Committee (2010), Numbering and Addressing in M2M Communications, ECC Report 153.
[E.164] ITU-T Recommendation E.164. The International Public Telecommunication Numbering Plan (2010).
[E.118] ITU-T Recommendation E.118. The International Telecommunication Charge Card (2006).
[23.040] 3GPP TS 23.040. Technical Realization of the Short Message Service (SMS) (first published 1999).
[23.041] 3GPP TS 23.041. Technical Realization of Cell Broadcast Service (CBS) (first published 1999).
[23.002] 3GPP TS 23.002, Network architecture http://www.3gpp.org/ftp/Specs/html-info/23002.
[23.003] 3GPP TS 23.003, Numbering, addressing and identification.

7

The Role of IP in M2M*

Laurent Toutain and Ana Minaburo
Telecom Bretagne, Cesson Sevigne Cedex, France

7.1 Introduction

Very few computer network protocols lasted as long as the IPv4 protocol. Even if, at the outset, IP was never designed to become the universal protocol we know nowadays, it was made versatile enough to evolve and cover an ever-increasing number of applications. At first glance, the success can only be linked to Metcalfe's law stating that the value of the network (i.e., the benefit that users get from the technology) is the square of the number of users. Even if [1] tempers this formula to *n.log (n)* (where *n* is the number of users), this leads to a mitigated virtuous circle: on the one hand, the network becomes more attractive, while on the other, resistance to deploying new protocols or new forms of behavior also increases.

For the general public, the Internet is synonymous with services offered by the network, but for network engineers, the Internet can be viewed as a protocol stack defined by IETF with RFC 791 [2] at layer 3, some transport protocols such as TCP [3], UDP [4], and more recently SCTP [5] or DCCP [6] and finally some applications such as DNS [7].

Another definition can be based on the actual name itself since it is composed of *interconnection* and *network*, and sometimes the motto "the network of networks" is used. This more general definition does not take into account the fact that any protocol leads to a more "philosophical" approach where the network design and architecture become more important than protocols. This definition is more accurate for M2M communications. Protocols developed for the current Internet do not fit well when exposed to M2M

* The authors want to thank Dominique Barthel from Orange Lab and the ANR ARESA2 project for their contribution to the chapter.

M2M Communications: A Systems Approach, First Edition.
Edited by David Boswarthick, Omar Elloumi and Olivier Hersent.

constraints. Energy constraints are not taken into account. Internet protocols are sometimes very talkative or require states that are not available in an upgradable environment. Auto-configuration for regular services or for bootstrapping are needed since these items of equipment do not necessarily have a keyboard or a massive deployment of devices, nor do they allow manual or managed configuration. Internet evolution for supporting M2M and wireless sensor network (WSN) requires more than just a few simple changes in the routing protocols. It leads to a drastic redefinition of header formats, protocols, routing algorithms, and application design with a revised role for addresses. On the other hand, interoperability with existing items of equipment and current applications is needed, since otherwise the effort to include Internet protocol will have been in vain.

The Internet is currently undergoing a drastic protocol evolution. Packet format is changing to include a 128-bit rather than the 32-bit address used initially with version 4 of IP (IPv4). Current forecasts envisage an exhaustion of IPv4 addressing space at the middle of 2012. The new version of the Internet standard (IPv6) has been standardized for a while, but even if some major players have included IPv6 in their network devices, actual deployment of IPv6 remains very limited. Metcalfe's law works against IPv6, since most of the services or content are only available in IPv4 and there is no real need to use the new version of the protocol. Since there is no urgent need, new applications are deployed, perhaps with greater difficulty, on version 4 of the protocol, and the need for version 6 remains low. The scenario foreseen by the IETF, to simultaneously run both versions of the protocol, has failed and the most likely scenario will be that providers will adopt IPv6 addresses in the core network, with legacy applications remaining on IPv4.

If IPv4 remains the main protocol for legacy applications, new areas of telecommunication leading to a massive deployment cannot be supported by this IP version. The 32 bits dedicated to addresses in IPv4 theoretically only allow up to 4 billion devices if the constraints due to routing protocol and allocation policies are not taken into account. This number is relatively low considering the number of cars, machines, and sensors that can be connected to the Internet. IPv6 appeared to be an attractive solution for services that will be deployed in a near future. Certain new network infrastructures, such as LTE, also made the choice to mandate IPv6.

7.1.1 IPv6 in Brief

IPv6 protocol is a simplification of the IPv4 packet and retains the same principle. Fields that proved to be unused after years of exploitation have been removed. The first of these, the option field leading to a variable header size, has been withdrawn and replaced by extensions. An extension can be viewed as an intermediary protocol between Layer 3 (IP) and Layer 4 (UDP, TCP), except for the hop-by-hop extension that must be processed by each router on the path. Other extensions are invisible from the core network and only processed by the final destination. Nowadays, very few extensions are really used apart for those defined for IPsec. They are mainly used to manage mobility with mobile IP and multi-homing with shim6.

Fragmentation has been removed from the main header and is now viewed as an extension. This leads to a simpler packet with more stable and predictable fields. Even if the size of the addresses is four times larger, the size of the header is only twice the size of an IPv4 header, as shown in Figure 7.1.

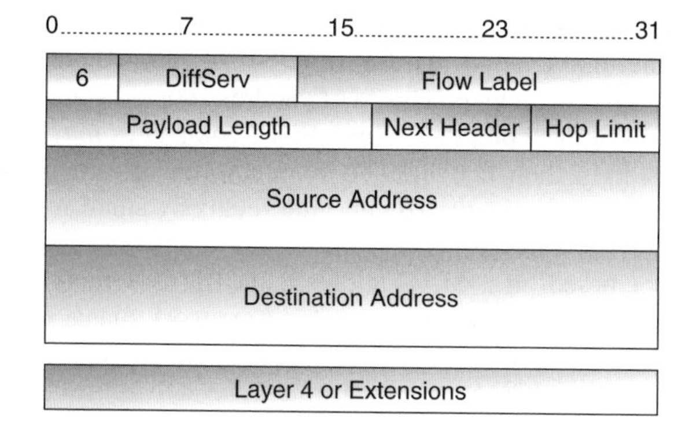

Figure 7.1 IPv6 header.

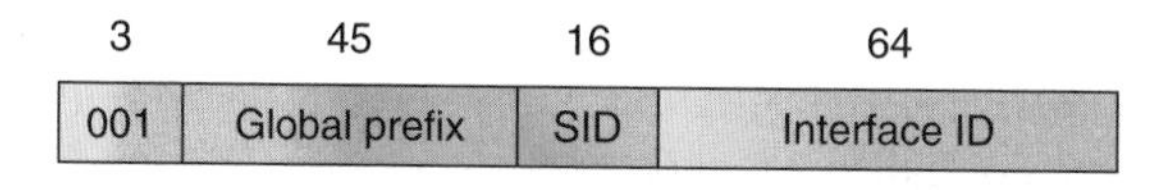

Figure 7.2 IPv6 header.

The only new field introduced by IPv6 is flow label. In the early days of IPv6 design, flow label attempted to allow easy flow identification within the core network, since Layer 4 port numbers cannot be as easily accessible as in IPv4 due to the use of extensions. As multiprotocol label switching (MPLS) has since redefined the concept of flows, the usage of flow labels became less important.

The packet format modifications create connectivity problems between both versions of the IP protocol leading to the use of application level gateways (ALGs) or complex header translations to guarantee compatibility between IPv4 and IPv6 applications. Nevertheless, the way the network is managed remains the same. Addresses still follow CIDR rules to be allocated and routed. Most of the protocols associated with IP must evolve to take into account larger addresses, but the algorithms are not changed.

The large address space provided by IPv6 not only allows a larger number of devices to be connected to the Internet, it also simplifies the constraints on the network management. More clearly than in IPv4, the IPv6 address is divided into three parts, as shown in Figure 7.2. The first part, global prefix (GP), is allocated by the ISP and is used to route the IPv6 packet in the Internet core. The site network manager assigns the second part – the subnet identifier (SID) that is allocated to forward the packet internally to the appropriate network. GP and SID must fill the first 64 bits of the address. Generally, the GP is 48 bits long and the SID 16 bits long, but depending on the site size and the provider, these values can be different. The remaining 64 bits of the address are dedicated to the interface identification (IID). Address notation can be represented with 16-bit words in hexadecimal, separated by colons ":". A colon repeated twice represents a single long sequence of zeros, but this can be done once in an address.

There are several different ways of allocating IID. Originally, the IID was derived from the MAC address (MAC-48 or EUI-64) of the interface. This simple way of ensuring a single allocation on the link had certain drawbacks: even if the device is moved from one network to another and changes the GP:IID::/64 part, the IID remains the same and some servers may follow the device. The MAC address contains an organization unique identifier (OUI) indicating the computer brand. Some companies may be reluctant to send this information outside of their network. To counteract these criticisms, the IID can be randomly drawn; the large size of the IID and the relatively small number of devices on a link makes conflicts very rare.

Whether random or based on a MAC address, the IID is sometimes very difficult to manage since it is unpredictable and difficult for a network manager to remember. IID can be manually assigned with a small value that can be easily remembered. IID can also be derived from a certificate (CGA: cryptographic generated address); this way, the host is authenticated in the network.

IPv6 defines other prefixes:

- FE80::/64 is used with random or MAC IID-generated to create link-local (LL) addresses. Packets using those addresses cannot be routed outside of the link, but are used during the bootstrap period to allow the starting hosts to build a configuration at Layer 3 with a protocol called neighbor discovery (see Section 7.3.3). LL addresses are also found in the routing table.
- ULA (unique local address) addresses are composed of a prefix randomly generated for the site. The packet carrying these addresses cannot leave a site, but may be used if the site is not connected to the IPv6 or to isolate some devices from the global Internet, as was the case for private addresses in IPv4.
- Multicast prefixes starting with FF00::/8. The four bits following the prefix contain some flags used to manage large multicast networks. The next four bits give the scope of the multicast (one stands for multicast limited to communications inside the host, two for communications on the link, five limits to an administrative entity, etc.).
- Some special groups are defined: for instance, FF02::1 is used to send a packet to all the devices connected to the sender's link and FF02::2 represents the group of routers connected to the sender's link.

7.1.2 Neighbor Discovery Protocol

Neighbor discovery protocol (NDP) is a new way of automatically inserting a host into an IP network. For IPv4, a client/server approach was possible using DHCP. IPv6 redefines DHCP but is mainly used to distribute static parameters, such as DNS servers, although DHCPv6 can also be used to allocate addresses to a host, or prefixes to a router. IPv6 promotes stateless auto address configuration (SLAAC) mechanisms. NDP uses four specific ICMPv6 messages – RS: router solicitation; RA: router advertisement; NS: neighbor solicitation; and NA: neighbor announcement. The first pair of messages are used for an exchange between a node and a router and the latter pair between two nodes and are mainly used to resolve mapping between the IPv6 addresses and a MAC address, as was done through ARP in IPv4.

When a node is initializing, it creates its LL address using the well-known FE80::/64 prefix and concatenates its IID, before testing that no other host has this address by sending an NS request, mapping its own LL address. If it does not receive a response, the node assumes the uniqueness of its LL address and that it can use it for communication with the router. This phase is known as DAD (duplicate address detection) and several NS probes can be sent to guarantee that neither the request nor the answer has been lost on the link.

If the DAD succeeds, the node can query the router for configuration parameters using RS messages. The router answers with several parameters, such as:

- the prefix used on the link and its length;
- whether the node can construct the global address by concatenating the received prefix with its IID, otherwise, a DHCPv6 request must be sent to obtain the global address;
- whether some static parameters can be retrieved through DHCPv6;
- whether the link is a broadcast link where nodes can directly talk to each other or a non-broadcast multi-access (NBMA) link where all communications must go though the router, in which case, there is no need for address resolution (NS/NA exchange) since the router knows this mapping;
- maximum transmission unit (MTU) and hop limit (HL) used on the link.

A node configuring its address with a prefix contained in an RA message will use the replying router as the default router. GP uniqueness is also checked using DAD. To avoid having to use DHCPv6 queries to obtain the DNS resolver, the RA message can also optionally contain the resolver's IPv6 address.

7.2 IPv6 for M2M

In an M2M network, not all devices have the same capabilities to run over the networks that are initially defined for IPv6. Energy and bandwidth consumption are not taken into account. IPv6 also supposes that the transmission support has functionalities close to those of the Ethernet. Wireless sensor networks (WSN) is the first area that has shown that these constraints were not realistic. They cover only one part of M2M networks (which can include 3G networks) but are representative of these new areas for networking. Not so long ago, applications for WSN were directly designed over a Layer 2 protocol such as IEEE 802.15.4, giving limited flexibility, since the entire networking part, such as routing, has to be developed for each case. This also gives rise to interrupt applications that cannot directly talk with other devices on the Internet. The ZigBee forum defines a protocol stack and profiles for sensors and actuators with better interoperability. Networks can also be used in different ways to carry alarm system messages and lighting control. Nevertheless, the ZigBee network is a walled garden, and communication with the outside world has to go through application gateways. This can make the deployment of new services more complex since the gateway has to be modified to take new applications into account.

As presented before, IP offers a more flexible architecture, since routers do not take applications into account and are able to forward packets from one interface to another and may adapt the packet delivery to reflect the nature of the media. The 6LoWPAN working group at the IETF works on the adaptation layer that allows IPv6 to be carried on LoWPAN networks. Even if a LoWPAN network definition is more generic than IEEE 802.15.4, the work is mainly targeted at working in the low-powered environment.

If some home automation applications, such as lightning control, can support the ZigBee closed model, some others such as ZigBee Smart Energy require interconnection with the outside world, since some messages have to be exchanged with energy providers to take actions based on the energy price, so global interconnection is needed. The ZigBee forum has selected IPv6 for this profile. In other domains, they have made a similar choice, as ISA is the defining standard for sensors in the industry. They have also selected IPv6, not for its interconnection capabilities, since industrial processes are mainly controlled locally, but in order to benefit from the cost reductions generated by deploying equipment that has been developed for the mass market.

The main constraints concerning IPv6 and related protocols on LoWPAN networks are the following:

- In the standard defining IPv6 [8], a link layer must support datagram with a minimum of 1280 bytes for payload. This limitation stems from NDP, and NDP messages, especially RA, should not be fragmented. This size can also allow several encapsulations of an IP header to support tunneling. The standard also states that if a link layer cannot support this minimum size, an adaptation layer must be used to hide these constraints. This was the case with ATM and the adaptation layer AAL5.
- For WSN, a new adaptation layer has been developed supporting IPv6 header compression and a more accurate support of the Layer 2 protocols allowing broadcast and routing facilities.
- As was shown in the previous chapter, IPv6 adapts the NDP protocols to WSN in order to generate a background multicast traffic on the links. Furthermore, in LoWPAN, a link is not well-defined since the frame scopes vary very rapidly over time. The adaptation will consist of drastically reducing the number of messages produced by the protocol and of centralizing the information on some well-known and reachable items of equipment.

The IETF working group for routing over low-power and lossy networks (Roll) is currently developing a routing protocol that considers the nature of the underlying link and the traffic patterns. They are working on RPL routing protocol for WSN. RPL takes into account the evolution over time of links between nodes, mainly due to the weakness of the radio signal. Finally, a Layer 4 protocol, such as TCP used for control, is not always appropriate for communication with sensors. The opening and closing phases of connection-oriented protocols are too heavy for simple query/response exchanges. TCP is also a very complex protocol requiring a large footprint in sensor memory to handle connections efficiently. Some simple query/response schemes can be deployed over UDP. The IETF CoRE working group is currently focusing on that topic.

7.3 6LoWPAN

6LoWPAN working group aims to adapt IPv6 protocols that can be supported on LoWPAN networks. The main goal is to ensure interconnection between the Internet and the WSNs in order to adapt the IPv6 protocol, and associated protocols such as NDP, to LoWPAN network specificities. The 6LoWPAN protocol is implemented in sensor nodes and one or several LoWPAN border routers (LBSs) connected to the public Internet. The LBR device contains router functionalities and may also compress and decompress packets. Several constraints require an adaptation of IPv6 to LoWPAN environment:

- Layer 2 frame size may be limited compared to the IPv6 specifications with a minimum mandatory MTU of 1280 bytes. Fragmentation is required to provide the frame lengths specified by IPv6 RFCs.
- Energy can be drastically limited in some nodes. For instance, some metering devices may run for several years (up to 15) on the same battery. Unnecessary exchanges and multicast traffic must be avoided in order to save power.
- Neighbor discovery messages should not be fragmented. If some fragments are lost, the protocol becomes inefficient.
- Radio range may be reduced and the IPv6 packet will have to be relayed from node to node to reach the destination or an LBR.
- The neighborhood is not as well-defined as in wired or WiFi networks and radio range variation causes the neighbor list to change.

A 6LoWPAN network can be organized around three topologies:

- **Star topology:** All sensor nodes can reach and are reachable from the LBR.
- **Meshed:** Nodes are organized at Layer 2 in order to relay frames toward the destination. Algorithms to organize the interconnection are not defined by 6LoWPAN, which simply offers generic support to manage broadcast and hop-by-hop bridging. From the point of view of the Internet, a meshed network is similar to an Ethernet network where every node shares the same prefix. 6LoWPAN refers to that technology as mesh-under (MU).
- **Routed:** Nodes act as routers and forward packets toward the destination. A routing protocol must be running on some nodes in order to construct forwarding information. Nodes acting as a router inside the LoWPAN network and not directly connected to the Internet are called LoWPAN routers (LRs). 6LoWPAN refers to that technology as route-over (RO). The best candidate is the RPL protocol, as defined by the Roll working group.

7.3.1 Framework

The work is organized into two aspects. The first is the compression of the IPv6 header and the above-mentioned protocols in order to minimize the overhead and allow simple queries and NDP messages to be held in a single limited-size frame. The second task is to adapt NDP to the specificities of the link, especially when NDP has to interact with routing protocols.

7.3.2 Header Compression

6LoWPAN is not linked to a specific form of technology. However, current implementations are mainly based on IEEE 802.15.4. Four types of frame are defined in this MAC protocol network: beacon, data, acknowledgment, and control, as shown in Figure 7.3. Each one starts with 32 bits of preamble and a start frame delimiter to synchronize the

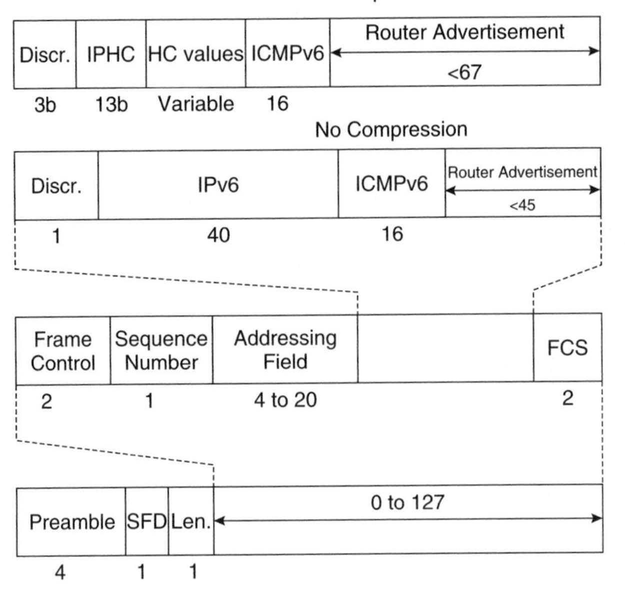

Figure 7.3 IEEE 802.15.4 RA encapsulation.

receivers. This is followed by the length coded on seven bits allowing up to 127 bytes of data. Thereafter, a frame-control field gives some information, such as the type of frame, the presence and size of the source and destination addresses (16 or 64 bits), the use of a PAN (personal area network) identifier (PADid). A sequence number is used to identify and acknowledge frames. The acknowledgments are not used when frames carry an address field composed of the PANid and the addresses for both destination and the source. Only the acknowledgment frame is empty, while the others contain data. All of the four frame types end with a CRC.

The protocol allows a maximum data size of 127 bytes, but the IPv6 standard [8] imposes a minimum value of 1280 bytes. 6LoWPAN [9] contains a fragmentation protocol to adapt to this constraint. 6LoWPAN also offers a header-compression mechanism that reduces the size of the IPv6 header, thus avoiding the need to send well-known information or information found at the sensor network layer. The compression technique defined in the standard will be soon obsolete and replaced with a more efficient version called IP header compression (IPHC).[1] In this instance, we will only describe this new header-compression algorithm. The protocol also carries some fields that can be used for mesh networks or broadcast support, but there are no protocols defined, since Layer 2 protocols are out of the scope of IETF specifications. 6LoWPAN may also define L4 header compression. Currently, a mechanism only exists for UDP, but other protocols, such as ICMPv6 or TCP, should also soon be covered.

[1] RFC 6282.

6LoWPAN defines the first byte of the data field as a discriminator, which indicates how the data is structured, for instance:

- 01000001 indicates that the remaining bits contain a non-compressed IPv6 header;
- 011 indicates that the header is compressed using the new compression scheme;
- 10 indicates a mesh encapsulation, that is, source and destination addresses;
- 01010000 carries a tag limit propagation of broadcast frames in a mesh topology;
- 11000xxx and 11100xxx indicate a fragmentation header, that is, the first and remaining fragments, respectively.

The way in which these values are placed into the frame is fixed in order to simplify the decoding. If they are present, the frame must start with the mesh header and the broadcast header, followed by the fragmentation header and finally the L3 header (compressed or not) and L4 header (compressed or not).

Even if the discriminator value is selected in order to avoid collisions with ZigBee data frames, there is no reliable way of distinguishing between 6LoWPAN packets and ZigBee packets, so only one of these protocols should run on a WSN at any given moment.

Figure 7.3 gives an example of an ICMPv6 message with and without IPHC compression, where, in the best cases, the values carried after the IPHC discriminator can be reduced to two bytes, compared to the 40 bytes of the original header. For more details on the compression result, see Section 7.3.2.5. The following sections describe the discriminator behavior.

7.3.2.1 Interface ID

6LoWPAN benefits from the IID derived from MAC addresses. Since the MAC address is supposed to be unique, the DAD algorithm will not be required to guarantee that the IPv6 address is unique. This will also help the compression process developed by 6LoWPAN. If the address is carried at Layer 2, then it is not necessary to send it on to Layer 3.

If the Layer 2 address is coded on 16 bits, then the IID is built concatenating 0000:00ff:fe00 before it.

7.3.2.2 Mesh Discriminator

In a radio network, every node located in a radio range can receive the frames. If all the receivers copy the frame toward the destination, the destination will receive several copies of the original message. To avoid this, the sender must specify which node will relay the frame toward the destination. In a mesh network, the addresses contained in the frame header are used to design hop-by-hop nodes, that is, the node that really sends the frame is its neighbor that should then relay the frame toward the destination. In the mesh header, as described in Figure 7.4, 4 carries the end-to-end addresses. The V and F bits are used to indicate the address length of the originator (Very first address) and the Final destination; if it is set to 0, the length is four bytes and if it is set to 1, the length is 16 bytes. The address values are given after the discriminator. The hop left field is used to limit the impact of infinite loop in the mesh network.

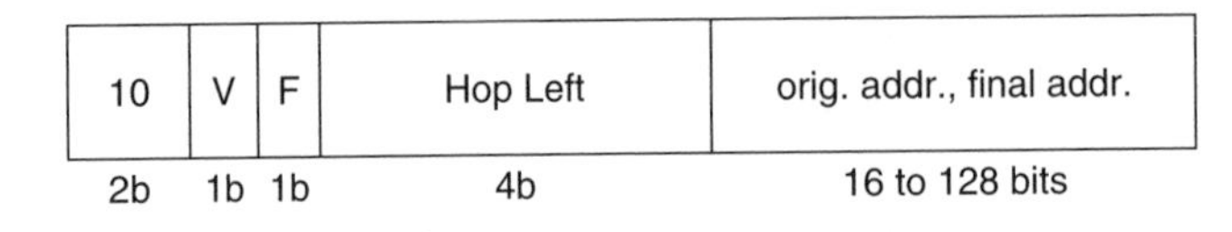

Figure 7.4 Mesh discriminator.

6LoWPAN does not define any specific algorithm for discovering the next hop address regarding a specific destination. The discriminator used for MU forwarding is not widely used. RO, with each node acting as a router, is preferred. In that case, the discriminator is not required. The packet is forwarded by looking at the destination address, while the routing table gives the next hop. Routing protocols, such as RPL, are used to fill this routing table.

7.3.2.3 Broadcast Discriminator

If a broadcast frame is needed to reach all the nodes in the WSN or to easily but inefficiently implement multicast, a flooding mechanism has to be implemented. The broadcast discriminator carries information to avoid loops. After the eight-bit discriminator, an eight-bit unique sequence value is selected by the source. Even if the RFC does not specify how it should be used, the algorithm is relatively simple. When forwarding a message with the broadcast discriminator, the node memorizes the source address and this sequence value for a period of time. If during this interval a new frame is received with the same source address and sequence number, the frame is simply discarded.

7.3.2.4 Fragmentation Discriminators

These two discriminators manage fragmentation as shown in Figure 7.5. The fragmentation process is very similar to the one used by IP. A tag selected by the router performing the fragmentation is used to distinguish between fragments belonging to different datagrams.

All fragments carry the original datagram length, and all fragments, except the first, carry the offset giving the location in the original datagram.

7.3.2.5 Header Compression Discriminator

IPv6 header compression is the most important part of the 6LoWPAN protocol. This compression is stateless, which means that no context information is kept to compress or reconstruct a packet header. In this way, the rerouting can occur and the datagram may leave through different LBRs. The compression scheme is based on the fact that most of the fields of the IPv6 header are predictable, and therefore the length can be measured; the flow label is often set to 0 and if IPv6 addresses are built including EUI-16 or EUI-64, it is not necessary to send this value since it can be found in the Layer 2 frame. The prefix part is more complex to compress, since in [9] compression was only optimal for link-layer addresses and then it can only be efficient for NDP. The RFC 6282 generalizes the compression mechanism to any kind of prefix and also multicast addresses. Since the prefixes are not predictable, they can be stored in some tables, and the compression mechanism will use an index in order to refer to a specific prefix.

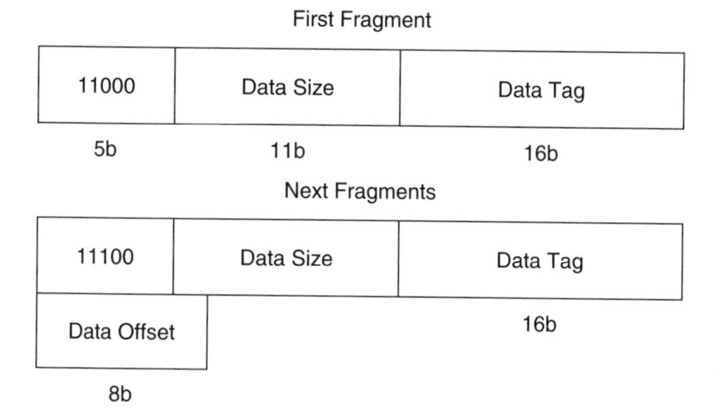

Figure 7.5 Fragmentation discriminator.

011	TF	NH	HLIM	CID	SAC	SAM	M	DAC	DAM
3b	2b	1b	2b	1b	1b	2b	1b	1b	2b

Figure 7.6 IPHC discriminator.

The packets start with a three-bit discriminator followed by a 13-bit bitmap describing which parts of the original IPv6 header must be sent and which parts can be avoided. Figure 7.6 gives the values of the different flags.

In keeping with coding convention, when all the bits of a flag are set to 0, the field is sent online after the discriminator, and if they are set to 1 the field is omitted. Since the length of all fields is known, it is easy for the destination to recover the value.

The TF (traffic class, flow label) covers the DiffServ Field (DSCP: six bits), the explicit congestion notification (ECN: two bits) and the flow label (20 bits). Flag values avoid either the DSCP or the flow label.

Next header (NH) flags specify how the Layer 4 protocol is processed. If the flags are set to 0, the NH field is sent unmodified into IPv6 header. If they are set to 1, another tag indicating the upper compression mechanism can be defined after the IPHC header values. Currently, the standard defines two possible tags: one for IPv6, which can be used to carry IPv6 extensions or IPv6 tunnels, and also a simple compression scheme for UDP. The other mechanism may be defined in the future for other Layer 4 protocols, such as TCP.

HLIM (hop limit) flags define some well-known values in addition to the online value.

Context refers to a prefix known by every sensor. This avoids sending the prefix value in each IPHC header. A single context (0x) is, by default, defined. This context can be different from the source address and the destination address. When the CID flags are set to 1, this indicates that 16 contexts can be used. In that case, an extra byte of information is added in the value field after the discriminator. The first four bits of the context field refer to the source prefix values and the remaining four bits to the destination

Table 7.1 SAC and SAM bits

SAC/SAM	00	01	10	11
IID		64 first prefix bits are not sent, IID is fully sent	112 first prefix bits are not sent, last 16 IID bits are sent	128 bits are not sent
0	Address is sent completely (Link-Local and Global)	Prefix is FE80::/64	Prefix is FE80::0:ff:fe00 /112	Prefix is: FE80::/64 and IID is taken from L2 source address
1	Unspecified address (::/0 (not sent))	Prefix is given by the context	Prefix is given by the context, IID starts with 0000:00ff:fe00: and 16 bits online	Prefix is given by the context and IID is taken from the L2 source address

prefix values. Synchronization between context numbers and values can be performed by enhanced NDP for WSN (see Section 7.3.3).

The SAC (source address compression) flag gives the nature of the prefix for the source address. If the flag is set to 0, the prefix is mostly a LL prefix. Table 7.1 gives an overview of possible values.

Compression will differ in terms of topology. SAC/SAM (source address mode) = 0/00 can be used in any situation since it does not lead to any compression at all. SAM = 11 is mainly defined for star topology and the MU network, since the L2 address must be present in the header, the IEEE 802.15.4 header, or in the mesh header, respectively. SAM = 01 or SAM = 10 can be used in the RO topology since hop-by-hop forwarding renders the information contained in the L2 header strictly local between two hops. These two values are necessary for distinguishing between the lengths of the source codes if they are addressed on 64 or 16 bits.

The principle remains the same for the destination address. A discovery address confirmation (DAC) flag indicates whether the prefix is link-local or from a context. The destination address can also be a multicast address, so an M flag is added to distinguish from unicast addresses. Table 7.2 gives an overview of possible values.

7.3.3 Neighbor Discovery

In a LoWPAN network, the radio range is limited, and not all of the nodes can talk directly. Moreover, the radio range may vary drastically due to, for instance, the air humidity level or disruption generated by other networks or devices. Bidirectional traffic cannot always be guaranteed. The link definition on which the NDP protocol relies, as defined in [10], is not as clear as the definition used with Ethernet, or even WiFi. The NDP has to be adapted not only to take into account the specificities of the link, but also to reduce the energy required to assure the functionalities that allow the node to talk with its neighbors.

NDP relies extensively on Layer 2 multicast to obtain network parameters, run DAD, or resolve the mapping between the IPv6 address and the MAC address of a remote

Table 7.2 M, DAC, and DAM bits

M-SAC/ SAM	00	01	10	11
00	Address is sent completely (link-local and global)	Prefix is FE80::/64	Prefix is FE80::0:ff:fe00 /112	Prefix is: FE80::/64 and IID is taken from L2 source address.
01	Reserved	Prefix is given by the context	Prefix is given by the context, IID starts with 0000:00ff:fe00: and 16 bits online	Prefix is given by the context and IID is taken from the L2 source address
10	Address is sent completely	48 bits are sent online and are spread in a multicast address as: FFXX::00XX: XXXX XXXX	32 bits are sent online and are spread in a multicast address as: FFXX::00XX: XXXX	8 bits are sent online and are spread in a multicast address as: FF02::00XX
11	48 bits are sent. They are used for large-scale multicast as defined in [11]. Context value contains the rendezvous point address	Reserved	Reserved	Reserved

node, since Ethernet, WiFi, or bridged networks easily support this kind of traffic. On a LoWPAN, multicast implies a more complex implementation:

- In MU models, a specific dispatch value has been defined to allow broadcast. An L3 multicast address can be easily mapped into an L2 broadcast address but this leads to network flooding and all nodes will have to process the message.
- RO networks can be viewed as being derived from works on NBMA[2] network where multicast may be done by a specific server. In the adaptation of NDP to RO LoWPAN, multicast will only be allowed when a node needs to discover neighboring routers. Once the address of a router has been determined, the traffic will then be sent using unicast.
- Furthermore, the RO network is mainly composed of nodes acting as routers to relay the information to the destination, but from the addressing perspective, it appears as a single link, that is, the same prefix may be assigned to all nodes. NDP is not, by construction,

[2] Non-broadcast multiple access: any node can be reached, but only one at a time, as with telephony networks.

supposed to cross routers, since the specification was originally intended for use with a physical link. NDP for 6LoWPAN introduces the concept of multi-hop prefixes. NDP should not be viewed as a routing protocol, but will to allow communication between physically reachable neighbors. A routing protocol inside the LoWPAN, such as RPL, is needed to build connectivity between all the nodes belonging to a LoWPAN.

Another way of reducing multicast traffic is to avoid periodic RA as a way of informing the neighboring nodes of the router's status and the LoWPAN prefix values. In a LoWPAN network, each node has to verify, before the expiration of its address, that the prefix and network parameters are still valid.

DAD usage is also fully reviewed. DAD is based on a time-out mechanism that is used to guarantee that the address has not been assigned to another node. In a LoWPAN, the time-out may be exceeded due to the neighboring node being temporarily unreachable. This means that the DAD process will be inefficient.

In a LoWPAN network, it will be assumed that the EUI-64 is uniquely assigned, which is the case if the IEEE process has been correctly respected. The LL addresses derived from IEU-64 are, by construction, unique and do not need to be verified in order to guarantee uniqueness. The same assumption can be made for addresses determined through DHCPv6. For other addresses, such as random values or those issued from a certificate, verification will be carried out by some central elements, which will maintain a list of addresses used in their area and the MAC address of the sender.

The goal is to reduce the traffic and avoid the use of multicast messages. Periodic router advertisement messages are suppressed and RA can only be sent using unicast when a node requests it.

7.3.3.1 Link-Local Addresses

LL addresses are supposed to be on-link. Since the IID is based on the MAC address and by definition they share the same media (i.e., there is no router between the source and the destination) there is no need for address resolution. In the MU model, the LL address may be used to reach any host on the LoWPAN. Layer 2 maintains a type of routing protocol (not specified by IETF) to reach any other node. In the RO model, the LL address can only be used to directly contact neighbors that are reachable by the radio interface.

Figure 7.7 provides an example of neighbor discovery. Each node has a Layer 2 EUI-64 that is unique by design. Neighboring nodes create their LL addresses by concatenating FE80::/64 with their respective EUI-64 values. If at Layer 3, node 1 has to send a packet to node 2, the MAC address of node 2 will be extracted from its IPv6 address.

7.3.3.2 Global Addresses

Contrary to LL addresses, there is no way for global addresses to distinguish between an IID based on an IUE-64 and, for example, a random value. For this reason, this kind of address can be managed in the off-link mode. IPv6 defines this mode (carried in RA messages) that mandates that all packets are sent to the default router Layer 2 address, which has a better vision of surrounding hosts and will be able to resolve the mapping

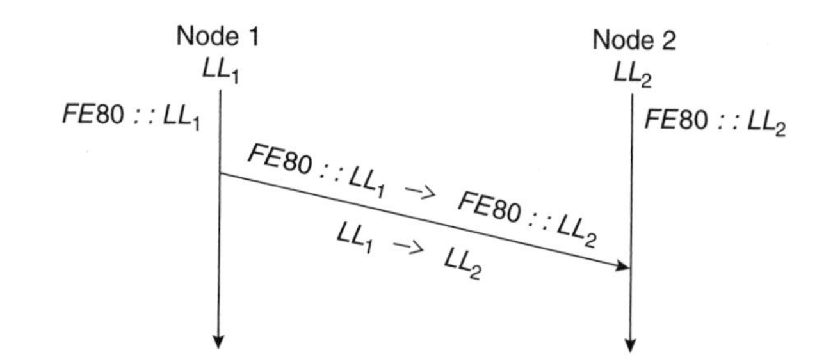

Figure 7.7 Neighbor discovery.

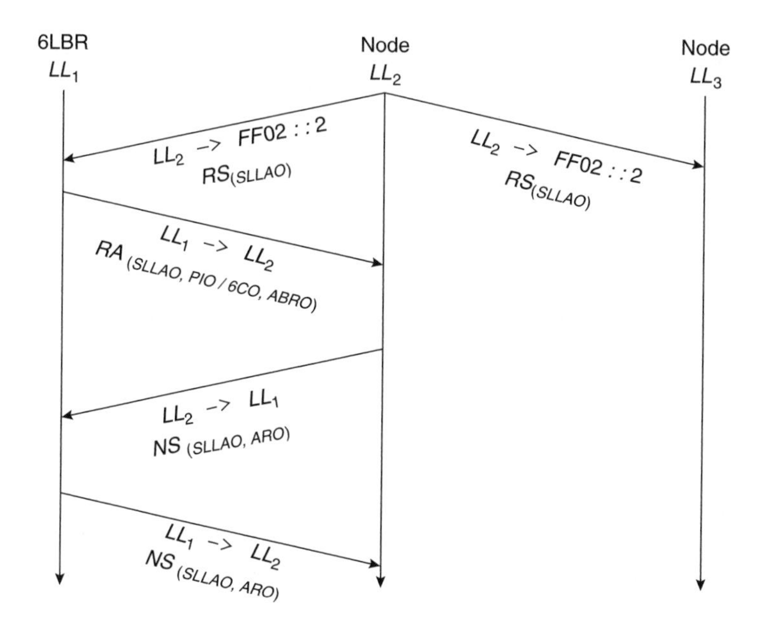

Figure 7.8 Neighbor discovery.

between the address and MAC address.[3] To enable the router to understand the mapping between the IP and the MAC address, nodes have to register their addresses. Registration also provides a simple mechanism for verifying the uniqueness of the address.

Figure 7.8 gives an overview of NDP for global addresses with a scenario that includes one router and two nodes in the RO topology. As in the previous scenario, all kinds of nodes create their LL addresses based on their EUI-64. In the simplest case, all sensor nodes are able to send and receive information from an LBR, with no requirement for relaying.

[3] Off-link models can allow the router to send ICMPv6 redirect messages to inform the sender of the MAC address of the destination. This provides a more direct transmission for the remaining packets.

When a node enters the network, several phases are required. The first phase is to determine all routers and to receive the network parameters:

- The node sends a multicast message to determine the neighboring routers. As with any RS (router solicitation), this message contains an option giving the node MAC address.
- The border routers (6LBR) and the other nodes receive the RS, but only the 6LBR routers answer with a router advertisement (RA) containing their MAC address and prefix(es). The prefixes are managed by the 6LBR (PIO) and/or a specific option for 6LoWPAN that maps a context identifier to the announced prefix (6CO) allowing prefix compression in 6LoWPAN packets. Bit L is not set to indicate an off-link mode.
- Compared to the standard specification of NDP, the maximum prefix lifetime can be set up to the maximum value (0xFFFF; about 18 hours) The border router also adds a specific option (ABRO: authoritative border router option) that will be relayed by other 6LRs. In this way, the nodes will know the IPv6 address of the border router that is announcing the prefix and will be able to directly send all packets using unicast traffic.

The second phase is to register the global address:

- The node registers its address and delegates it to the 6LBR in order to check its uniqueness by sending a unicast NS message that contains both the MAC address () and the tentative IPv6 address (ARO: address registration option).
  If the node has received several RAs, it is recommended to register its address(es) to all 6LBR to enhance reliability.
- The router maintains an active neighbor cache, and checks if there is a conflict between the pair composed of MAC address contained in the SLLAO and the tentative IPv6 address included in the ARO. If the entry is new or it already exists, then there is no conflict, and the 6LBR answers with an NA containing an ARO with status set to 0. If a conflict is detected – the same IPv6 address but a different MAC address – then status 1 is returned. If the neighbor cache is full, status 2 is returned.
  If the node has set the address lifetime to 0, then the routers remove the entry from the neighbor cache.
- In case of conflict, the node may choose another address – for example, by drawing a random number – and restarts the registration process.
- The second node proceeds in exactly the same way to discover and register its address. When node 1 wants to send a packet to node 2, node 1 sends the IP packet to the default router.
- The router checks the MAC address of the destination in its neighbor cache and sends the packet to the destination.

In the RO mode, the node may not directly reach the 6LBR, and the message should be relayed by 6LR. The 6LR keeps a copy of information gleaned when the RA, coming from other 6LBRs or 6LRs, is received. As each of them is tagged with the IPv6 address of the 6LR that issued the original messages, only a single copy of the original announcement is kept.

As in the previous case, a node sends a multicast RS message and the surrounding 6LR or 6LBR answers with a point-to-point RA message. The node registers its address with some of the routers, following the process explained earlier. At this point, it is only possible to check the address against nodes registered on the same router. To check the uniqueness on the whole LoWPAN, the router has to send the request to the 6LBR assigning the prefix. This is done via two new messages, DAR (discovery address request) and DAC, that are used to tunnel the NS sent by the host and the NR status given by the 6LBR, respectively.

The neighbor cache maintained in routers is not for routing purposes; it merely allows connectivity on a link. In an RO meshed network, a routing protocol must be launched to learn addresses (or prefixes, if aggregation is possible) known by other routers.

7.4 Routing Protocol for Low-Power and Lossy Networks (RPL)

The LoWPAN network introduced in the previous chapter can be implemented on different topologies. The star topology is the simplest to manage but will not be the most common, since the LoWPAN has to cover a larger area than the radio transmission range. Therefore, in most cases, the node has to relay the information. As shown before, 6LoWPAN architecture covers two scenarios:

- The first, called *mesh-under*, supposes that an L2 relaying strategy is defined, resulting in the mesh network acting as a link for the IP layer.
- The second, called *route-over*, supposes that some nodes have routing capabilities and are able to forward packets toward the destination.

In this chapter, we will focus on the RO architecture, which is better integrated within IP architecture.

Several routing protocols have already been defined for fixed and ad-hoc IP networks. In the series of RFCs [12–15], the Roll[4] working group has defined the requirement for different network architectures:

- urban networks;
- industrial networks;
- home networks;
- building networks.

Nowadays, none of the existing protocols meet the specific requirements of the study architectures. Protocols based on algorithms of link state (such as open shortest path first (OSPF) or intermediate system to intermediate system (IS-IS)) generate significant flooding traffic when connectivity flapping and protocols for ad-hoc networks, such as ad hoc on-demand distance vector (AODV) or optimized link state routing (OLSR), may not scale correctly with a large number of sensor nodes.

The protocol defined by the Roll working group is called RPL. It is based on distance vector algorithms, such as routing information protocol (RIP), and is designed to correct

[4] The work done by this working group is orthogonal to the 6LoWPAN working group, so vocabulary will be different. A LoWPAN network will be called a low-power and lossy network. Devices will also have another name derived from graph theory, that is, 6LBR will be called *root*, 6LR will be a *node* and 6LN will be a *leaf* if it cannot forward packets.

performance issues due to bad loop detection. RPL must also remain flexible in the definition of routing strategies. An urban network will not have the same constraints as a home network, and also, within each network type, the applications do not always share the same objectives. An alarm system may try to reduce propagation delays, and an application generating periodic messages will try to select routers with the highest level of energy (powered).

Figure 7.9 shows the relationship between the different specifications set up by the Roll working group. The RFC standards [12–15] show the diversity of constraints for different application cases. RPL will consider the routing protocol making its routing metrics as agnostic. It is based on the *rank* notion, which represents the distance to the root. Several instances of RPL may run on a single node, each of them being used to take into account different routing topologies. The RPL messages can contain different metrics or constraints to further refine the graph. The method for transforming the metrics into a rank is defined in an objective function. Currently, two objective functions are defined. The first one, OF0, is the default objective function and can be used when RPL messages do not contain any metrics, as routing is based on hop counts. The second one, minimum rank with hysteresis objective function (MRHOF), is used to transform metrics. A hysteresis is used to maintain relative stability in the routing, even if metrics change rapidly due to the poor link quality.

The primary focus of the working group is to process upward traffic: leaves to roots – also called MP2P (multi-point to point) – and downward traffic: roots to leaves, called P2MP (point to multi-point). MP2P traffic is generated when the low power and lossy network (LLN) is connected to a traditional Internet, and packets are forwarded to/from outside of the LLN. Since M2M traffic is primarily between the gateway and multiple devices, MP2P and P2MP are the most important aspects to be covered. This corresponds to ZigBee Smart Energy and ETSI M2M architectures.

P2P traffic – that is, traffic between two nodes located inside the same LLN – is not optimized in the first standard versions. P2P packets are sent to the root until they reach

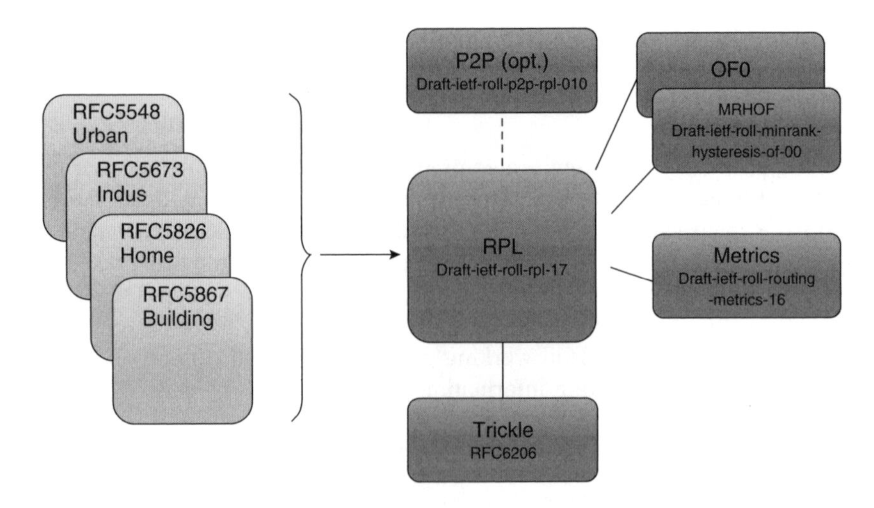

Figure 7.9 RPL galaxy.

a node able to forward the packet toward the destination. Optimizations are covered by another set of drafts.[5]

To reduce the number of messages sent on the network, a trickle algorithm may limit the number of periodic messages that are sent. A node regularly sending the same message will increase the period until reaching a defined maximum value. If the announcement changes, the period is set to the defined minimum value. The principal difference compared with RIP, where a lack of three periodic messages is viewed as an indication that the sending device is dead, is that the trickle algorithm will delay the detection. Nodes will have to deploy other strategies to detect dead routers, such as neighbor unreachability detection (NUD) or Layer 2 acknowledgments.

7.4.1 RPL Topology

The RPL topology is based on a direction oriented directed acyclic graph (DODAG). The DODAG is used to allow the upward traffic to leave the LLN, but can also be used afterwards to build a reverse path toward the nodes and leaves. One difference compared with RIP, which maintains a single path toward a destination, is that the DODAG may contain more loop-free paths toward the destination, which increases reliability and speeds up recovery in the event of a route failure.

The DODAG construction adheres to the generic rules defined by RPL protocol specifications and also respects more specific patterns of behavior defined in separate objective functions documents (see Section 7.4.1.2).

A single DODAG may cover the entire LLN, but several DAGs may compose a DODAG. Also, different root nodes may be available on the LLN as shown in Figure 7.10 where node 11 and 51 are root nodes.

Nodes can run different instances of the RPL protocol, which will lead to an equivalent of VLAN, and each instance may use different metrics and constraints to build a DODAG. If it is relatively easy to build different forwarding tables based on the different instances of DODAG, packets will have to carry a hop-by-hop extension to indicate which instance of the DODAG has to be used.

Roots can change the DODAG version number. This can be done, for instance, if some inconsistencies are found in the DODAG, perhaps due to routing message losses. In that case, a new DODAG is built.

A DODAG can be grounded, meaning the DAG's root is a border router. During the DODAG building phase or when connectivity is lost with a parent, a subDAG's root may not be a border router, in which case the DAG is said to be floating.

7.4.1.1 Neighbors and Parents

All nodes except "leaves" generate DIO (DODAG information object) ICMPv6 messages. The message format shown in Figure 7.11 provides an opportunity to understand how the DODAG is constructed. DIO messages are periodically sent where the period is driven by the trickle algorithm in order to reduce traffic when the network is demonstrated as being

[5] Draft-ietf-roll-p2p-rpl, draft-ietf-roll-p2p-measurement.

Root
11
Preferred
parent
21
22
31
32
33
41
42
43
44
51
52
53
54
55

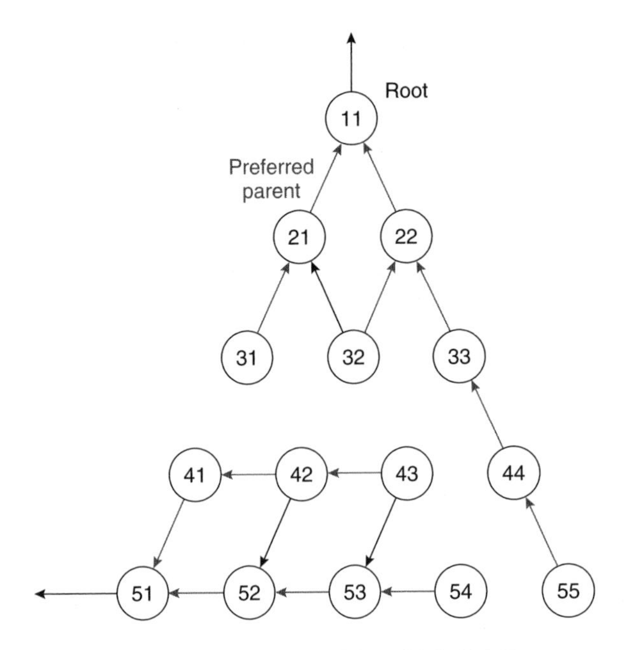

Figure 7.10 DODAG with multiple DAGs.

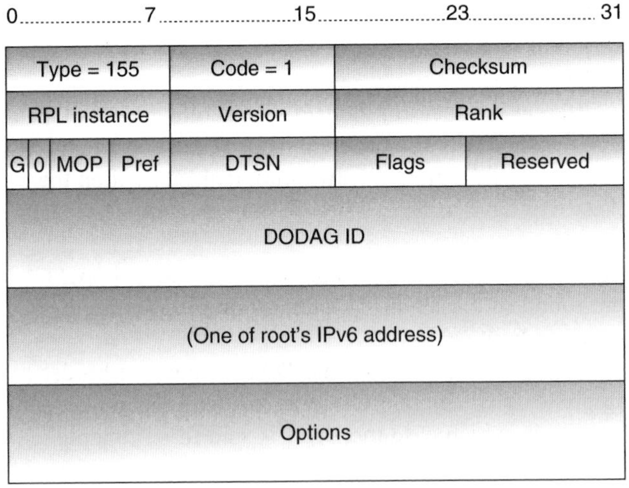

0
7
15
23
31
Type = 155
Code = 1
Checksum
RPL instance
Version
Rank
G
0
MOP
Pref
DTSN
Flags
Reserved
DODAG ID
(One of root's IPv6 address)
Options

Figure 7.11 DIO format.

stable. DIO can also be sent in response to DIS (DODAG information solicitation) sent by other nodes. DIS may contain options to limit the number of answering neighbors.

A node uses the DIO messages received from its neighbor to determine their rank and, optionally, some additional metrics and constraints. Based on that information and rules defined in the objective function, the node will select a set of possible parents and a preferred parent to which the upward traffic will be sent. A node may more easily move closer to the root if some better rank announcement is received.

If a node has to increase its rank, for example, when all possible parents have disappeared, the operation is made more complex in order to minimize the risk of routing loops. The node poisons its neighborhood by setting INFINITE_RANK in the DIO messages that, in turn, breaks the sub-DODAG root and becomes the root of a floating DODAG. Nodes receiving this message prefer to be attached to another parent, if the rank does not increase, or they themselves become floating and only re-attach to other floating nodes.

In the reverse direction, the node (or leaf) can send DAOs (DODAG advertisement objects) to inform some of its parents of the addresses (or prefixes) that are accessible through this node. By going up to the root, this node will fill the routing tables. Nevertheless, some nodes may not have the possibility of maintaining a routing table due to memory limitations. RPL implements two forms of behavior:

- storing mode, where all nodes have enough resources to memorize downward addresses so the routers forward the packet to the next hop.
- non-storing mode, where only the root keeps track of the network topology. Nodes send their DAO message containing their address and that of their parents to the root. The root is able to find the path to a destination and will send the packet using, for example, IPv6 routing extensions containing the list of routers that have to be crossed.

7.4.1.2 Objective Functions

Objective functions adapt the network-routing behavior to the applications. Different forms of behavior are possible and can be run in parallel in different RPL instances. The root periodically floods a code point defining the objective functions that should be applied in the RPL instance.

Currently two objective functions are defined:

- **OF0:** This is a very simple form of behavior, which can be used as a last resort if the node implements neither the objective function announced by the root nor the metric containers included in DIO announcements. This leads to a topology that minimizes the number of hops, for example, RIP. This is not the best routing choice for an LLN.
- **MRHOF:** The rank computation is based on metrics contained in DIO messages. These metrics are more dynamic than the hop-count and can take into consideration, for instance, the link quality. The MRHOF works only for additive metrics (i.e., the value increases each time a node is crossed) and to avoid creating instability on the route. A hysteresis mechanism allows the selection of a better root only if it is significantly different from the current preferred parent's route.

7.4.1.3 Metrics and Rank Calculation

RPL intends to use different metrics and constraints to build the topology. The metrics used can be additive (i.e., path quality), can report a minimum or a maximum (i.e., the energy of node on the path) and finally can be multiplicative (i.e., error rate on the path). Additional constraints are used to extend the scope of the DODAG. The choice of one of the criteria used is independent of the metric itself. For instance, link quality can be viewed either as an additive metric or as a minimum metric. Metrics or constraints can be also related to the node:

- energy: how the node is powered and percentage of remaining energy;
- hop count: number of crossed hosts;
- throughput, latency;
- link quality level: from 1 (highest) to 7 (lowest);
- ETX (expected transmission): number of transmissions required to successfully send a packet.

7.5 CoRE

Previous chapters have shown how to bring about connectivity between devices or at least between a device and a gateway. The CoAP (constraint application protocol) protocol is a way of structuring the exchange of information based on the representational state transfer (REST) paradigm but optimized for M2M applications.

The HTTP protocol is the most famous example of REST architecture. Some simple commands, such as GET and PUT, allow for a document to be accessed from or pushed to a server. The REST architecture can also be applied to the sensor world. A GET could be used to access a value, and a PUT to store a value or trigger an activator. Unfortunately, HTTP and TCP are considered too heavy for devices such as sensors. UDP is preferred for light queries and HTTP can be simplified to make the parsing of data messages easier and also to reduce its overhead.

CoAP can be used for compressing simple HTTP interfaces but also offers built-in discovery, multicast support, and asynchronous transactions. The CoAP defines:

- a simple message format easily processed by devices with limited processing resources.
- a simple transport protocol to detect and correct packet losses.
- a way to query or store information on a node using the REST paradigm.
- a way to discover resources inside a LoWPAN network.

7.5.1 Message Formats

CoAP messages are carried over the UDP under any port number, but to optimize 6LoWPAN UDP compression, port numbers between 61616 and 61631 are used. Figure 7.12 gives the basic message format. CoAP can be secured by using TLS (transport layer security) over TCP but DTLS (datagram transport layer security) [rfc4347] is preferred in order to better encrypt the information.

The message contains the following fields:

- version number in two bits, current value is 1.
- a T field (two bits) giving the nature of the message:
  - 0: indicates a confirmable message (CON); the receiver will have to acknowledge it.
  - 1: indicates a non-confirmable message (NON); the receiver will not acknowledge it.

  - 2: is the acknowledgment message (ACK).
  - 3: is a reset message (RST).
- OC (option count) is a four-bit field containing the number of options contained in the header. As shown in Figure 7.12, the mandatory fields are limited and optional fields are needed to extend the message header and extra fields stored as TLV (type length value). In this way, the CoAP header can carry only the information appropriate for a specific action.
- Code indicates the nature of the message. Code is encoded on one byte, which can be viewed in two four-bit nibbles. In HTTP [16], the status codes are encoded on three-digit numbers. 1xx is informational, does not notify an error except that the request be taken into account and continue to be processed. 2xx indicates a success, 3xx includes a redirection, 4xx is sent by a client to notify an error, and 5xx is sent by a server. Since, in any case, xx are smaller than 15, the code can be compressed in a byte that be denoted as S.XX (where S is a value between 1 and 5):
  - Since no status code is equal to 0, values below 31 are used for requests (O for nothing, 1 for GET, 2 for POST, 3 for PUT, and 4 for DELETE). GET allows a client to obtain the content of a resource located on a server, POST creates a new resource on a server with a specific content, PUT stores content on an existing resource, and DELETE removes a resource.
  - Values above 32 are for responses. Table 7.3 gives the error codes. Note that HTTP status 200 (OK) is not converted into a CoAP code, but 2.05 may be used instead.
- Message ID is used in the acknowledgment process. CONfirmable messages contain a unique value selected by the sender. ACKnowledgment messages or ReSeT messages must copy this value.

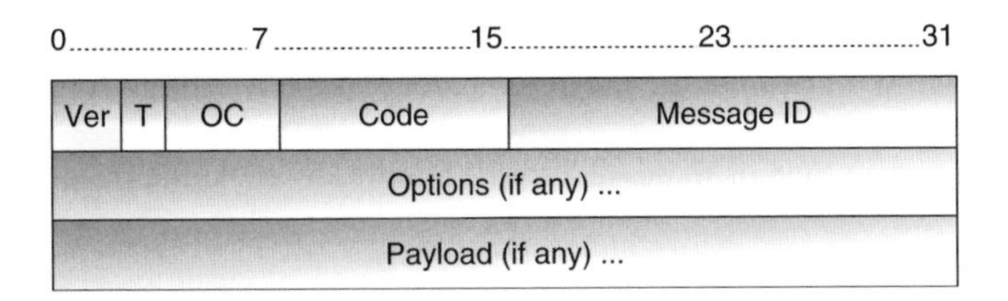

Figure 7.12 CoAP header format.

Table 7.3 Response values

Success notification codes			Client error notification code			Server error notification code		
65	2.01	Created	128	4.00	Bad request	160	5.00	Internal Server error
66	2.02	Deleted	129	4.01	Unauthorized	161	5.01	Not implemented
67	2.03	Valid	130	4.02	Bad option	162	5.02	Bad gateway
68	2.04	Changed	131	4.03	Forbidden	163	5.03	Service unavailable
69	2.05	Content	132	4.04	Not found	164	5.04	Gateway timeout
			133	4.05	Method not allowed	165	5.05	Proxing not supported
			141	4.13	Request entity too large			
			143	4.15	Unsupported media type			

CoAP makes use of asynchronous transactions using a simple binary header format with a base part followed by options in TLV format. The datagram length implies the length of the message payload.

To simplify parsing, the options are always sent in the same numerical order, based on the type. To force this in implementation and in many cases to reduce the size of the type field, only the delta between two consecutive types is sent. For instance, if a message contains option types 1, 5, 6, 7, and 11, the delta type will be 1, 4, 1, 1, 3. Currently, the defined types are as follows:

- 1 (Content-Type) is the value referring to a mime value describing the syntax of the payload, for example, 0: text/plain, 44: application/soap+xml.
- 2 (Max-Age) gives the maximum duration in seconds for which the answer may be cached.
- 3 (Proxy-Uri) contains the URI that will be processed by a proxy.
- 4 (ETag) is used to check whether the document version in a cache and in a server are the same.
- 5 (Uri-Host), 6 (Location-Path), 7 (Uri-Port), 8 (Location-Query), 9 (Uri-Path), and 15 (Uri-Query) contain the different elements of the URI to simplify the parsing by the server.
- 10 (Observe) is used to receive regularly updated values from the server.
- 11 (Token) is used to match a response with a query.

Odd numbers indicate a critical option, which means that the receiver should know how to process it. If this is not the case, an error code is generated. Even numbers indicate elective options, which mean that the receivers will ignore it if they do not know how to process it.

Some options may appear several times in the header, for example, the path in the URL will be split into elements separated by /. Each of them will be stored in a different URI-path option.

The definition and usage of these TLVs will be introduced in the following chapters.

7.5.2 Transport Protocol

In "traditional" Internet, TCP is used to detect and correct packet losses and adapt the throughput to the network and destination capacity. However, TCP is a highly complex protocol leading to a large footprint. Certain operating systems, such as contiki,[6] have implemented a restricted version of TCP, where the performance is limited, since the transmission window is set to one segment. So, to avoid such additional complexity, CoAP relies on UDP but implements its own "transmission" protocol to correct the errors or duplications.

The basic protocol uses the message ID field to keep track of messages. The source selects a unique value when sending CONfirmable messages, for example, by incrementing a counter whenever a new message is sent, where the first value should be a random number. The receiver either sends an ACK message, if it can process the message, or in other cases sends an RST message containing the same message ID.

[6] See http://www.sics.se/contiki/.

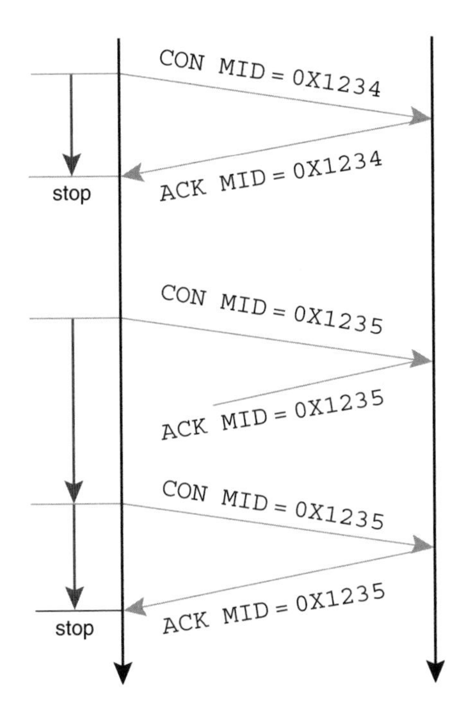

Figure 7.13 Simplified examples of CoAP transmisson flows.

The sender keeps track of the messages it has sent and triggers a timer. When the sender receives an ACK or RST message, it can stop the timer and remove the message associated with the message ID from its memory. ACK messages can also contain information. In other cases, when the timer expires, the sender will resend the message with the same message ID.

Figure 7.13 shows some simple examples.

In the first case, the sender sends a CON message with message ID 0x1234 and the receiver immediately acknowledges the message. In the second case, the ACK message is lost and then the sender resends the information. The receiver should notice that the message is a duplicate, since the message ID is the same, and does not process this message but acknowledges it to stop the sender's timer.

The timer-value setting is the tricky part, since the node relaying the message may sleep. The propagation delays may be high and have some large variations. It is difficult to define a standard value for all kind of networks. The default value is 2 seconds. This value is doubled whenever a message is resent. After several attempts (the default is four), the sender discards the message.

7.5.2.1 Asynchronous Exchanges

The scenario described in Figure 7.13 shows that the ACK is immediately sent after receiving a CON message. It is difficult to delay the answer, since it can lead to the triggering of the retransmission timer on the sender side. However, the CON message

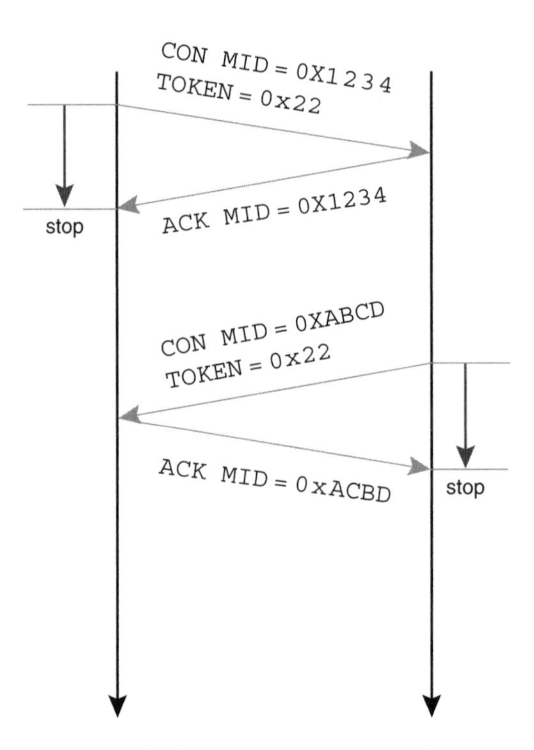

Figure 7.14 Example of token option.

may ask the receiver for a value that may not be immediately available. A token option (11 critical) may be used to map answers with queries. Figure 7.14 shows an example of token option usage.

The sender sends a request with a token option set and the receiver immediately acknowledges the request. When the receiver is able to reply to the query, it sends a CON containing the answer and waits for the acknowledgment.

7.5.2.2 Periodic Messages

Occasionally, it is necessary to regularly probe a sensor node to obtain up-to-date state information. This may be done by endlessly repeating a query, and waiting for an answer. CoAP defines the elective option observe to periodically send a response to a single query. The option observe is linked to the use of a token, which establishes the link between the query and the answer. The observe option contains an increasing value, allowing the sender to reorder the messages, if necessary. Figure 7.15 gives an example.

The sender includes an observe option in the request. If the receiver recognizes the option, it will return periodic notifications. In this example, they will be carried into NON, since the periodic sending losses are less important. However, CONs can also be used. In that case, where a new value has to be sent, retransmission attempts from the previous case have to stop.

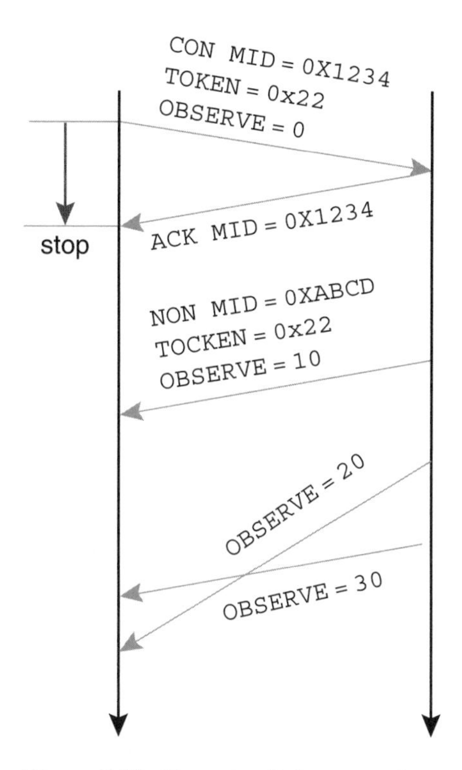

Figure 7.15 Example of observe option.

This example shows that the receiver of notification can choose to ignore the notification with an observe value equal to 20, since it receives this notification before the one with a value set to 30.

To complete an observe sequence, the device that initiated the observation can either send a RST message in the event of observation being sent in CONs or perform another query without the observe option. The device sending the notification can terminate with an error message, or stop if CONs cannot be acknowledged.

7.5.2.3 Large Block Transfer

A request or its result can sometimes not be sent in a single message. The sender of large files can use the block option (value 13 critical) to break it down into smaller parts. The option can be one, two or three bytes long and is composed of a NUM field (4, 12, and 20 bits) indicating the block number in the file. An M bit when equal to 1 indicates that more blocks have to be sent, and when equal to 0 indicates that it is the last block. The last three bits contain the block size (SZX) that may be negotiated between both entities. The block size is always indicated to the power of two and is represented by the formula $block\_size = 2^{\wedge}(SZX + 4)$. As such, the block can be between 16 and 2048 bytes long.

Block numbering always starts at 0, and blocks are acknowledged the same way as messages.

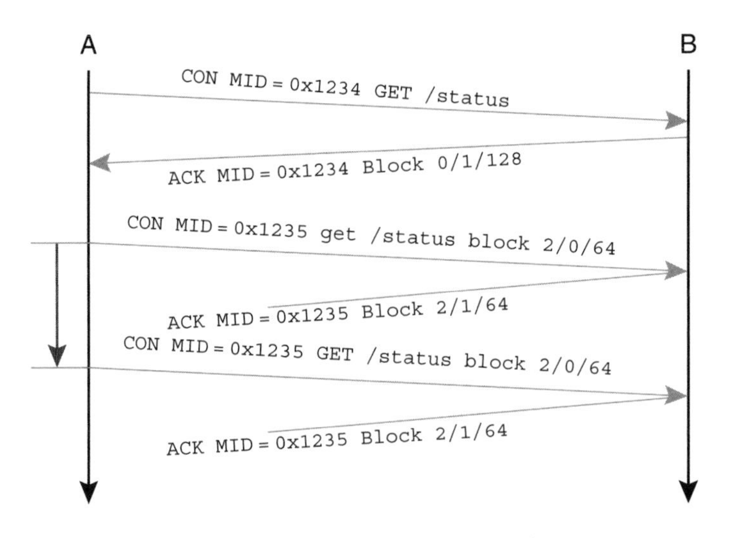

Figure 7.16 Example of block transfer.

In a GET method, the client is unaware of the size of the corresponding resource, so the request does not contain a block option. The server reply uses the block option, and the server will then use this option to request the subsequent blocks.

In PUT or POST methods, the client will use the block option directly.

Figure 7.16 gives an example of the block transfer. Host A (client in REST terminology) sends a GET request to Host B (server) to get the resource/status. Ignoring the size of the answer, the block option is not included in the request. B answers with the first block of the requested document. B sends the first block (0) and M bit is set to 1 since other blocks will follow. B also selects a block size of 128 bytes (SZX = 3). Note that data is included in the ACK message. Exchanges using the token option as shown before are also possible.

Server A accepts the information and requests the next block, but the size chosen by B is too high in terms of its memory constraints, so A requests a smaller block of 64 bytes. The block number is changed to adapt the new block size and M bit is set to 0, since it does not have any more information to send.

In this example, the message is lost. A has triggered a timer for each of its requests, in which case the timer expires and A resends its request.

7.5.2.4 Multicast

Multicast may be a useful feature, for example, when switching on several lights in a building at the same time. RPL routing protocol may be used to implement multicast in a LoWPAN, but the group management may require significant amounts of energy to flood the whole network. The CoRE working group is currently defining a method for implementing multicast at the application level in a proxy gateway. Any node wishing to belong to a multicast group will send a CoAP message to register its address in the gateway. When the gateway sends the message to that group, it will send it in unicast to all the members.

7.5.3 *REST Architecture*

REST is widely known as being the architectural model of the web through the HTTP protocol. As for the web, the RESTful architecture is well-suited to most types of M2M applications. It is based on the client-server paradigm. The server contains information that can be saved or retrieved by clients. In this instance, a sensor or an actuator can be viewed as a server and the application or the gateway is treated as the client.

Requests are stateless, which means that one request is always independent of the other. Nevertheless, the server is stateful and can be updated by requests. This allows requests to be cached or proxied by certain gateways between the client and the server.

REST defines four interactions between the client and the server:

- **GET (code 1 in the CoAP header):** The client requests the information located on the server.
- **POST (code 2):** The client creates and stores the information on a resource located on the server.
- **PUT (code 3):** The client stores information on a resource based on the server.
- **DELETE (code 4):** The client removes a resource from the server.

Resources on the server can be represented using URI, which contains the protocol used to access the resource, followed by the server address or name, and finally the path within that server to access the resource.

The CoAP URIs are comparable to the HTTP URI. The “coap” URI is used to identify and locate the CoAP resources within the host and the port. The resources are organized in a hierarchical way and are governed by a CoAP server, which is identified via a generic syntax’s component authority with a host identifier and an optional UDP port number. The remainder of the URI identifies a resource. The path consists of a sequence of path segments separated by the slash character “/”.

The syntax of the CoAP URI is:

coap-URI=“coap:” “//” host[“:” port] path-abempty [“?” query]

Application designers need to make a trade-off between shortness and descriptiveness to take into account the CoAP environments constrained by bandwidth and energy.

For instance, if a client has to query a server with this URI:

coap://sensor1.example.com:61620/external/temperature?max_value

it will generate a message to the IP address of sensor1.example.com on port 61620. The CoAP message will contain:

- the URI-host option with sensor1.example.com.
- two URI-path options with external on the first one and temperature in the second.
- the URI-query option with max_value.

7.5.3.1 HTTP Mapping

In the current version, a proxy can be used to interconnect an HTTP to a CoAP, or vice versa. The mapping function transforms HTTP fields into CoAP options. Since this

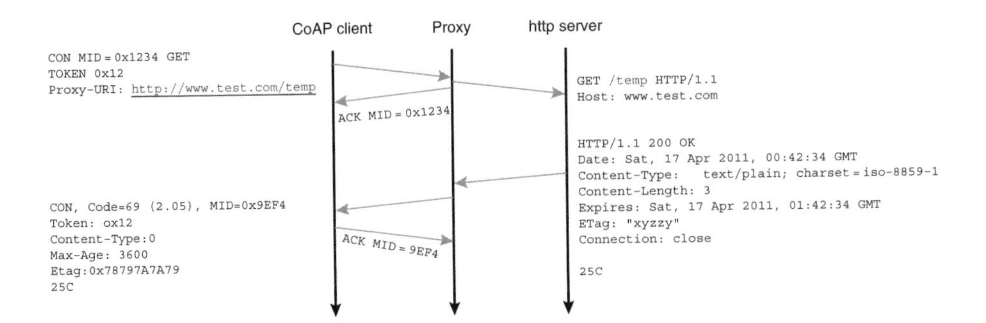

Figure 7.17 Proxy behavior for CoAP to HTTP.

conversion is performed at the application level and the device may be designated through a DNS name, it is possible to have HTTP requests on IPv4 and CoAP in IPv6.

The process is initiated by using the proxy-URI option with the "http" URI in the CoAP request to the CoAP-HTTP proxy or by sending a CoAP request to a reverse proxy mapping CoAP to HTTP. The CoAP method or response code, content-type and options are converted to the corresponding HTTP features. The payload does not need any conversion since it is carried in a similar way by both protocols.

Figure 7.17 gives an example of the proxy behavior for CoAP to HTTP.

A client sends a CoAP message to a CoAP/HTTP proxy with the proxy-URI option containing the target URL http://www.test.com/temp. This message also includes a token to match with the answer.

The proxy resolves the server name contained in the proxy-URI option to obtain the IP address of the server and opens a TCP connection with it. The proxy then sends a GET message containing the absolute path (/temp) and specifying the HTTP protocol version. The Host: header field allows virtualization on the server side.

The server answers with an HTTP message containing a header and content separated by an empty line. The header contains a content-type announcing that the content is coded with the Latin alphabet (ISO-8859-1). The proxy will transform the answer into UTF-8, which is more generic since it contains ASCII and Unicode coding. On a CoAP message, the content-type value 0 corresponds to the default coding and can thus be omitted. The expires HTTP header field is transformed into a max-age option, indicating how long (in seconds) it will be before the content is cached and an ETag contains the information tag for caching.

An alternative scenario sees the node on the Internet directly contacting a proxy in order to access the resource in a CoAP server. Figure 7.18 gives an example of this conversion.

The HTTP client opens a TCP connection with the proxy and sends a GET request containing the hostname of the CoAP device that it wants to probe. The proxy resolves the name and sends a CoAP message with the path split into two URI-path options. The CoAP server answers by the value and the proxy converts the answer into HTTP content.

7.5.3.2 Caching

Nodes can keep responses on a cache in order to reduce latency, network bandwidth usage, and response time, but they need a corresponding freshness mechanism. The CoAP

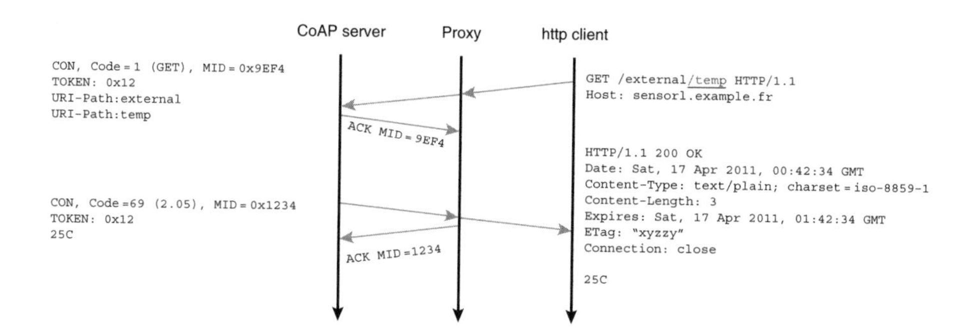

Figure 7.18 Proxy behavior for HTTP to CoAP.

cacheability does not depend on the request method but on the response code. A stored response is not used unless:

- the method used for the actual request and the one used to get the stored one is the same;
- the options are the same in both requests, except for the token, max-age or ETag request options; and
- the stored response is fresh or successfully validated.

The freshness mechanism uses the max-age option to determine whether a request can be used for subsequent requests without contacting the server. The max-age option specifies in a matter of seconds the time during which a request can be used, where the default value is 60 seconds. If a server does not want to use caching, it must include a max-age option with a value of zero.

When there is more than one response in the cache for a GET request, the end-point can use the ETag option to select one of the stored responses and to validate the freshness. A code 2.03 (valid) in the response determines that the stored response can be used and it refreshes the max-age option. Any other code will replace the stored response by the one sent.

7.5.3.3 Resource Discovery

The resource discovery is important for M2M interactions where there are no human beings in the loop. A well-known URI as been defined and the client can use it to determine what the server capabilities are. When a client sends a GET message to a node with a /.well-know/core path, the response payload is based on the RFC5988 describing link message formats between resources. The CoAP response contains the following attributes:

- A path surrounded by < > characters indicates the way to obtain the resource on the server.
- rt (resource type) describes the type of resource.
- if assigns a name to the resource.

- ct indicates the content-type of the resource. It is the same code as that found in CoAP options.
- sz returns the maximum estimated size of the resource.

To identify resources inside a network, the query can be in multicast. To limit the number of answers, a query string can be added after the path to allow servers implementing this functionality to limit their response.

References

1. Briscoe, B., Odlyzko, A., and Tilly, B. (2006) *IEEE Spectrum*, IEEE, pp. 34–39, July 2006.
2. Postel, J. (1981) Internet Protocol. RFC 791 (Standard). Updated by RFC 1349, September 1981.
3. Postel, J. (1981) Transmission Control Protocol. RFC 793 (Standard). Updated by RFCs 1122,3168, September 1981.
4. Postel, J. (1980) User Datagram Protocol. RFC 768 (Standard), August 1980.
5. Tuexen, M. and Stewart, R. (2011) Stream Control Transmission Protocol (SCTP) Chunk Flags Registration. RFC 6096 (Proposed Standard), January 2011.
6. Kohler, E., Handley, M., and Floyd, S. (2006) Datagram Congestion Control Protocol (DCCP). RFC 4340 (Proposed Standard). Updated by RFCs 5595, 5596, March 2006.
7. Mockapetris, P.V. (1987) Domain Names – Implementation and Specification. RFC 1035 (Standard). Updated by RFCs 1101, 1183, 1348, 1876, 1982, 1995, 1996, 2065, 2136, 2181, 2137, 2308, 2535, 2845, 3425, 3658, 4033, 4034, 4035, 4343, 5936, 5966, November 1987.
8. Deering, S. and Hinden, R. (1998) Internet Protocol, Version 6 (IPv6) Specification. RFC 2460 (Draft Standard). Updated by RFCs 5095, 5722, 5871, December 1998.
9. Montenegro, G., Kushalnagar, N., Hui, J., and Culler, D. (2007) Transmission of IPv6 Packets Over IEEE 802.15.4 Networks. RFC 4944 (Proposed Standard), September 2007.
10. Narten, T., Nordmark, E., and Simpson, W. (1998) Neighbor Discovery for IP Version 6 (IPv6). RFC 2461 (Draft Standard). Obsoleted by RFC 4861, updated by RFC 4311, December 1998.
11. Savola, P. and Haberman, B. (2004) Embedding the Rendezvous Point (RP) Address in an IPv6 Multicast Address. RFC 3956 (Proposed Standard), November 2004.
12. Dohler, M., Watteyne, T., Winter, T., and Barthel, D. (2009) Routing Requirements for Urban Low-Power and Lossy Networks. RFC 5548 (Informational), May 2009.
13. Pister, K., Thubert, P., Dwars, S., and Phinney, T. (2009) Industrial Routing Requirements in Low-Power and Lossy Networks. RFC 5673 (Informational), October 2009.
14. Brandt, A., Buron, J., and Porcu, G. (2010) Home Automation Routing Requirements in Low-Power and Lossy Networks. RFC 5826 (Informational), April 2010.
15. Martocci, J., De Mil, P., Riou, N., and Ver-meylen, W. (2010) Building Automation Routing Requirements in Low-Power and Lossy Networks. RFC 5867 (Informational), June 2010.
16. Fielding, R., Irvine, U.C., Gettys, J. *et al*. (1999) Hypertext Transfer Protocol – HTTP/1.1. RFC 2616, June 1999.

8

M2M Security

Ioannis Broustis, Ganesh Sundaram,
Simon Mizikovsky and Harish Viswanathan
Alcatel-Lucent, New Jersey, USA

Security is an important function in any digital communications environment, and the role of security is even more critical with machine-to-machine (M2M) communications. In several existing service infrastructures, such as cellular systems, typically the service, network, and device distribution are tightly coupled and managed by a single entity: the network operator. On the other hand, M2M solutions typically involve multiple entities, such as application providers, network operators, and numerous device manufacturers. All of these entities may be related in very diverse ways, while certain entities may not be related at all. In other words, complex trust relationships exist in M2M environments, since certain players may not directly interact to establish formal business relationships (and hence trust). This fundamental issue necessitates novel, scalable, and automated methods for security association establishment. Such methods should be able to deal with a potential explosion in the number of M2M devices and hundreds of applications provided by a few M2M operators utilizing multiple access network technologies.

In this chapter, the complexity of trust relationships among the various M2M players is explained in some detail. Such complexity provides guidelines for designing security strategies and solutions for M2M, as well as for avoiding design pitfalls. In addition, various security policies that need to be incorporated in M2M systems to guard against potential threats are discussed. Following this, the chapter provides an overview of ongoing standardization efforts.

M2M Communications: A Systems Approach, First Edition.
Edited by David Boswarthick, Omar Elloumi and Olivier Hersent.

8.1 Introduction

The M2M market is highly fragmented with many players across numerous vertical domains. On the other hand, the industry has reached a point where the value of M2M services for improving productivity in the enterprise, as well as convenience and comfort for consumers, is broadly understood. Given the increased focus on M2M applications by large network operators with a view toward leveraging their existing assets, coupled with the opening up of their networks and devices, M2M communications is poised to take off in an unprecedented fashion. The communication security is a critical enabler for the mass market adoption of M2M. End-users demand that M2M communications achieve at least the same level of security as traditional human communications. In a digital context, confidentiality, integrity protection, privacy, authentication, and authorization are some of the key elements of security that need to be addressed in an end-to-end fashion.

Some M2M service systems have already been deployed in a number of different scenarios. In particular, cellular networks are presently being used for many M2M solutions. In this chapter, we focus on the use case of cellular M2M communications for assessing the applicability of potential security solutions. The following section discusses the inherent properties of cellular M2M systems, which render certain security strategies inapplicable.

8.1.1 Security Characteristics of Cellular M2M

Cellular M2M differs from current cellular networks in three important ways, which suggests that the current security mechanisms in place today for mobile devices in such networks are not appropriate for M2M.

First, cellular network services today are typically provided by a single service provider that typically owns the device distribution along with the SIM card distribution, device provisioning, network infrastructure, and service delivery for voice and data services. In some cases, the device distribution, device provisioning, and service delivery are owned by a mobile virtual network operator (MVNO), while the underlying network infrastructure is owned by a different network provider. On the other hand with cellular M2M, multiple players exist with limited business relationships among them.

8.1.1.1 Use Case

As an example, let us consider a scenario where a utility company plans to deploy smart meters that employ cellular wireless modems for connectivity. In one approach to implementing this solution that is typical of M2M services, the utility company will order meters from a meter manufacturer, and subscribe to the M2M service of an M2M MVNO that has negotiated network access and related services with multiple cellular operators. The utility company obtains and deploys the meters while provisioning them in their database and in the database of the MVNO providing the connectivity service. The entities involved in providing this solution are thus: the utility company, which is the application provider in this case, the M2M operator that they enlist for this purpose, the cellular access network provider, the meter manufacturer, and the end-user. Note that there is no relationship between the utility company that deploys the devices and the cellular access network provider; in other words, these two entities do not necessarily trust each other. Therefore, adopted security solutions should be workable in such an open and disconnected business environment.

The second important difference is due to the fact in many M2M applications there is a very large number of devices each contributing only a small amount of data and thereby generating relatively low revenue streams from each device to the various players involved. The economics of M2M do not allow a security provisioning process similar to that of cellular handsets to be adopted. A simpler automated process with the capability of efficiently dealing with a large number of device manufacturers is desirable. For example, in the use case described above, one desired deployment process would be to simply ship the meter directly from the manufacturer to the end-user. In general, M2M devices are typically not customizable prior to delivery to the site. They are shrink-wrapped clones that potentially differ only by their unique medium access control (MAC) address and electronic serial numbers. The devices are often installed over multiple time frames and should seamlessly integrate into the network. Furthermore, the installation is often done by end-users or contractors with little communication engineering experience.

Finally, unlike cellular phones, smartphones, or wireless-enabled laptops, M2M devices are often unattended and are subject to a higher risk of vandalism and misuse. In particular, the threat of using a cellular M2M device for other applications such as web-browsing is a real risk and such tampering may involve users who do not have the authorization to access the device. Applicable security solutions should be designed to protect against such inappropriate use of the device and subsequently protect the network operator and the M2M application provider.

Furthermore, in the cellular M2M context, we are dealing with an explosion in the number of devices (potentially running into the billions) and hundreds of applications provided by a few virtual operators using a mix of access network technologies. Therefore, we should expect a multitude of device vendors participating in a relatively open ecosystem selling low-cost devices directly to consumers or through an application provider. For instance, device manufacturer and M2M service providers may not have a business relationship (and hence there is no predefined mutual trust) where subscribers buy devices in an open market. Alternatively, in residential applications such as metering, the end-user (home-owner) may not even be aware of the existence of the virtual network operator who is providing the service on behalf of the owner of the organization which owns the application. However, securing the application is critical to all players in the chain.

8.1.1.2 Sample Attacks

In the following examples, we discuss two possible M2M device-hijacking attacks, taking into consideration the inherent properties of M2M system deployments. We specifically consider cellular network connectivity, although the attacks are easily applicable to other network technologies.

Example 1: Theft of Removable Credentials

In this case, the adversary extracts the (removable) universal integrated circuit card (UICC) module from an M2M health monitoring M2M cellular device and installs it into a smartphone. Given that the UICC is legitimate, the mobile access network will successfully register and authorize access to the smartphone to the network services. This is because all the necessary information for registering a mobile terminal with the network is located within the UICC. In other words, the network has no way of explicitly identifying the device that carries the UICC. As a result of this attack, the adversary could use the smartphone to browse the web at a billing rate that corresponds to the M2M device subscription and which could be quite inexpensive (or even free, depending on the relations between the subscriber, M2M operator, and network operator). More importantly, access network operators carefully plan their networks to deliver a certain amount of traffic based on the number and types of subscriptions. Any malicious use of M2M devices to access unauthorized services completely falsifies such network planning which leads to inferior services for legitimate end-users. Along similar lines, in scenarios where the M2M device is not physically accessible, the adversary would have to overhear the credentials of the M2M device, such as the IP address and the temporary identifier assigned to the device by the network during registration, for example, the unicast access terminal identifier (UATI) for the case of code division multiple access (CDMA) 1xEV-DO networks [1] or its equivalent in UMTS/HSPA networks. In this scenario, the attacker could potentially manage to inject data packets into the legitimate data session that was set up by the network in place of the victim M2M device.

The physical removal and use of removable credentials (such as in cases of removable UICCs) has the following effects:

- The mutual network authentication procedure verifies the credentials contained in the UICC. Hence, irrespective of the device that is hosting the UICC, the registration process will verify the UICC validity and successfully authorize the device into using the network services. Note that this problem exists with any ordinary mobile device that carries a UICC, so it is not specific to M2M.
- All temporary identity credentials and associated network configuration parameters will be generated for the device that carries the legitimate UICC, that is, the adversary's smartphone device. In other words, currently, access networks have no way of differentiating a smartphone from an M2M device. As long as the device carries a legitimate UICC, the device will be granted access to network services.
- The adversary may even own the abused M2M device that carries a UICC. In that case, the adversary need not attempt to physically access an M2M device belonging to someone else. For example, an adversary could detach the UICC from his/her health monitoring device and install it into a smartphone. An incentive for such an action could be that the access network operator billing rates corresponding to M2M operations are much lower than for regular 3G data connectivity. In addition, it is difficult for their affiliated M2M servers to detect that they are being abused since traffic requests to the Internet will not be routed through M2M servers.

- The use of a foreign, legitimate M2M UICC can clearly go undetected even if the access network provider has activated encryption and integrity-protection operations. This is because the keys used for encryption and integrity protection are renewed at every new registration. Hence, the adversarial device will be able to generate such keys, since it carries a legitimate UICC. Packets originating from this device will be normally encrypted and potentially integrity-protected using those keys.

Example 2: Hijacking through Passive Medium Overhearing

An adversary can hijack M2M cellular network connections even in the absence of UICC, by simply overhearing control and data traffic between M2M device and access network. The goal of the attacker is to use legitimate device credentials in order to freely access network services. As an example, let us assume that the attacker wishes to establish a UDP connection with a remote host, and stream data (e.g., video traffic) on the uplink direction for the case of an EVDO-based access network (the attack is similarly applicable to other technologies as well). The adversary would be able to freely use the uplink traffic channel as follows:

- The adversary overhears all packet exchanges between M2M equipment (M2ME) and the radio network controller (RNC) during registration. Recall that all these messages are not encrypted. Hence, the radio-channel-specific identity of the user equipment assigned by the serving system for operating on the air interface, the so-called UATI, and the IP address are known to the adversary.
- The application layer on the adversarial client device generates video frames, for example, through a webcam application. Each such frame is forwarded down to the IP layer, where it is encapsulated into an IP layer packet. The IP encapsulation function has been modified by the attacker to include: (i) the overheard IP address as the IP source and (ii) the IP address of the attacker's desired remote host, as the IP destination.
- Every fragment is forwarded to the MAC layer, where a MAC header is added. This header includes the overheard UATI24 (UATI-coded to 24 bits) value. The physical layer (PHY) transmits every MAC frame to the base station using the long code that corresponds to the assigned UATI. Since the UATI is known to the attacker, the latter is able to reconstruct the long code corresponding to the overheard UATI. For each fragment transmission, the uplink traffic channel that was allocated to M2ME by RNC during connection set-up is used.
- The base station forwards each fragment to the RNC, which merges fragments in order to reconstruct the IP layer packet. As is typical with RNC implementations, the RNC checks the MAC header of each fragment in order to identify the originating M2ME, through the UATI24 value. This value is then used for querying the locally maintained ⟨UATI24 – generic routing encapsulation (GRE) key⟩ association in order to tunnel the IP packet to the packet data serving node (PDSN), using the A10

interface. The GRE key identifies a specific A10 channel (or GRE tunnel) associated with a particular user session. Note here that the RNC receives fragments containing a legitimate UATI value and thus, the RNC is left to "believe" that the fragments were sent by the device that initially requested the specific UATI assignment.

- The PDSN receives the IP packet from the GRE tunnel and consults its locally maintained look-up table to verify that the source IP address actually corresponds to the GRE key used. As long as the entry is verified, the packet is routed to the adversary's intended destination. For this, the destination IP address of the packet is used. Clearly, the PDSN has no further reason to challenge the validity of this packet. The source IP is a legitimate address that was allocated to the M2ME during registration, and corresponds to the matching GRE key.
- Assuming that the adversary can always overhear signaling messages and commands from the radio access network (RAN) toward the abused M2ME, the adversary can always respond to such commands either by using its own credentials or the overheard ones. Clearly, the adversary need not be always located next to the M2ME, but alternatively needs to be close to the base station.

From this second example, it is evident that the adversary simply needs to overhear the UATI and the IP addresses that are assigned to the M2ME by the access network operator. Using these values, the attacker can generate packet headers which "mislead" the access network into believing that the packets are originating from a legitimate client. Note that the adversary needs to use both the specific overheard UATI and IP addresses (assigned to the same legitimate M2ME). Only one of these values is not sufficient, since:

- if a different UATI is used, it will correspond to a different long code and a different GRE key. Assuming that this UATI actually exists, it will lead the RNC into using a different GRE tunnel. The PDSN will then reject the packet, since there will be no match between the used IP address and the GRE key.
- if a different IP address is used, the packet will be rejected by the PDSN for the exact same reason: the PDSN will not identify any valid <GRE key – IP address> tuples in its locally stored look-up table.

Therefore, both the UATI and the IP address (assigned to the same M2ME) need to be overheard in order for the attack to be successful.

It is evident that currently the access network will authorize access to any device carrying a legitimate UICC. In other words, cellular access networks have no way of detecting this type of hijacking attack.

Throughout this chapter, we elaborate on the cellular M2M ecosystem, identifying all the different business entities involved as well as the security trust relationships among them. Moreover, we take a look at standardization efforts that are related to M2M security. In addition, we discuss the security requirements for M2M services, and further assess the suitability of existing security schemes for M2M.

8.2 Trust Relationships in the M2M Ecosystem

Traditional network services, such as cellular voice and data services, are provided by operators that deploy and operate the network, market the services, certify and channel the devices to end-users, and manage the end-user subscriptions and billing. Essentially, a single entity is involved in all aspects of each service. On the other hand, M2M services typically involve multiple entities in the delivery of each service. The entities most commonly involved are the end-user, the network provider, the M2M service provider, the application provider, and the device manufacturer. M2M operators negotiate network services in bulk from multiple network providers and manage the network access subscriptions on behalf of many application providers. Application providers, as the name suggests, are the entities that provide the end-user application, such as collecting and processing the data. An end-user may buy a device that incorporates the network connectivity module directly from the manufacturer (through a retail distribution network) or from the application provider. Hence, the ecosystem is much more complex than a traditional network service. This renders the design of efficient end-to-end security solutions quite challenging.

The M2M service delivery ecosystem is more complex for multiple reasons:

- Given the scope of M2M communications, many device manufacturers are participating in an open ecosystem. This implies that the application provider may or may not have a relationship with the manufacturer. In such a case, there is no predefined trust relationship between them and therefore the application provider cannot rely purely on security solutions offered by the manufacturer. Especially for mass-market applications, where individual machines communicate sparingly, maintaining the device distribution supply chain would result in extra costs to the application provider. Conversely, for high-end, low-volume, high-value applications, such as acute health condition monitoring, it makes sense for the application provider to own the distribution of the individual devices to the end-users.
- An important requirement for many M2M services is the need for guaranteed network connectivity across many regions, since the customers for a given application may span a wide area even though a given network provider may be regional. This necessitates the use of multiple access networks, belonging to independent network providers. Furthermore, some of the processes, such as provisioning, billing, and device management, are so substantially different for M2M devices, compared to those of human devices, such as handsets, that traditional network providers are ill-equipped to economically perform these processes. This has resulted in the need for an M2M service provider that bridges the relationship between the application provider and the network operator. Clearly, the application provider always maintains business relationships with the customer and the M2M service provider. Similarly, the M2M service provider has relationships with one or more access network operators. While the access network operator may itself be the M2M service provider, we focus on the case where the M2M service provider is an independent entity from the perspective of application providers as customers.
- The device may be used by an end-user customer, which may not have ownership of the device. Moreover, due to the inherent effort of keeping the device cost low, the device may not always have a user interface for customer interaction.

The following use cases illustrate the business relationships involved in M2M service delivery. The first is that of smart metering. in this example, an electric utility company will buy smart meters from a meter-manufacturing company and distribute them to the end-users that have subscribed to the utility service. The utility company will own and deploy the meters in the premises of the end-consumers. The utility company subscribes to the M2M service from an M2M service provider for meter data collection and device management. The M2M service provider, in turn, has business relationships with network providers for use of the bandwidth for transmission of data. Table 8.1 below shows the various entities and the relationships between them.

For comparison, Table 8.2 shows the relationships involved in the case of the traditional cellular voice service. A comparison of the table shows that many of the entries in Table 8.2 are void, indicating a simpler ecosystem.

The second use case, fleet management, is illustrated in Table 8.3. In this service, an application provider is providing fleet-tracking and scheduling services to small and medium-sized companies. This service differs from the smart metering case, primarily in that the devices are bought by the end-customers from retail outlets, and are owned and deployed by themselves, as opposed to by the application provider.

In summary, the M2M ecosystem involves a situation where potentially a billion customers (each customer typically having more than one device) communicate and are served by thousands of application providers with which they typically maintain trust relationships. These application providers work with a handful of M2M service providers, who in turn may potentially maintain business relationships with multiple network operators. The device ecosystem consists of many manufacturers reaching customers through retail outlets, or leveraging application provider-owned distribution channels.

Given the above summary, one may note that the complexity of the M2M ecosystem is strongly characterized by diverse business (and thereby trust) relationships, which cannot be accurately predicted during the design of security solutions. Therefore, security protocol design for M2M systems has to inherently assume that the M2M service provider may not have trust relationships with other stakeholders in the ecosystem. Throughout the following sections, we will assess the suitability of various security strategies for the case of M2M and provide design recommendations for applicable security solutions.

8.3 Security Requirements

The M2M security requirements are drawn from the need to protect the various entities in the end-to-end solution from perceived threats. We will provide examples of the threats experienced by each of the players identified in the previous section and then summarize the security requirements.

8.3.1 Customer/M2M Device User

Devices can be hijacked for the data they provide or for their actuation capabilities by adversaries pretending to be the back-end application servers. A scenario where such an attack could be of some economic value to the adversary is the remote control of home automation devices, such as alarms and garage door openers. By pretending to be the network-based home automation server, attackers can deactivate the alarm,

Table 8.1 Business relationships between the main entities in smart metering service

	Role	Consumer	Device	Network provider	M2M operator	Application provider
Consumer	Subscribes to utility	–	–	–	–	–
Device	Smart meter with wireless module	Device installed in consumer's premises	–	–	–	–
Network provider	Provides cellular communication	None	Certifies the wireless module in the devices	–	–	–
M2M operator	Virtual network operator that serves as the home network for devices and provides connectivity over multiple transport network providers	None	None	Roaming relationship	–	–
Application provider	Utility company that provides the smart metering application and services	Consumer subscribes to the utility service	Device owned and deployed by utility	None	AP subscribes for connectivity services with the M2M operator	–
Device manufacturer	Makes the meter and integrates wireless module	None	Makes the device	None	None	Manufacturer provides the devices to the utility for deployment

Table 8.2 Business relationships between the main entities for voice service

	Role	Consumer	Device	Network provider	M2M operator	Application provider
Consumer	Subscribes to voice service	–	–	–	–	–
Device	Cell phone	Device owned by consumer	–	–	–	–
Network provider	Provides cellular communication	Provides the device	Certifies the devices	–	–	–
M2M operator	Does not exist	None	None	None	–	–
Application provider	Same as network provider	None	None	None	None	–
Device manufacturer	Makes the device	None	Makes the device	Wholesale supplier	None	None

Table 8.3 Business relationships between the main entities in the fleet management service

	Role	Consumer	Device	Network provider	M2M operator	Application provider
Consumer	Subscribes to tracking service from the AP	–	–	–	–	–
Device	Tracking device	Consumer buys off-the-shelf device from retailer	–	–	–	–
Network provider	Provides cellular communication	None	Certifies the wireless module in the devices	–	–	–
M2M operator	Virtual network operator that serves as the home network for devices and provides connectivity over multiple transport network providers	None	None	Roaming relationship	–	–
Application provider	Provides fleet management services to small and medium enterprises	Consumer subscribes to the tracking service	None	None	Ap subscribes for connectivity services with the M2M operator	–
Device manufacturer	Makes the device	None	Makes the device and sells through retail outlets	None	None	None

and open doors to enter the house. Devices must be protected from unauthorized entities trying to establish communications to and from the devices.

In many M2M applications the data collected from the M2M devices is sensitive in nature. For example, in a child-tracking application it should not be possible for an unauthorized person to acquire information about the location of the child. Thus, the M2M security solution must be such that it is not possible to acquire information about the stored data by eavesdropping at any point in the network.

Identity is a valuable piece of information as it can be correlated with other data such as the location of network elements from which this identity information is retrieved to discern some patterns. In some applications, it is important that the identity of the end-customer is not available. Thus, unencrypted transmission of the actual identity of the device is unacceptable, since the device and or its usage can be tracked by adversaries eavesdropping on the network.

8.3.2 Access Network Provider

Unlike the case of consumer electronic devices, in many cases M2M devices are owned by the application providers and deployed in premises that are not constantly physically monitored or protected. For example, in the case of smart metering, the meters are typically owned by the utility companies and deployed in homes and small business locations. These devices are thus more susceptible to theft. Misuse of stolen communication modules found in these devices for the purpose of regular Internet communication, such as web-browsing, should not be permitted.

The idea behind M2M hijacking, where an adversary makes use of legitimate device credentials in order to freely access network services, also threatens access network operators. While such attacks can be launched even on non-M2M devices (such as regular smartphones, PDAs, etc.) using intelligent malicious software, they are much more prominent in M2M deployments owing to two inherent properties that make M2M devices prone to such attacks:

- **Inexpensive equipment:** A low-cost heart-monitoring device would simply measure the average number of pulses per minute and send this information to a remote server through the attached cellular network. Such a device would typically have no further capabilities or tasks. Similarly, a home smart meter would measure the amount of water that is consumed during a measurement period, and report this information to the water company. In both these examples, a single data session would always have to be established between the M2M device and a predefined server; the device would not need to contact any other entity. Therefore, such equipment will be naturally inexpensive. This will enable clever adversaries to easily hack into such devices and compromise them.
- **Ease of accessibility:** Potential M2M deployments are mainly comprised of static (or perhaps very-low-mobility) devices, placed in visible and physically accessible locations. From the security point of view, this is expected to have two complications. First, adversaries will be able to physically access the device, remove/replace the credentials if they are stored on a removable card (e.g., UICC), install sniffers, and perform other potential malicious actions. Second, in a static device installation where there are

no hand-offs to different base stations or RANs, an attacker will be able to easily identify the affiliated access network identity as well as the assigned temporary M2M device identity.

8.3.3 M2M Service Provider

The threats to an M2M service provider include all those applicable to a network operator. In addition, it is the responsibility of the M2M operator to ensure service availability to application providers and guarantee that the data sent to application-provider data centers is not tampered with.

8.3.4 Application Provider

Adversaries may benefit from modifying the data transferred from the device to the back-end application server or vice versa through man-in-the-middle attacks. It should thus be possible to verify the integrity of data transferred between the entities.

It is possible for some devices to masquerade as other devices, that is, to impersonate other devices in the network and upload data to the back-end server. For example, in the smart metering service where meter data is collected from the various meters, a dishonest home owner might alter the meter identification so that it appears as a neighbor's meter to the application server, thereby avoiding paying for use of electricity. Another possible threat is for an adversary to send false information to the server by pretending to be a legitimate device. For example, in a prisoner-tracking service, another device could act like the prisoner's device, and falsely report the location, while the legitimate device is compromised or even destroyed.

Based on the above threats, we can derive the following common requirements for any security solution.

- **Mutual authentication:** Only authenticated M2M devices can access the network and M2M system, and conversely any M2M device should authenticate the server before accepting any data, such as commands or management-related updates. This will ensure that data collected by the back-end-server is coming from a legitimate device, and the device is assured that it is communicating with a legitimate server. The mutual authentication procedures have to be completed between the M2M device and the M2M operator network before initiating data transfer.
- **Confidentiality:** Data transfer between the M2M device and the application server should be protected from unauthorized eavesdropping.
- **Integrity:** Application servers should be able to verify the integrity of the data originating from an associated device. Similarly, devices should be able to verify the integrity of the data sent by the affiliated servers. This protects the data from unauthorized modifications or manipulations.
- **Exclusivity:** Communication devices in the machine module should not be capable of being used for other applications. When the network provider entity is separate from the application provider entity, this requirement implies that the network provider has to deny service to a tampered device.
- **Anonymity:** The identity of the device should not be revealed to an eavesdropper in the network.

Additionally, the following requirements on bootstrapping the security solution exist, which are very specific to M2M communications.

8.3.5 Bootstrapping Requirements

Within the M2M ecosystem, we are dealing with an explosion in the number of devices (perhaps running into the billions, if widely adopted) and hundreds of applications provided by a few virtual operators using a mix of access networks. Therefore, several constraints are imposed on the bootstrapping of security keys:

- **Complex ecosystem:** We should expect a multitude of device vendors participating in a relatively open ecosystem, selling low-cost devices directly to consumers.
  - For instance, as described earlier, the complex ecosystem may involve device manufacturers selling on an open market while service providers may not have a business relationship with the device vendor or reseller.
  - Alternatively, in residential applications such as metering, the end-user (homeowner) may not even know the virtual network operator who is providing the service on behalf of the owner of the organization which owns the application, but yet securing the application is critical to all players in the chain.
- **Scalability:** M2M services typically involve a large number of devices, each sending a small amount of data. The revenue per device is generally small for any of the players in the value chain. Thus, it is important to minimize expenses incurred in deploying and maintaining these devices. One of the key steps in the deployment of devices is the provisioning of the device in the appropriate databases of the network and application provider. In particular, provisioning of keys or other information for the security solution should be simple and scalable. The provisioning process cannot assume that the device will be activated at a known, prearranged specific time. Devices may be turned on temporarily at the manufacturing site for testing but then subsequently turned off until deployment. It can be installed at one time but then the communication may only be turned on later. Thus it is not possible to predict a particular time for a provisioning process to be executed. The bootstrapping application should be as automated as possible, given the volume of devices involved in the ecosystem. In fact, in many cases, any manual intervention (offline or otherwise) should be limited to consumers or installers entering a few keystrokes of data into an online database (if need be) or using a command line interface on the device, if one is available.
- **Bootstrapping requires network authentication:** More importantly, it is highly likely that the network operator (who provides bandwidth to the M2M operator) will not be able to tell when a device is executing a bootstrapping application. For instance, this constraint requires that a device authenticate itself with the mobile data network using a network access identity, which allows the network operator to "recognize" the device and "authorize the data transaction," despite the device not yet possessing security keys or other access credentials. In other words, bootstrapping protocols for M2M devices should assume that the M2M system is an overlay over existing (and deployed) networking systems, thereby requiring devices to successfully register with the access network before the bootstrapping application is invoked, and such registration procedures should clearly comply with existing networking standards.

- **Flexibility in operator selection:** In addition to the requirements expressed earlier in this section, there clearly exist cases where the application provider may choose to switch M2M operators at some point of time (for various business reasons). This, in turn, requires that any bootstrapping procedure has to provide perfect forwards and backwards secrecy, with no leakage of key material across operator boundaries. Additionally, this requirement has to be met with little or no human intervention.

Given these constraints, the overall bootstrapping issue is to ensure that security requirements are met and there are no compromises at the device, link, network, or application layer.

8.4 Which Types of Solutions are Suitable?

Various attack detection and prevention schemes for other types of deployments have already been proposed. We develop them further below.

8.4.1 Approaches Against Hijacking

8.4.1.1 Encryption and Integrity Protection

A large body of research and standardization studies have proposed the use of cryptographic operations in order to detect and prevent hijacking attacks [2–15]. In summary, the overriding idea proposed in these studies is to integrity-protect the control and data messages by: (i) hashing the packet contents using a secret key known only to the participants of the session and (ii) appending the result of the hash operation at the end of the message. The recipient can then use the same secret key to produce the hash result, and compare it against the one that is appended in the received message. If they match, then the recipient is certain about the authenticity of the message. With this, hijacking is only possible when the adversary manages to obtain the common secret key in addition to the device identity. As well as offering integrity-protection, the encryption of control and data traffic ensures that the device identity is also protected from spoofing. However, in certain cases the device identity has to be sent unprotected. Furthermore, many of these cryptographic operations are extremely overhead-intensive [16] in terms of processing operations, and such processing may be prohibitive for inexpensive M2M devices. Studies have proposed that the user-data traffic packets are not themselves integrity-protected. In practice, integrity-protection of user-traffic adds tremendous overhead onto the RAN and thus is not supported by cellular network standards [2], while encryption of user data is widely used. Control signaling, on the other hand, is typically integrity-protected. Considering the limited amount of signaling messages and their sporadic nature, even the large overhead imposed by integrity-protection does not severely impact network performance, and is therefore widely accepted.

8.4.1.2 IP Address-Based Filtering

The idea here is to check the source IP address of packets at intermediate routers along a routing path in order to detect impersonation, reflection, and hiding attacks. Such techniques involve checking incoming packets using routing tables and looking up the return path for the source IP address, as well as comparing the IP address that has been allocated to a specific MAC address, potentially through IPCP or DHCP [17]. Such techniques are, however, inadequate in the context of our study, since: (i) the adversarial device may be located in very close proximity to the victim device; (ii) the identity (either temporary or permanent) of the source node can also be easily overheard during the device-registration procedure, as we demonstrate in the following section; and (iii) depending on the RAN implementation, it is possible that packets originating from client devices do not include any device identities. Another body of work uses IP address information to perform traffic-pattern monitoring and classification. The main goal of such mechanisms is to detect network anomalies by observing whether certain client devices have been compromised and, as a consequence, whether their associated traffic patterns have been recently changed – for example, whether new destination hosts are being used and different types of traffic are being introduced. Such mechanisms have been mainly proposed for detecting worm attacks. However, such techniques are less efficient in detecting M2M hijacking, since an astute adversary may use a very large data set of overheard IP addresses and temporary identities in a random fashion. With this, the traffic monitor/classifier cannot effectively identify any malicious traffic, since this is uniformly distributed across an apparent multitude of legitimate source nodes.

In the following sections, we will review the currently available solutions for M2M security bootstrapping and secure session establishment in order to assess their suitability for M2M, in terms of meeting the security requirements discussed in the previous section.

8.4.2 Public Key Solutions

One well-known method for bootstrapping keys involves provisioning a private-public key pair along with a certificate of a public key [18]. This pair can be individually generated during manufacture and can be pre-installed along with the corresponding certificate. Following this, during installation, the device could execute a well-known public-key-based key-agreement protocol (such as internet key exchange – IKE [19]) to bootstrap a root key. While this sounds straightforward (and uses known techniques), some of the issues with this method include the need for a large-scale public key infrastructure (PKI) that can potentially handle a billion devices.

PKI is a set of hardware, software, people, policies, and procedures needed to create, manage, distribute, use, store, and revoke digital certificates [18]. It is an arrangement that binds public keys with user identities by means of a certificate authority (CA). The binding is established through the registration and issuance process. The PKI function that assures this binding is called the registration authority (RA). For each device, the device identity, the public key, their binding, validity conditions, and other attributes are included in public key certificates, which are issued by the CA. The primary role of the CA is to publish the key bound to a given identity. This is done using the CA's own key, so that trust in the device's public key relies on people's trust in the validity of the CA's key.

8.4.2.1 Digital Certificates and Signatures

The digital certificate certifies the ownership of a public key. This allows other entities to be able to rely upon signatures or assertions made by the private key that corresponds to the public key that is certified. In this model of trust relationships, a CA is a *trusted third party* that is trusted by both the owner of the certificate and the party relying upon the certificate. In other words, a certificate is a digital "document" that contains both an identity and a public key, binding them together by a digital signature, which is constructed by the CA. This organization guarantees that upon creating the digital signature it has checked the identity of the device that has the corresponding public key, and that it has checked that this device is in possession of the corresponding private key. Any entity in possession of the CA's public key can verify the CA's signature on the presented certificate. With this, the (trusted) CA guarantees that the public key in the certificate belongs to the entity whose identity is in the same certificate.

The globally accepted standard for digital certificates is called X.509 [18]. An X.509 certificate contains the information that is signed and, therefore, is guaranteed to be true and unmodified. This information may contain the serial number, validity period, issuer name (this is the identity of the issuing CA), subject name (this is the identity of the party being certified; this can be a device, operator's authentication server, etc.), subject public key (the public key of the party being certified), and a path-length field indicating the maximum number of intermediate CAs that are allowed between the "root CA" and the end-user. The digital signature is produced using the issuing CA's private key.

8.4.2.2 Use of Certificates for M2M Service Bootstrapping

In the following section, the different possible approaches for bootstrapping using certificates are discussed, including the approach followed by the OMA DM (open mobile alliance device management) standard.

OMA DM Procedures

With OMA DM, a DM enabler is generally implemented using a DM server that stores the management objects (MOs) to be transferred to a DM client using the DM protocol. With OMA DM, devices would be bootstrapped in any of the following three ways:

- at the factory during manufacture;
- with a smart card;
- via a DM server, whereby after initial handshaking between the device and the DM server, the latter sends the root key to the device.

In cases where DM server provides (pushes) material to the device, at which time the device can initiate a management session with the DM server, the following steps take place:

- The DM server and the device are mutually authenticated using either a shared secret, or an out-of-band delivery of authentication information. The shared secret may be based on a user PIN or a network PIN. Certificates may also be used for this.
- The DM server sends out the bootstrap message with a push mechanism. The server must know the device address or some other mechanism for communicating with the device before initiating the bootstrap. The bootstrap message contains enough

information for the device to be authorized and to initiate a session with the DM server that sent out the bootstrap message. It is a one-time transfer of information and not part of an ongoing session between the device and the DM server.

Upon bootstrapping, and for regular registration of the device, the device and the DM server may authenticate each other in various ways, such as: (i) transport layer security pre-shared key (TLS-PSK), where the pre-shared key is provided either by the manufacturer or the DM server, (ii) TLS with server certificate and digest-based client authentication, where a client password is used to authenticate the client, or (iii) TLS with both client and server certificates, where the client certificate is provisioned either by the manufacturer or by the DM server.

Bootstrapping Using Certificates

Potential ways of bootstrapping using certificates could involve the following approaches (which can be included in the OMA DM scope):

- The device is pre-provisioned with *one* certificate by a CA that is trusted by all candidate M2M service providers. Such a CA may be a single trusted CA, operated by an entity that maintains business relationships with all possible service providers. Such a CA needs to also maintain business relationships with, or be trusted by, all device manufacturers. The device is also pre-provisioned with one or more public keys of CAs that are affiliated with service providers.
- The device is pre-provisioned with as many certificates as the total number of CAs that are trusted by all candidate service providers at the time of factory manufacturing. In cases where such CAs are owned by the service providers, the device needs to be provisioned with a certificate from each CA. The device is also pre-provisioned with one or more public keys of CAs that are affiliated with service providers.
- The device is provisioned with a certificate by a CA during bootstrapping. Such a CA must be trusted by the service provider (e.g., the latter owns the CA). In such a case, the device needs to be pre-provisioned with credentials that will be used for device authentication prior to reception of the certificate (potentially accompanied by the reception of an RSA private/public key pair, issued by a CA that offers such services).

In the above cases, whenever a certificate is to be provisioned to the device, one or more trusted certification authorities need to be contacted. More specifically, in the case where the manufacturer pre-provisions the certificates, a device needs to be provisioned during manufacture with at least as many certificates as the number of root CAs that are trusted by the different existing M2M service providers (assuming that the manufacturer is not a priori aware of the particular service provider that the device will be bootstrapped with). If the manufacturer owns a root CA that is trusted by all different candidate service providers, then all service providers need to trust all manufacturers; in such a case, a single certificate (issued by the manufacturer's CA) could be pre-provisioned to the device at the factory. In addition, the device would also need to be pre-provisioned with the public keys of all the (trusted) CAs that are responsible for issuing service provider certificates. Similarly, if certificate provisioning is performed by the service provider, the latter needs to provision a certificate issued by a CA that is trusted (e.g., owned by the service provider). Clearly for all the above cases, such certificate provisioning would have to take place for all potential types of authentication-serving entities, that is, not just for DM servers in OMA DM.

Revocation of certificates is based either on certificate revocation list (CRL) queries or on certain revocation protocols such as online certificate status protocol (OCSP) [20]. CRL queries require extra traffic initiated by the device or the M2M service provider. Given that billions of M2M devices are to be deployed, this is translated to increased bandwidth occupancy (generally, increased network resource occupancy) across multiple access networks and therefore to increased complexity. CRL databases need to be always online in order to serve revocation-related queries. Given that the CRL database only contains the list of revoked certificates, it often returns an inconclusive result that the "certificate in question is not in a database."

Given this discussion, in what follows we list specific concerns which indicate that public key cryptographic methods may not be ideal for addressing the M2M bootstrapping problem.

- For cases of cellular networks, public key methods are not part of major existing cellular standards, and almost all registration and related link-layer authentication mechanisms are addressed using symmetric key protocols [3GPP, 3GPP2, etc.]. Any use of public key methods to bootstrap keys over the air would then imply that devices cannot register with a mobile network prior to bootstrap. Hence, public key methods for bootstrap can be used only after a standards change and hence M2M operations over cellular will not be an overlay technology.
- The need for a PKI to manage certificates (including provisioning, maintenance, and revocation) is critical for secure functioning of public key methods. In the M2M context, any public key method is limited to bootstrapping "one symmetric key" per device – which in turn implies that transaction costs would be very high; that is, the use of existing public key methods with an extensive PKI for certificate management for "one transaction per device" is very high. Recall that public key systems are intended for long-term use and hence the cost of the PKI (especially revocation), while high, when amortized over several transactions per key ends up being reasonable. Furthermore, extending existing PKIs to manage billions of keys is not a scalable task given the complex ecosystem. Recall that M2M devices can be fairly inexpensive (e.g., metering applications). Hence, the use of public key methods to bootstrap security associations will have the unintended effect of increasing device costs.
- Additionally, private keys in a public key setting have to be stored safely. Given the cost-constrained nature of M2M, physically tamper-resistant security cannot be assumed to provide long-term protection of certificates and private keys.
- Moreover, public key methods are attractive only when multiple nodes share keys with each other, reducing the problem of managing $O(n)$ keys as opposed to $O(n^2)$ keys. The problem we address involves multiple devices sharing a security association with one server.

8.4.3 Smart Card-Based Solutions

The use of smart cards (SIM cards for GSM systems or UICC in general) is a very attractive option for scenarios with M2M deployments over cellular networks. Since the late 1990s, SIM cards have been widely used by cellular network operators in mobile phones and other mobile devices that attach to cellular base stations. Hence, for certain cellular M2M cases where devices are by default equipped with smart cards, such cards could provide the necessary key material for use by the M2M service or the application layer, based on policies and relationships.

More specifically, certain network standards mostly defined by the 3GPP mandate the use of specific types of smart cards, such as UICC with SIM or USIM applications, to any device that is attached to the network. In such settings, the use of smart cards is mandated for M2M devices. Smart cards contain a pre-provisioned permanent secret key, which is hard-coded by the smart-card manufacturer and delivered to the network operator. Such a key can be used for generating session-protection keys for use by the M2M service layer or application layer, and various techniques can be used for such a key derivation, such as the generic bootstrapping architecture (GBA) framework.

Generating and providing keys for use by higher layers (service, application) suggests that recipients of the new keys are entities that trust the owner of the smart card (key generator), that is, the cellular network operator. Hence, if a generated key is to be used for establishing secure sessions between the device and an M2M service provider, the latter should have trust relationships with the network operator. This is especially convenient when the network operator offers M2M services by performing M2M service provider tasks. Similarly, if the application provider trusts the network operator, the latter can provide a key to the former for use in establishing end-to-end application-layer secure sessions. Once again, the network operator may be offering application-layer services to devices – such as operating the application server – and hence M2M applications may use keys extracted by protocols that use the permanent key stored in the smart card as well as in the operator's secure key database. Clearly, if the cellular network operator is not trusted by the application or the service provider, an independent method for key provisioning that does not rely on the smart card needs to be used, unless such a method can derive a key in a way that the network operator (as a non-trusted entity by the key user) cannot obtain it.

Smart cards are not appropriate in cases where they are exclusively used for service or application-layer security. This is because smart cards inherently add to the cost of devices and moreover the problem of provisioning root keys into the smart card is often left to the card manufacturer and may be a proprietary procedure. Typical methods involve manual provisioning when a customer purchases services. Such methods, however, are not scalable in cellular M2M deployments that may be comprised of billions of devices. More importantly, the fundamental use of smart cards in cellular telephony is to enable subscription portability, that is, to enable the end-user to use any arbitrary device and connect to the network via a valid subscription without operator intervention. This feature has little or no value in the M2M context, and in fact any removable identity module opens the possibility of malicious end-users using these cards in other devices and potentially stealing service (in those cases where the M2M application service is paid for by the application provider). Additionally, when application providers switch between operators, the smart cards need to be replaced. This may require the replacement of millions of smart cards and can potentially be very expensive. Moreover, this cannot be achieved without human intervention.

8.4.4 Methods Based on Pre-Provisioned Symmetric Keys

An M2M device can mutually authenticate with the M2M service provider in cases where both of them are pre-provisioned with the same key, which can be used for performing symmetric cryptographic operations during every device registration. Such a key may be a permanently stored key in the device, within a secured environment. On the service provider's side, such a key is typically stored in a secure database, which interfaces with an authentication server, such as an HSS or AAA server.

Given the diversity of scenarios discussed above regarding M2M device deployments, it is not possible to a priori assume that the M2M service provider is involved in the device-deployment process and distribution to the end-customer. Moreover, due to the cost of M2M devices, in most cases such devices are not equipped with a user interface that can be used for manually provisioning a key upon purchase. Thus, for pre-provisioned key solutions, it is the device manufacturer's responsibility to perform the following two tasks:

- Install a key into the device at the factory. Typically, such a key is stored within a secure environment in the device.
- Optionally (depending on the device distribution scenario) provision the same key to the service provider. Such provisioning may utilize back-office servers that interface with the manufacturers' secure databases. This is typically the case where the application or the service provider orders and purchases devices directly from the manufacturer.

Nevertheless, a potential scenario where the customer purchases the device from a retail outlet, while not maintaining any business relations with the operator or the manufacturer needs to be accommodated. In such a case, the manufacturer may not maintain any business/trust relationships with any other player in the M2M ecosystem. Indeed, in this case, manufacturers simply make and sell M2M devices to the retail outlet – not even directly to end-users or application providers. Due to the absence of trust relationships involving the manufacturer, other players such as the device user/owner and the M2M service provider cannot assume that adversarial actions involving the manufacturer (implicitly or explicitly) will never take place. Therefore, pre-provisioned keys by manufacturers generally cannot be used as permanent keys for establishing secure service-layer sessions. However, such pre-provisioned keys can be used for two purposes: (i) for initial mutual authentication between device and service provider or device and application provider and (ii) for use in a secure procedure, which generates a new shared permanent key, which should not be implicitly or explicitly obtained by non-trusted entities, such as the manufacturer, through passive attacks.

As an example of such a procedure, in CDMA1x-based cellular voice and narrow-band data systems, a general Over-The-Air Service Provisioning (OTASP) procedure has been specified. This OTASP procedure includes the optional use of password-authenticated key exchange (PAK) [21] to provision a long-term secret used for link-layer authentication and privacy. PAK involves the use of a "password" in the context of a Diffie-Hellman key exchange. The crucial step in this procedure is to provision individual passwords in every device during manufacture, following which these passwords are shared offline through back-office infrastructures that are accessible to manufacturers as well as to M2M service providers or application providers.

8.4.5 Protocol for Automated Bootstrapping Based on Identity-Based Encryption

In Section 8.4.2, we discussed the challenges that are introduced by a potential adoption of PKI solutions in M2M. identity based encryption (IBE) protocols have been proposed as alternative methods to public key protocols requiring the existence of PKI. The idea behind IBE is that the public key of a user-device is a mathematical function of the identity that is associated with this key [22]. Therefore, there is no need to bind an identity of an entity with a public key through the use of certificates, since the public key is inherently derived from the identity, using a known algorithm. Note that with IBE, messages are encrypted with the IBE public key of the recipient. The latter is the only entity that can decrypt these messages, using an associated (to its identity) private key that is known only to the recipient. Such a private key is issued by an IBE key generation function (KGF), through the use of a domain secret and publicly known cryptographic functions through which the requirement for identity verification using certificates that are managed by large PKIs becomes obsolete. The only cryptographic material that is required by the message sender for IBE-encrypting a message is a set of publicly known cryptographic parameters that were used for generating the public key of the recipient [3].

IBE is based on the concept of elliptic curve cryptography (ECC) according to which the encryption/decryption procedure is dependent upon certain inherent properties of elliptic curves as well as a set of parameters that are a priori known to the entities that participate in bootstrapping. IBE can be used for establishing a shared key between two entities. In what follows, a protocol employing IBE for use in bootstrapping, while at the same time performing mutual authentication, is discussed. The protocol is called identity-based authenticated key exchange (IBAKE) [23]. IBAKE may be used in cases where key bootstrapping is not facilitated by a third party due to the lack of business/trust relationships. For example, it can be used for bootstrapping a permanent shared secret between an M2M device and an M2M operator's authentication server in cases where the M2M operator does not have trust relationships with the device manufacturer or the access network operator.

The IBAKE protocol involves the exchange of three messages between the two participants, with each message being IBE encrypted with the public key of the corresponding recipient. Since each message is IBE encrypted, only the intended recipient will be able to decrypt it. By the end of the three-way handshake, the participants are mutually authenticated and can establish (bootstrap) a permanent, shared secret key, which is known only to the two of them.

In order for bootstrapping to be performed, a set of parameters need to be provisioned to the entities that participate in bootstrapping. In what follows, we consider specifically the scenario where an M2M device and an M2M service provider generate a shared root key, which is then securely provisioned to an authentication server (e.g., an HSS or a AAA server). Clearly the protocol can be followed in the case where the M2M operator's responsibilities are performed by a different stakeholder, such as a cellular network service provider.

- **Temporary ID and password for device:** During manufacture, the device is provisioned with a public identity (e.g., in the form of MAC_address@realm) as well as with a temporary password. This password is used for temporary authentication with the M2M service provider (prior to bootstrapping) through which certain material is "pushed" to the device. More specifically, prior to bootstrapping, the device needs to ensure that it bootstraps with the correct M2M operator. Similarly, the operator wants to verify that a legitimate device will be bootstrapped. The use of a temporary password allows mutual authentication between the device and the M2M service provider in order to obtain initial connectivity to the M2M operator resources. Once this connectivity has been obtained, the permanent subscription-related credentials are established between the device and the M2M operator, as described below.
- **IBE-related material for the device:** Upon initial device activation and for the purposes of bootstrapping, the device needs to be provisioned with an IBE private key. For this, the device contacts a trusted KGF, which makes use of a "domain secret" as well as a set of publicly available cryptographic parameters in order to generate an IBE private key for the device. The communication between the device and the KGF is facilitated by the M2M service provider's infrastructure. Therefore, in order for the device to contact the KGF, mutual authentication between the device and the M2M core takes place first, using the temporary password that is pre-provisioned in the device. The domain secret used for generating the private key is known only to the KGF, and therefore the KGF needs to be a trusted entity in terms of not committing (in isolation or in collusion) active attacks. The generated private key is provisioned securely from the KGF to the device through the M2M operator, that is, it is pushed to the device using an appropriate interface. Note that the IBE private key is generated using date information (included in the public key) and thus these keys are automatically revoked (invalid) upon expiration of the used date. Due to such automatic invalidation, the IBE private key becomes useless upon expiry; it cannot be used to decrypt content beyond the time frame for which it was issued. The device is also provisioned with the public cryptographic parameters of the KGF that is associated with the M2M operator; such provisioning takes place in conjunction with private key provisioning. Note here that the KGF does not need to store any private/public key information upon device provisioning, since these keys can be re-generated upon request.
- **IBE material for operator:** The operator is similarly provisioned with an IBE private/public key pair by the associated trusted KGF. The operator is also provisioned with the identity (and thereby the public key) of the device, as well as the public parameters of the KGF associated with the device. Note also that the device and the operator may use the same KGF, which may be either external or owned by the M2M service provider.

8.4.5.1 IBAKE Protocol Execution

As soon as the device and the operator are provisioned with the aforementioned cryptographic material and are allowed to exchange messages, they may proceed with executing

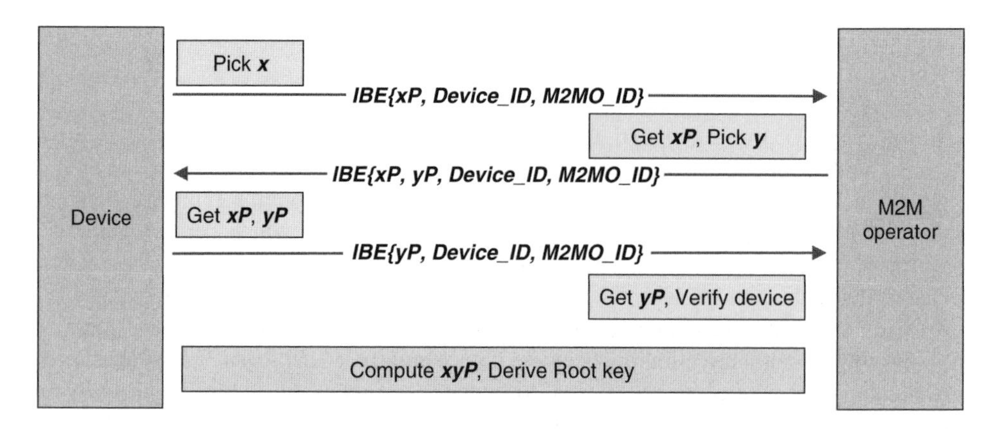

Figure 8.1 Root key bootstrapping using IBAKE.

the IBAKE handshake procedure. As explained in [23], IBAKE consists of three messages between device and operator. This message exchange is shown in Figure 8.1 and is described below.

- The device picks a random number x, calculates xP using ECC, where P is a known point on a known elliptic curve. The device encrypts xP with the IBE public key of the operator and sends the encrypted message to the operator (MESSAGE_1). As mentioned above, only the operator can decrypt this message.
- The operator receives and decrypts MESSAGE_1 from the device, thereby obtaining xP. The operator then generates a random number y and calculates yP. Furthermore, the operator encrypts both xP, yP with the IBE public key of the device, and sends a message (MESSAGE_2) to the device containing the IBE-encrypted xP and yP.
- The device receives and IBE-decrypts MESSAGE_2 from the operator, thereby obtaining xP, yP. If the received xP value matches the one sent in the initial message from the device to M2M Service Bootstrap Function (MSBF), the device has authenticated the operator (specifically the serving entity in the operator's infrastructure that participates in bootstrapping), since the latter is the only entity that can decrypt MESSAGE_1. Subsequently, the device IBE encrypts yP and sends it back to the operator (MESSAGE_3). The operator receives and decrypts MESSAGE_3. If the obtained yP value matches the one in MESSAGE_2, the operator has authenticated the device.

As such, both the device and operator have been mutually authenticated and are in possession of xP and yP values. Finally, each of them can use ECC functions to calculate xyP. They further use xyP to derive the permanent shared key using a publicly known ECC function. The European Telecommunications Standards Institute (ETSI) M2M functional architecture specification provides details of how the IBAKE protocol can be used for automated bootstrapping of M2M devices and gateways.

8.4.5.2 Advantages with IBAKE

Leveraging IBE in the afore-mentioned manner has some advantages:

- **Avoiding key escrow points:** All online steps in the protocol exchange are encrypted using IBE. Hence, clearly the KGF is able to decrypt all the exchanges. However, the KGF cannot compute the session key. This is because of the hardness of the elliptic curve Diffie-Hellman problem. In other words, even if xP and yP are obtained by a third party, it is computationally hard to compute xyP. Hence, even if the KGF were to be compromised by a passive attacker (who would obtain the domain secret), the latter would not be able to derive the established (root) key. Therefore, the KGF needs to be trusted only in terms of not launching or facilitating active attacks. Furthermore, although the device manufacturer has pre-provisioned the temporary password into the device, it is not possible for the manufacturer to passively obtain the root key, even though the manufacturer can obtain the IBE private key of the device by overhearing the communication between device and operator;
- **Mutually authenticating device and operator during bootstrap:** As explained above, IBAKE is a mutually authenticating protocol; the device verifies the identity of the operator upon receipt of MESSAGE_2, while the operator verifies the identity of the device upon receipt of MESSAGE_3;
- **Achieving perfect forwards-and-backwards secrecy:** Since x and y are random, xyP is always fresh and unrelated to any past or future sessions between the device and the operator. Hence, a new M2M operator cannot derive old root keys, while old M2M operators cannot obtain future root keys; and
- **KGF availability:** KGFs are not contacted during IBAKE protocol exchange, which dramatically reduces availability requirements on an external entity. Moreover, a KGF needs to be online only if the operator is expecting to perform bootstrapping for devices. In other words, if the KGF is owned by the M2M service provider, the provider may set KGF offline if there are no devices expected for bootstrapping.

IBAKE is currently published as an Internet Engineering Task Force (IETF) Request For Comments (RFCs) in the form of MIKEY-IBAKE (Multimedia Internet Keying) and under consideration by various standards bodies, such as ETSI and 3GPP.

8.4.6 Security for Groups of M2M Devices

Traditional secure registration protocols rely on clients (or devices) authenticating with a server (or network). For instance, a user may log on to a server, or a mobile device might register with a network. The registration protocol often includes authentication of the client/device to the server/network or a mutual authentication protocol. More recently, with a view toward securing the entire session (not just access), these authentication protocols have been augmented to include Key Agreement which allows the client/device and the server/network to agree on a set of keys to secure the entire session. Examples of one-way authentication protocols include Challenge Authentication Protocol (CHAP) [1], Password Authentication Protocol (PAP) [1], GSM triplet-based authentication protocols, and so on, and examples of mutual authentication protocols include: 3GPP's Authentication and Key Agreement (AKA) protocol [2], various Extensible Authentication Protocol (EAP)-based mutual authentication protocols such as EAP-TLS [24], EAP-GPSK (Generalized Pre-shared Key), and so on.

While the above mechanisms are applicable in the M2M context, they do not scale well in scenarios where hundreds or thousands of devices need to be authenticated by an access network operator, M2M operator, or application provider within a short period of time. Thus, recently the notion of *device-group authentication* has been discussed in the various standards bodies. Groups of M2M devices are being considered in standards in a broader perspective, that is, not only for the purposes of security, as described in [25]. Devices are grouped together based on a particular policy and are aware of the groups that they belong to. The policy-making procedures and the functionality of network entities that are involved in developing such a policy are performed by potentially different players, such as the M2M operator and the access network operator, either jointly or separately.

With group authentication, a group of M2M devices can securely register with the same system using a single registration procedure. This allows a dramatic reduction in the required signaling related to registration procedures, as we will discuss later. In other words, there is no need to establish individual sessions with each M2M device for the purposes of authenticating each of them individually. In contrast, devices belonging to a particular group are authenticated in bulk, thereby reducing to a large extent the complexity, and the bandwidth requirements of the authentication procedure. A group of N devices is represented by a group ID, which is known to all the network entities that are involved in the group authentication process; they are the same entities that are used for authenticating individual devices. Those entities are also aware of the identities of all the devices that belong to such a group. Such information can be stored locally in every entity in the form of a look-up table.

In order to provide an idea of how group authentication saves on signaling, we shall consider the following example. Let us assume a deployment of N devices and that an authentication entity (authenticator) needs to authenticate each of these devices before authorizing them to access one or more services. To date, the authenticator needs to establish an individual session and exchange a fixed number of authentication messages with each device. Hence, if the deployment consists of N devices, and if k number of messages need to be exchanged between the authenticator and every device, then a total of $N \times k$ messages needs to be transmitted over the air in order for all of these devices to be authenticated. As an example, let us assume that $N = 100\,000$, and that the authenticator uses the AKA protocol [2, 24]. As per this protocol, the authenticator sends a *user authentication request* to the client, which consists of a random challenge (RAND) and an authentication token (AUTN) [4, 7]; hence $k = 2$ in this case. As a result, 200 000 individual messages are required in total. At the end of this overall process, each of the N devices will generate a session key, that is, a total of 100 000 session keys will be mutually generated between the authenticator and each device. Note that instead of one authenticator, multiple authenticators may exist, each authorizing a different type of service. In such a case, every device will have to separately hold a distinct authentication process with each of the authenticators (fortunately for such scenarios, the notion of "single sign-on" can be used, wherein a device can use a single set of credentials to authenticate with the various different authenticators). In such a setting, a potential group authentication solution would save on signaling not by addressing each device individually, but by treating all N devices as a group, by challenging them simultaneously, using a broadcast downlink message. In this way, only one (instead of 100 000) messages would have to be sent to all N devices.

A key feature of such methods should ensure that the session keys that are generated as a result of registration are not the same across M2M devices, although all devices are authenticated as a group. Such a property would allow each M2M device to securely communicate with the server/network, without compromising on privacy within the group. More specifically, the following requirements should be accommodated:

- ensure the same strength of security as in the case of individual device authentication;
- establish individual, unique keys per device, such that device data is secured against potential attacks originating from other devices (even those belonging to the same group); and
- use a reduced number of authentication messages. Ideally, exchange as many messages with the authenticator, using the same bandwidth, as in the case of a single authentication process.

One example of such procedure is described below.

Assume that each device belonging to the group is provisioned with two secret keys, that is, the unique individual key Ku and the group key Kg. Assume further that the AKA authentication protocol is used for performing mutual AKA.

In the group-authentication scenario, the authenticator would request from the HSS the authentication vector specifically designed for group authentication. This vector would contain the Random Challenge RAND and the Network Authentication Token AUTN computed using the group key Kg. But it will also contain expected authentication results XRESi and sets of Session Keys $(CK|IK)_i$ individually computed for each device in the Group by using an XOR of Kg and $(Ku)_i$ of each device.

The authenticator will send the RAND and AUTN to all devices in a group as a single broadcasted message. Each device in the group will verify the validity of the AUTN, and respond with individual RES_i, while computing an individual device-specific $[CK, IK]_i$. The responses are sent back to the Authenticator, which verifies them and activates the corresponding device-specific session keys.

In this exchange, for 100 000 devices in the group only 100 001 messages are exchanged instead of 200 000 as would otherwise have been the case in a conventional AKA exchange case.

While the details of such a procedure are not known at the time of writing this book, the expected benefits from the application of a group-authentication solution are manifold.

8.5 Standardization Efforts on Securing M2M and MTC Communications

Recently, there has been an increasing interest in standardizing various aspects of M2M security. Such interest is aligned with parallel efforts on standardizing functional architecture settings. In what follows, we discuss the current state of such efforts for different standardization bodies.

8.5.1 ETSI M2M Security

As explained in Chapter 2, the ETSI M2M Standards Group is focused on designing the functional architecture of the M2M service provider. It considers that the M2M provider is a service-layer entity, which logically resides above the access layer, and which is independent of (but may be collaborating with) the access network operator(s), as well as the M2M application provider(s).

The establishment of secure data sessions between M2M devices (or gateways) and the ETSI M2M service capabilities is based on the existence of a key hierarchy.

Security originates from a permanently stored common root key (K_{mr}) in a safe location within the device (e.g., a device-trusted environment), as well as in a secure user database within the M2M service provider's premises. Such a key is securely bootstrapped into the device and provider, as we will explain later. The root key is used for mutual authentication between the gateway and the provider, as well as for generating of temporary registration session keys (known as). Given that K_{mr} is bootstrapped into the device during the service bootstrapping of the device with the M2M service provider, K_{mr} may be permanently stored in the device as long as the device is affiliated with the same service provider. Upon termination of any such affiliation, the stored root key is discarded. Note that a device may be simultaneously affiliated with more than one M2M service providers. In such a case, one or more root keys for each provider are securely stored in the device, in such a way that the different providers are not aware of each other's root key values. Note also that a root key is bound to a client identity; a device may be hosting multiple client identities, even if it is affiliated with a single service provider. In such a case, multiple root keys with one key per client identity are once again stored and used by the device.

A K_{mr} key is used for mutually authenticating a device (client identity residing in a device) with the M2M service provider's service capabilities whenever the device registers with the M2M core. Upon successful mutual authentication, a K_{mc} key is generated by K_{mr}, by both the device and the authentication server, using a commonly known KGF. K_{mc} is a session key that is valid for the duration of the device registration session. K_{mc} is generated by the secure authentication server and is delivered to the Network Security Capability (NSEC) capability in the M2M Core. In fact, the NSEC may participate in the mutual authentication procedure by serving as an authenticator. In such a case, the NSEC establishes a secure session with the M2M Authentication Server (MAS), which interfaces with the secure user database, where user root keys are stored. Authentication requests from devices are directed to NSEC, which in turn requests and receives authentication material from the MAS. K_{mc} may be further used to secure data exchanged between device/gateway and M2M operator, using TLS-PSK or other methods.

The complexity of the M2M ecosystem is a major consideration in the work of ETSI M2M. Security needs to be standardized in such a way that functional requirements related to potential trust relationships in the ecosystem are met. Given this, ETSI M2M clearly distinguishes two high-level use cases regarding security:

- the M2M service provider maintains business relationships with other players, such as the access network operator and the device manufacturer; and

- the M2M service provider does not trust other players in the ecosystem and therefore service layer functions need to independently establish security associations between devices and M2M core.

In the first case, the M2M service provider trusts other stakeholders, such as the access network provider and the device manufacturer. Existing trust relationships between M2M and access network providers suggests that key material can be provided by the access network provider on a permanent basis, given that the latter supports such mechanisms. As an example, an M2M service provider may trust an affiliated mobile network operator, which already performs certain security operations in order to protect traffic from/to mobile devices. In such a trust scenario, the access network operator may generate security keys from the permanent secret that is stored in HSS/HLR (Home Location Register) and the corresponding UICC, and provide them to the M2M service layer for use in establishing secure service layer sessions. One method of performing such a key material generation could be the GBA method [26], which uses access network session keys to derive new keys for use by higher layers. Clearly, the access network operator would know the root key used at the service layer; however, this is not a problem for the M2M service provider due to the trust relationships maintained between the two operators. In a similar fashion, existing relationships between the device manufacturer and the M2M service provider allow the manufacturer to provide keys to the service layer for use in establishing security associations. In particular, the manufacturer pre-provisions a key into the device at the factory and also provisions the same key to one or more M2M service providers (depending on which and how many such providers the device will associate with). Such a key could potentially be used as the root key, given that the manufacturer is a trusted entity and provisions the key to the M2M service provider in a secure fashion. Typically, however, such trust relationships do not exist it the M2M ecosystem, unless the M2M service provisioning is offered directly by the access network operator or the manufacturer.

If no trust relationships exist between the M2M service provider and other entities, the M2M service provider needs to independently establish security association with affiliated devices, in such a way that non-trusted entities are not able to passively obtain root or session keys. In other words, bootstrapping and secure data session establishment between M2M devices and M2M service capabilities at the M2M core should be performed using methods that prohibit passive adversarial attacks involving the manufacturer and the access network operator from obtaining root or service keys. Note also that such methods need to be aligned with considerations regarding the complex M2M ecosystem, as well as the limited capabilities of M2M devices. Potential examples of such methods for root key bootstrapping have appeared in ETSI M2M standards, and can be found in [27].

8.5.2 3GPP Security Related to Network Improvements for Machine Type Communications

Standardizing machine-type communications (MTCs) has received a lot of interest from the 3GPP. The 3GPP security group, SA3, has been guided by the group requirements and the network architecture group (SA1 and SA2, respectively) toward designing and standardizing security procedures related to network improvements for MTC. Since 3GPP is predominantly interested in operations exclusively involving the deployed 3GPP network, architecture, and thereby security support for M2M communications is tied to the existing 3GPP network design.

The main requirement imposed for MTC security is that MTC optimizations must not degrade security compared to non-MTC communications. Given this broad requirement, SA3 has the task of designing security solutions for protecting the functionality of MTC features, as defined and standardized by SA2. Such features are related to group communications, time control, low mobility, monitoring, IP addressing, small data transmissions, device triggering, signalling congestion control, etc. SA2 provides updated draft versions of the designs of each of these MTC features. The task of SA3 is to determine whether already standardized security countermeasures are affected by the new designs, and further act toward eliminating identified security design flaws.

References

1. Mizikovsky, S., Wang, Z., and Zhu, H. (2007) CDMA 1x EV-DO security. *Bell Labs Technical Journal*, 11 (4), 291–305.
2. 3GPP TS 33.102. Technical Specification Group Services and System Aspects; 3G Security; Security Architecture (Release 9, 2009).
3. European Telecommunications Standards Institute (1997) GSM Technical Specification GSM 03.20 (ETS 300 534): Digital Cellular Telecommunication System (Phase 2); Security Related Network Functions, August 1997.
4. Aboba, B. and Simon, D. (1999) PPP EAP TLS Authentication Protocol. RFC 2716.
5. Cox, R., Grosse, E., and Pike, R. (2002) Security in plan 9. USENIX Security Symposium, 2002.
6. 3GPP TS 33.401. Technical Specification Group Services and System Aspects; 3GPP System Architecture Evolution (SAE); Security Architecture (Release 9, 2009).
7. Gallagher, M.D. and Snyder, R.A. (1997) *Mobile Telecommunications Networking with IS-41*, McGraw-Hill, ISBN: 0-07-063313-2.
8. Handley, B. (2000) Resource-efficient anonymous group authentication, *Financial Cryptography, 4th International Conference FC 2000 Anguilla*, British West Indies, Springer-Verlag.
9. Gohil, S. and Fleming, S. (2004) Attacks on Homage Anonymous Group Authentication Protocol. Technical Report OUCS-2004-16, Department of Computer Science, University of Otago.
10. Jaulmes, E. and Poupard, G. (2002) *Financial Cryptography 2001*, Lecture Notes in Computer Science, Vol. 2339, Springer-Verlag, pp. 106–116.
11. Hanaoka, G., Shikata, J., Hanaoka, Y., and Imai, H. (2005) *Unconditionally Secure Anonymous Encryption and Group Authentication*, Oxford University Press, December 2005.
12. Martucci, L.A., Carvalho, T.C.M.B., and Ruggiero, W.V. (2004) A lightweight distributed group authentication mechanism. Proceedings of the 4th International Network Conference, INC, July 2004.
13. Kent, S. (2005) Security Architecture for the Internet Protocol. RFC 2401.
14. Teofili, S., Mascolo, M.D., Bianchi, G., Salsano, S., and Zugenmaier, A. (2008) User plane security alternatives in the 3G evolved Multimedia Broadcast Multicast Service (e-MBMS). *Security and Communication Networks*, 1 (6), 473–485.
15. Arkko, J. and Haverinen, H. (2006) Extensible Authentication Protocol Method for 3rd Generation Authentication and Key Agreement (EAP-AKA).
16. Feng, Z., Ning, J., Broustis, I., Pelechrinis, K., Krishnamurthy, S.V., and Faloutsos, M. (2011) Coping with packet replay attacks in wireless networks. IEEE Communications Society Conference on Sensor, Mesh and Ad Hoc Communications and Networks (SECON), 2011.
17. Droms, R. (1997) Dynamic Host Configuration Protocol. RFC 2131.
18. Cooper, M., Dzambasow, Y., Hesse, P., Joseph, S., and Nicholas, R. Internet X.509 Public Key Infrastructure: Certification Path Building. RFC 4158.

19. IETF RFC 4306. Internet Key Exchange Protocol.
20. Myers, M., Ankney, R., Malpani, A., Galperin, S., and Adams, C. X.509 Internet Public Key Infrastructure Online Certificate Status Protocol – OCSP. RFC 2560.
21. Brusilovsky, A., Faynberg, I., Zeltsan, Z., and Patel, S. Password Authenticated Key (PAK) Diffie-Hellman Exchange. RFC 5683.
22. Boyen, X. and Martin, L. (2006) Identity-Based Cryptography Standard (IBCS) #1: Supersingular Curve Implementations of the BF and BB1 Cryptosystems. RFC 5091.
23. Cakulev, V. and Sundaram, G. (2011) IBAKE: Identity Based Authenticated Key Agreement, Internet draft, http://tools.ietf.org/html/draft-cakulev-ibake-03.
24. Simon, D., Aboba, B., and Hurst, R. (2008) The EAP-TLS Authentication Protocol. RFC 5216.
25. 3GPP TR 23.888 System improvements for Machine-Type Communications (MTC) (to be published in 2012).
26. 3GPP TS 33.220. Generic Authentication Architecture (GAA); Generic Bootstrapping Architecture (GBA) (2004).
27. ETSI TS 102 690. Machine-To-Machine Communications (M2M); Functional Architecture. (Release 1, 2010).

9

M2M Terminals and Modules

Gustav Vos
Sierra Wireless, British Columbia, Canada

This chapter focuses on M2M modules, their characteristics and the services they provide. However, before discussing these topics, it is important to provide a clear definition of the M2M module. An M2M module differs from an M2M terminal, insofar as an M2M terminal can be broken down into two logical components. The first is the application portion of the terminal that provides the specific hardware and software for the M2M application. For example, in a point of sales terminal, the application portion would be the keypad, LCD, and printer with the associated application layer software. The second logical component of the M2M terminal is the M2M module, which is mainly responsible for providing the connectivity services. The application portion is also sometimes simply referred to as the "host" to the M2M module. The application portion of the terminal is highly related to the M2M application, a subject that will be comprehensively covered in the second book in this series *The Internet of Things: Key Applications and Protocols*. Although M2M modules exist in wired and wireless variants (to be developed later in this chapter), more focus will be given to wireless M2M modules in this chapter, since the majority of deployed M2M modules use wireless networks.

9.1 M2M Module Categorization

With the highly diverse nature of problems that M2M applications resolve, it is no wonder that there is an equally diverse range of M2M module types. The following sections provide one method for classifying the various types of M2M modules and provide some key categories into which an M2M module may be classified. It should be borne in mind that these categories are not orthogonal and thus an M2M module may fall into

M2M Communications: A Systems Approach, First Edition.
Edited by David Boswarthick, Omar Elloumi and Olivier Hersent.

more than one category. These classifications should also assist an M2M application developer in determining which type of M2M module is best for the M2M application under development.

9.1.1 Access Technology

9.1.1.1 Wireless or Wired

The first and easiest way to categorize an M2M module is to look at the type of access technology that it supports. There are two main types of access technology: wireless and wired. Wired access technologies require a physical connection to a cable such as a telephone line (e.g., RJ11) or the cable company's coaxial cable (e.g., RG-6). Some of the more popular wired access technologies for M2M include powerline communications (PLC) (connection to AC mains), Ethernet (RJ-45), and xDSL. The use of wired-access technologies is not common for M2M for several reasons, namely the inability to support mobility, the high infrastructure costs of laying new cable (PLC excluded), and the comparative complexity of maintenance. Wireless or radio access technologies (RATs), on the other hand, do not require a physical connection and use radio waves to relay information. RATs are commonly used for M2M applications, so we will focus more on wireless M2M modules in the following sections.

9.1.1.2 Wireless Link Distance

One way to classify a wireless access technology is by the distance at which a wireless link can be maintained. These subcategories are called WPAN (wireless personal-area networks), WLAN (wireless local-area networks), and WWAN (wireless wide-area networks) (see Figure 9.1).

In general, a PAN link covers less than 10 m and includes technologies such as Bluetooth, ZigBee, ultrawideband, and sometimes infrared. A WLAN link is typically categorized as being less than 30 m, a distance that could cover an office building. The most popular WLAN technology is 802.11 (or WiFi). A WWAN link can traverse very long distances and is a very popular access technology for M2M modules. Therefore, we will focus on this for the remainder of the chapter. WWAN M2M modules are popular because they are the only wireless access method that can support the full mobility required by many M2M applications. WWAN technologies can be further broken down using the following access scheme:

- frequency division multiple access (FDMA), e.g., AMPS;
- time division multiple access (TDMA), e.g., GSM;
- code division multiple access (CDMA), e.g., UMTS;
- orthogonal frequency division multiple access (OFDMA); e.g., LTE (long-term evolution).

Each of these WWAN technologies offers different levels of performance in terms of throughput and latency. The decision to use a certain type of WWAN technology will depend, among other things, on the application's throughput and latency requirements.

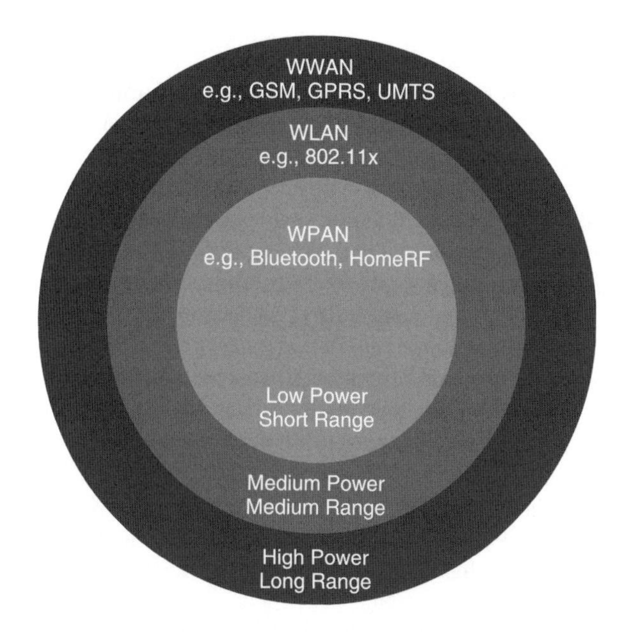

Figure 9.1 WPAN, WLAN, and WWAN.

9.1.1.3 Generation

A further method for classifying WWAN is by "generation" or "G." As with all technologies, WWAN has evolved over time. Sometimes WWAN improvements are incremental, but most often, and preferably, WWAN improvements are grouped together to create a significant leap in performance referred to as a generation (e.g., 2G or 3G). The ITU-R (International Telecommunications Union) has specifications for minimum performance for 3G and 4G, although marketing and advertising departments do not always adhere to the ITU-R definition, which causes inconsistency in terminology. In this chapter, the following categorization will apply:

- 1G (first generation) includes technologies such as AMPS, TACS and NTACS;
- 2G includes technologies such as IS-136, GSM, and most commercial deployments of 1xRTT;
- 3G includes UMTS, TD-SCDMA, and EVDO;
- 3.5G includes technologies such as HSPA (high-speed packet access);
- 4G includes UMB (ultra-mobile broadband), WiMAX (802.16e), and LTE.

Although the market place refers to LTE and WiMAX (802.16e) as 4G, strictly speaking, neither met the ITU-R's 4G performance requirements. In November 2010, the IU-R approved LTE-Advanced and WiMAX 802.16m access technologies as 4G technologies but it is uncertain what the market place will call these technologies, possibly 4.5G or 5G. Early indications suggest these 4.5G technologies will be commercialized by 2013, at the earliest.

9.1.1.4 2G Dominance

Although the retail channels for WWAN phones tend to be dominated by the newer access technology (e.g. 3G), M2M module sales do not follow this trend as the majority of WWAN M2M modules deployed in 2012 are 2G. Although the market trend is gradually evolving toward 3G and 4G modules, it will in all likelihood still be many years before 3G/4G-deployed M2M modules will outnumber 2G-deployed M2M modules. This is partly because the long life cycles of M2M applications; for example, automotive or utility metering can have a 10-year deployment lifetime. There are many reasons why 2G M2M modules are still being deployed in 2012 including:

- 2G has a significantly lower module cost. However, the difference in cost between a 2G and 2G + 3G module is reducing rapidly.
- 2G modules are seen as more mature, that is, more reliable, low power, and have more features and services (see the service section below).
- 2G coverage is seen as superior – since most 2G was initially deployed in the lower bands (cellular 800 and 900 MHz GSM band), and 3G in the higher bands (1800 MHz DCS and 1900 PCS), 2G has a natural advantage of better coverage, especially for indoor penetration.
- 2G is seen as good enough – since it is possible and even likely that the module will have to operate in 2G mode in rural and indoor areas (see the previous bullet-point on coverage), the M2M application needs to be designed for 2G throughputs and latencies.
- The slow-moving trend toward 3G or even 4G M2M modules can be very quickly accelerated by decisions made by the MNO (mobile network operator) or the telecoms regulators. For example, the MNO may forbid new 2G modules on their network as the MNO may have plans to re-farm the 2G spectrum to deploy more 3G or 4G technology or may even plan to decommission its 2G network entirely. Another possibility is if the MNO decides to charge less for services on their 3G and 4G network, thus swaying the TCO (total cost of ownership) in favor of the new technology. The MNO may also choose to subsidize the 3G and 4G device in order to make up for the difference in cost compared to providing a 2G device. The MNO decision as to when their 2G system should be decommissioned is very complex and has many variables. Some of these variables are geographically sensitive, such as spectrum and regulatory issues, and as such it is expected that the timing for the re-farming of 2G spectrum and decommissioning of 2G systems will be highly geographically dependent.

9.1.2 Physical Form Factors

The physical form factor is another method for M2M module classifications. The form factor typically relates to the level of integration that is required by the M2M application developer. Typically, the smaller the form factor, the higher the integration effort required by the M2M application developer. The form factor is also related to the M2M module's available hardware interfaces where the larger form factors typically have more physical interface options available.

9.1.2.1 Solder-Down Module

The first physical form factor category is the solder-down module that can be defined by the fact that it does not have a connector but only has solder pads or solder balls to allow it to be directly soldered to an M2M terminal's PCB. Solder pads are used with line grid array (LGA) interfaces and balls are used for ball grid arrays (BGAs). Commercial examples of such M2M modules using LGA and BGA interfaces are shown in Figure 9.2.

In general, solder-down modules are the smallest form factors available and in most cases this size reduction is achieved by using a multichip module (MCM), which is an integrated-circuit (IC) package where multiple semiconductor dies are packaged onto a substrate, facilitating their use as a single IC. As this technique is time-consuming and requires higher R&D costs, fewer M2M module vendors support this type of form factor, and the newer WWAN technology (e.g., 4G) is slightly lagging behind some other form factors, such as PCI express mini card. A solder-down module size varies but is in the range of 25×30 mm. Even with the constraints of size and I/O pins, these modules still often support many I/O options such as USB, I^2C, SPI bus, UARTs, and GPIOs, which typically give the M2M application developer enough flexibility. All I/O goes through the surface mount technology (SMT) interface, except in the rare cases where a separate antenna connection is provided. Beyond their small size, this form factor also has the advantage of being more reliable since it does not require a connector, which can become unreliable through vibration. As heat can be more easily transferred out of the M2M module and into the larger PCB, this form factor has superior head-dissipation properties making it suitable for high-temperature environments. Both the heat-dissipation advantage

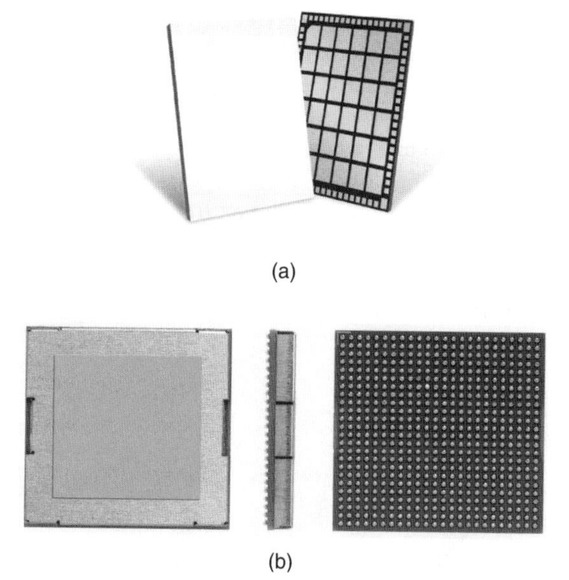

Figure 9.2 (a) LGA and (b) BGA. Images courtesy of Sierra Wireless.

and the reliability advantage make the solder-down form factor a suitable choice for automotive applications, but only a few M2M module vendors manufacture automotive-grade modules due to the large certification burden, such as the ISO TS-16949 production rules. There are also unfortunate downsides, the main disadvantage being that it requires more integration costs and effort from the M2M application developer. Although the integration effort is higher and more customized, the standalone module certification rules (see the certification section) will still apply, so the certification effort and costs should be the same as when using other module form factors. Other disadvantages include the absence of standards, so clients get locked into one vendor, maintenance is more difficult as it is not easy to replace a faulty M2M module, and it is more problematic to upgrade, for example, to a new access technology.

9.1.2.2 PCI Express Mini Card

The PCI express mini card form factor, or often simply referred to as the mini card form factor, was developed by the Peripheral Component Interconnect Special Interest Group (PCI-SIG) and was finalized by that group in June 2003. Although the mini card standard supports two options for host interface connectivity – PCI express and USB 2.0 – most M2M module vendors only support USB 2.0 connectivity. The physical dimensions of PCI express mini cards are defined ast 30×50.95 mm with a maximum thickness of 5 mm. PCI-SIG has also defined a half-length card that is specified at 30×26.8 mm (see Figure 9.3, which shows commercial examples of both the full- and half-sized mini cards).

With the exception of the antenna connectors, all the I/O is carried out via the 52-pin edge system connector. Although not explicitly defined by the PCI-SIG standard, the location of the antenna connectors are typically at the opposite end to the system connector in the PCI-SIG-defined I/O connector area.

The mini card standard was initially envisioned for use with laptops and PCs but has been expanded for use inside many types of M2M terminals. This form factor is available from most M2M module vendors and generally has the most complete and leading selection of access technologies and was the first form factor to support 4G technologies. As the system connector is designed for easy insertion, replacement due to failure, or technology upgrade is not an issue with this form factor.

The level of integration required by the M2M application developer is less than with the solder-down form factor but still includes mechanical housing, power supply, SIM card

Figure 9.3 Full- and half-sized PCI express mini card. Images courtesy of Sierra Wireless.

reader, antenna(s), certification, and software drivers. Although the physical form factor is standardized for the mini card, the software interface to the host is not. This implies that interoperability between vendors does not truly exist because M2M application developers still need to reintegrate to a new software interface when switching between vendors. This marginalizes the advantage that standardization offers M2M application developers.

9.1.2.3 Highly Integrated

Highly integrated M2M modules are typically categorized by having a commercial housing and a standardized host interface, such as USB, RS232, or RJ45 connector (see Figure 9.4 for some commercial examples of highly integrated modules).

There is no standard size but highly integrated M2M modules tend to be the largest of all the form factors. The key advantage to using a highly integrated module is that the M2M application developer has less integration and almost no certification testing to do (see Section 9.6 on certification). These types of M2M modules are also often called gateway terminals because they can include gateway-service functionality, thus allowing support for more than one host, which may be of great advantage to the M2M application.

9.1.2.4 Proprietary

The last form factor categorization is the proprietary form factor, which is very general and clearly not orthogonal to some of the other form factors mentioned above. Almost all M2M module vendors have some proprietary M2M modules in their portfolios (see Figure 9.5 for commercial examples of proprietary M2M modules).

The advantages here are highly varied, but in general, since the M2M module vendor is not constrained in any way, the vendors can often innovate beyond the other form factors to produce a module that is smaller or less costly or more reliable or has better heat-dissipation properties or is more specialized to a particular M2M application.

Figure 9.4 Commercial examples of highly integrated M2M modules. Images courtesy of Sierra Wireless.

Figure 9.5 Commercial examples of proprietary M2M modules. Images courtesy of Sierra Wireless.

9.2 Hardware Interfaces

Given the diverse nature of M2M applications, M2M module providers attempt to offer the M2M application developer several interface choices. This section will describe most of the available hardware interfaces.

9.2.1 Power Interface

Although a power supply interface is clearly mandatory for all modules, the range of the supported power supply voltages varies between 3.0 and 4.5 V. Some modules may have more than one required voltage for sleep modes, high-powered power amplifiers or real-time clocks.

9.2.2 USB (Universal Serial Bus) Interface

For higher-rate access technologies (i.e., 3G and above), the USB interface is the preferred communications interface. It can also be used for debugging but many modules will provide more than one USB interface for debugging purposes. Most legacy M2M modules only support the full-speed (12 Mbps) USB standard but due to the increase in the over-the-air (OTA) speed of the latest access technologies, M2M module vendors are also supporting the high-speed USB (480 Mbps) standard. The limited usefulness of the USB on-the-go (OTG) functions for M2M applications has led to very few M2M module vendors supporting this functionality.

Although M2M modules support the standard USB interface, the way in which the interface is configured – for example, end-points and the multiplying of control and data over the end-points – is not standardized. Therefore, each M2M module vendor requires specialized and customized USB drivers. The M2M module vendor typically supplies these drivers as long as a mainstream operating system (OS) – for example, Windows, Linux, iOS, Android, WinCE – is being used by the host.

9.2.3 UART (Universal Asynchronous Receiver/Transmitter) Interface

A UART interface is an extremely common host interface although it is slowly being replaced by the USB interface. The UART interface is mainly used for communications or debugging functions. The baud rate supported by the module is 1200–115,200 bps but ultimately this should not be a bottleneck in the communication and should therefore be greater than the maximum OTA data speed supported by the M2M module's access technology. Since a UART does not natively support multiplexing of data and control signals over one interface, two UARTs will typically be used; one for user plane data and the other for control signaling.

9.2.4 Antenna Interface

Every wireless M2M module must have an antenna interface that may be an antenna pad or an antenna connector. The location and design of the antenna is usually the responsibility of the M2M application developer, so that the correct trade-offs between performance, size, interference, and form factor can be made. Many of the newer access technologies require more than one antenna to perform receive-diversity or MIMO (multiple-input, multiple-output) signal processing. The second antenna is typically only used for receiving (not transmitting) and thus its design parameter trade-offs are different. If the module supports an embedded GPS (global position system), this will also require an antenna interface. However, this does not necessarily mean another physical connector, as the GPS antenna may be implemented in several ways:

- shared with primary WWAN antenna;
- shared with the secondary or diversity WWAN antenna;
- separate antenna connector.

9.2.5 UICC (Universal Integrated Circuit Card) Interface

M2M modules that will be used on GSM-evolved systems (i.e., GSM, UMTS, or LTE) or on some CDMA systems generally have a standard (ISO 7816-3-based) universal integrated circuit card (UICC) interface. Only the electrical interface is provided by the M2M module (except for highly integrated M2M modules), so the integration and position of the UICC reader is the responsibility of the M2M application developer. This allows the M2M application developer to choose a convenient location for the UICC to allow for easy access, should it need to be replaced, or a very secure place where the UICC cannot be easily stolen.

Some M2M module vendors are now supporting embedded UICCs within the M2M module. The embedded UICC may have a proprietary form factor or a standardized form factor, such as the MFF2 defined by the ETSI SCP group in ETSI TS 102 671. Since the embedded UICC cannot be reprogrammed or replaced, most M2M module vendors recommended that an additional external UICC reader be supported to provide future flexibility. If in the future the UICC becomes reprogrammable, the need for the M2M module to support a UICC interface and the need for the M2M application developer to integrate the UICC reader will be eliminated.

9.2.6 GPIO (General-Purpose Input/Output Port) Interface

The GPIO interface in M2M modules is needed mainly for controlling external peripherals and receiving digital input.

9.2.7 *SPI (Serial Peripheral Interface) Interface*

The serial peripheral interface bus, also called a "four-wire" serial bus, is a synchronous serial data link that operates in full duplex mode where the devices are configured as either a master or a slave. This interface can support clock rates of up to 70 MHz. This is a general-purpose interface that can be used to control numerous peripheral devices but is commonly used for controlling external display equipment.

9.2.8 *I^2C (Inter-Integrated Circuit Bus) Interface*

Similar to the SPI interface but slower, the Inter-Integrated Circuit (I^2C) interface, also called the "two-wire" interface, is a synchronous serial data link with a multimaster bus. The transfer rate of the I^2C bus can be as high as 400 Kbps. This is a general-purpose interface that can be used to control numerous peripheral devices but is commonly used for controlling external displays, reading sensors or storage peripherals.

9.2.9 *ADC (Analog-to-Digital Converter) Interface*

The analog-to-digital converter (ADC) interface in M2M modules is mainly used for sampling external voltages supplied by external sensors.

9.2.10 *PCM (Pulse Code Modulation) Interface*

The pulse code modulation (PCM) interface in M2M modules is used for digital audio transfer, usually voice.

9.2.11 *PWM (Pulse Width Modulation) Interface*

The pulse width modulation (PWM) interface provides a pseudo-analog output for M2M modules and is needed mainly for controlling lights, buzzers, and analog power output, but is very versatile.

9.2.12 *Analog Audio Interface*

M2M modules that support voice services will have at least one analog audio interface, including one audio input and one audio output interface.

9.3 Temperature and Durability

Most modules provide an operating temperature range of -30 to $+70°C$, which is satisfactory for most industrial applications, but the automotive, metering, and some other industrial applications are demanding higher operating ranges of -40 to $+85°C$. The thermal shock cycle is another important parameter, and module vendors typically provide as many as 1000 thermal shock cycles on some modules.

9.4 Services

The pre-integrated and supported services that an M2M module provides can often be the most important decision that an M2M application developer needs to consider when choosing an M2M module as it can affect costs, time-to-market (TTM) and reliability. The application design and development cost represents a very large percentage of the TCO. The GSMA published a white paper entitled "Embedded Module Guidelines," which found that only 3–7% of the TCO is related to the M2M module costs whereas 20–40% is related to non-recurring engineering (NRE) costs, that is, the large range is due to the varied nature of the M2M applications. These figures are also supported by similar findings in a report commissioned by the GSMA and published in November 2010 by Analysis Mason and entitled "The TCO for embedded mobile devices." The following sections describe some of the most common services that an M2M module could provide.

9.4.1 Application Execution Environment

Most of the major M2M module vendors support an application execution environment on the M2M module. In this case, the split between the M2M module and the host becomes blurred and only logical at best. Application execution environments are usually only available on the more mature access technologies, such as 2G and the older 3G access technology – for example, not as yet on LTE modules – as this feature requires more R&D on the part of the M2M module vendors. To create the application execution environment, the M2M module may have a dedicated application processor integrated into the M2M module or may use a form of virtualization where the processor is shared with protocol functions. The M2M module vendors have made every effort to use industry norms and flexibility wherever possible to make the development environment as familiar as possible for the M2M application developer, but there is likely to be some initial and specialized training required. Here is a list of the many features that are supported in the more established M2M modules:

- standardized integrated development environments (IDEs), e.g., Eclipse;
- choice of programming languages, e.g., JAVA, LUA, C++, C;
- real-time programming, e.g., < 1msec latency to external interruptions;
- pre-emptive multitasking, e.g., prioritized tasks;
- event-driven programming, e.g., "call-back" functional processing;
- OTA application download/update and debugging;
- memory access protection (prevents a rogue memory access from crashing the terminal);
- debugging tools, e.g., terminal emulator, remote traces monitoring.

The amount of processing power that is available for the M2M application varies between M2M module vendors but is typically more than 50 MIPS which is enough to run many M2M applications, such as metering, point of sale, automotive diagnostics, or fleet management. There can be a significant bill of materials (BOM) saving and simplification of the PCB to using an M2M module that supports an application execution environment, as it can eliminate the need for an application-specific CPU and the associated peripheral hardware, for example, power management, RAM, or Flash. Since the application execution environment includes an OS and development tools, software licensing costs can also be reduced.

TTM and development costs may also be reduced, since many prebuilt and pre-integrated services, such as TCP/IP, SMTP (email), FTP, and HTTP, are available. Although, in 2011, application execution environments are rarely found on 4G modules, given the benefits in respect of BOM, TTM, development, licensing, and maintenance costs, this feature will also migrate rapidly to M2M modules with the latest access technologies.

9.4.2 *Connectivity Services*

It is assumed that the M2M modules provide basic packet-switch connectivity, but it is also very likely to find an M2M module supporting additional services related to WWAN connectivity. The following section describes some of the additional connectivity services that an M2M module vendor may provide.

9.4.2.1 SMS Wakeup Service

Although in theory a packet switch (PS) WWAN system should provide the M2M module with an always-on connection so that it can receive mobile-terminated or push data from an M2M application server, in practice this service is not readily available in most commercially deployed WWAN systems. Some of the reasons for this deficiency include a limited WWAN public data network (PDN) gateway, such as gateway GPRS support node (GGSN) capacity, public IPv4 address exhaustion, the use of NATs (network address translators), and additional M2M module power consumption (e.g., keep-a-lives). Therefore, in practice, the M2M module is forced to operate in a pseudo-circuit-switched (CS) manner where the M2M module is always connected to the CS domain of the WWAN system and only intermittently connected to the PS domain. Being always connected to the CS domain allows the M2M module to constantly be able to receive an SMS (short message service). If an M2M application has push data for an M2M module, the M2M application would first send a special wake-up SMS to the M2M module, upon receipt of which the M2M module would decode this and initiate a PS domain connection through which the M2M application could then deliver the push data. This SMS wake-up service is often predefined and pre-integrated with available application programming interfaces (APIs) in most M2M modules. In the future, 4G systems will likely be able to cost effectively provide a scalable always-on PS connection, but the SMS wake-up mechanism is deeply entrenched within many M2M applications and services such as device management (DM) and SIM management. Therefore, this SMS wake-up method may still be used in a 4G system to allow application-layer backward compatibility.

9.4.2.2 Packet-Switched Services

Although providing a reliable PS data connection is the main responsibility of the M2M module, there is a considerable amount of complexity in the protocols above the IP layer that has to be dealt with by the M2M application developer. Although M2M application back-end servers often use standardized transport and application protocols to communicate to the M2M terminal, the integration and/or development of these protocols can be lengthy and costly for the M2M application developer. For this reason, mature M2M module vendors provide pre-integrated service plug-ins for the most commonly used packet-based protocols, and often do so at no cost. Some of the more common PS service plug-ins that may be available include:

- TCP/IP stack;
- SMTP;
- POP – post office protocol (Internet email protocol);
- FTP;
- HTTP (worldwide web protocol);
- XML – extensible mark-up language;
- SSL – secure sockets layer.

These types of services are more commonly available when the M2M module also provides an application execution environment. Since the M2M module vendor is an expert in its own environment, the performance of these plug-ins often surpasses what could be achieved by the M2M application developer.

9.4.2.3 Circuit-Switched Data

Before PS data services were offered by WWAN systems, CS data connections were the only connection option available. A CS data connection requires a call set-up procedure before any data can be sent, whereby dedicated resources are assigned and are not released until a call teardown procedure is initiated. Although most modules support some form of CS data, it is rarely used commercially anymore due to the many drawbacks of CS data. If there are gaps in the traffic, the dedicated resources assigned in a CS data call cannot be reused by others during this idle period, which is very inefficient. For CS data calls, the WWAN system operator may also have to provide a terminating modem to the public-switched telephone system, making the service expensive and not scalable. The M2M back-end application may also have to provide a set of modems. Mobile-terminated data is not well-supported with CS data but an SMS wake-up service could be used, as described in the section above to initiate a CS data call. However, CS data has the advantage of having a very constant delay, for example, no jitter.

9.4.2.4 Voice/Fax

Although voice and fax support is available on some M2M modules, this is very much a legacy service similar to CS data. There are very few M2M applications that require this service. One issue is that the M2M host interfaces used do not lend themselves well to an analog voice interface.

9.4.3 Management Services

There are many management services that M2M module vendors are offering with their modules, which are typically supported via pre-integrated software clients. The following section provides some example clients and services that may be support by the module.

9.4.3.1 Device Management Clients

One of the most common clients that module vendors include is a tamper-resistant secure DM (Device Management) client. DM is useful for mobile operators, service providers, and end-customers/companies. DM functionality typically includes OTA distribution of applications, data, and configuration settings. One of the more popular DM clients in the industry is based on the OMA-DM (open mobile alliance device management) standard but the TR-069 standard, as defined by the BroadBand forum group, is also popular. Some vendors add functionality over and above the standard offering by extending the mandatory node support to include monitoring nodes, which will automatically trigger the device to send an update (e.g., an received signal strength indicator (RSSI) node that will trigger an update when it goes below a defined value). The module provider may also offer the ability to create nodes dynamically and to manage nodes that are related to the M2M application and not just the module, for example, the temperature of a pump.

Some considerations with respect to DM include:

- support for device and RAM/Flash memory integrity checks;
- support for self-diagnostics and fault management (more on this below);
- support for M2M application and non-application software/firmware-over-the-air (FOTA) update;
- bootstrap provisioning, i.e., initial configuration;
- protocol security and device security, e.g., remote lock and wipe mechanisms;
- performance, e.g., network usage.

9.4.3.2 FW Update

Most module vendors provide some level of FOTA upgrade support. This is usually done as part of the DM services, for example, OMA-DM. However, not all vendors provide M2M application-layer software updating (application running on the module or running external to them module). The M2M application-layer software may be written in a variety of languages, for example, Java, Python, Lua, or compiled C.

Before initiating a module FOTA session, care must be taken to ensure that all the essential approvals (e.g., PTCRB (PCS terminal certification review board) and GCF (global certification forum)) and re-certification have been completed on the firmware. However, updates to the M2M applications, such as Java, Lua, or Python applications, running on the module or external to the module generally do not require re-certification.

9.4.3.3 Self-Diagnostics and Fault Management

Self-diagnostics and fault management are very useful services that many vendors offer through a pre-integrated software client on the module and a back-end portal to access and control the service. The main purpose of these services is to find device and application-design issues faster and to enable corrections in earlier releases, improving quality at the first market introduction. They are also useful in discovering faulty hardware in the field and even help diagnose mobile network field issues. Diagnostics related to the OTA performance can also be used to streamline the device field trial process that is sometimes required for carrier certification. Some of the basic services that a self-diagnostics and fault management service offer may include:

- watchdog protection – avoid locked software;
- report network usage – CS, data, and SMS
- respond to "ping" queries via ICMP;
- initiate PDP context activation via an SMS command;
- report device identification information, e.g., module/device/subscription;
- report current serving cell and neighbor cell ID, received signal level and other radio link quality data;
- report stored history of radio link quality data;
- report and execute device integrity checks of the HW/SW/configuration files;
- report battery charge level;
- report key events (e.g., reboots, resets, or crashes);
- start and stop event storage via remote commands;
- report assigned IP addresses;
- report location (assuming GPS support);
- check status of peripheral devices attached to module;
- re-boot module via remote command.

The vendor should provide some level of security to prevent unauthorized access to module diagnostic services.

Most of these services use proprietary protocols and clients but 3GPP has standardized some of these services in their "minimization of drive tests" feature standardized in Release 10. This service concentrates more on automating measurement collection of network link quality information for the purpose of measuring and optimizing the performance of the network and the module.

9.4.4 Application Services

Beyond management services, the module vendor may provide other application services such as security and location.

9.4.4.1 Security

The vendor may provide a library that provides an additional security support. Features may include RF jamming detection, secured sockets layer, secured Internet protocols (HTTPS), crypto libraries, and secured data storage.

9.4.4.2 Location

The vendor can provide pretested and pre-integrated libraries that drive certain GPS hardware, thus minimizing the integration required to be performed by the customer.

9.4.4.3 Other Services

Other services that the vendors may provide include:

- audio filtering;
- dual tone multi-frequency (DTMF) generation;
- gateway services – dynamic host configuration protocol (DHCP), network address port translation (NAPT), domain name system (DNS) proxy, port forwarding, simple network management protocol (SNMP);
- low-power service libraries and specialized CPUs;
- SIM services – EAP/SIM or EAP/AKA proxy for authentication purposes;
- email support;
- eCall.

In addition, some service-layer capabilities are being standardized in ETSI TC M2M, such as ETSI TS 102 689 and TS 102 690, and in TIA TR-50, which someday may be supported by the module vendor.

9.5 Software Interface

The above-described services provided to applications by the modules are presented to the application using a software interface. In general, the interface gives the application control to make, break, and monitor mobile connections and to send and receive data through it. Richer interfaces provide the ability to control peripherals and other services available to the module. The service-specific protocols and control interfaces supported by modules need to reflect the total set of services and interfaces available, which, as shown in the above sections, is very diverse. This leads to a very diverse set of commercially available software interfaces. This section will cover some of the types of module software interfaces.

9.5.1 AT Commands

The AT command interface carries serializing seven-bit ASCII character string commands. The interface was originally designed to be sent over a serial link but can also be commonly used over USB. The AT command interface is the oldest, best-known, and thus most commonly used interface by M2M applications today. In general, it is very well-suited for use with smaller microcontrollers but can be used for more complex implementation as well. Almost all module vendors support the AT command interface. 3GPP has standardized a broad set of AT commands in the TS 27.707 specification. These standardized AT commands are often complemented by proprietary extensions to support unique service and interfaces, such as network driver interface specification (NDIS) (vs. dial-up), GPS, and installation.

Even though the use of AT commands is very popular, this interface has some drawbacks. One of the disadvantages is that it only allows one outstanding AT command at a time, that is, it is synchronous. Although theoretically supported, asynchronous event-based notifications are not well-supported. In addition, raw binary data cannot be sent natively and requires hex-encoding. The AT command interface may also have some restrictions on the number of applications that can directly access the module at the same time. However, multiple AT command interfaces may be supplied using standardized multiplexing protocols.

9.5.2 SDK Interface

An SDK interface provides many function calls, methods, classes, and objects. The use of SDKs corresponds well with more powerful OS-based environments and can often be found in handheld mobile computing devices. Typically, the SDK runs on an external processor except in the case where the module supports an internal execution environment. One example of a more well-known SDK is the mobile broadband SDK first standardized in Windows7. There are varying levels of richness that an SDK may support. A simple SDK will only provide basic call-connection support, whereas a rich SDK will support many hardware peripherals, services, and higher-layer protocols.

The module vendor typically supports several SDKs for the same hardware, as an SDK is OS-dependent (e.g., Windows, Linux, Android, WinCE) and programming-language-dependent (e.g., C, Lua, Java). The SDK is linked to a particular programming language because at some point the application will be complied and bound to the SDK function calls. Some laptop vendors and MNOs have specific SDKs that also need to be supported by the vendor. All this yields many SDKs that a module vendor may need to support, thus most vendors do not support all OSs, only support one programming language and may not have SDK support for all their modules.

An SDK provides many advantages over the AT command interface such as:

- excellent event-notification support;
- it encourages the use of standard programing practices;
- concurrent access: the single common API can be used by multiple concurrent applications;
- SDKs tend to be more responsive;
- binary data support: binary data can be sent natively without first hex-encoding.

One of the issues with SDKs is that they are currently all proprietary. This makes it very difficult for customers to move from one module to another, even if they are from the same vendor. However, with this clear need in mind, there are some standardization groups, for example, GSMA or OMA, talking about creating such a standardized open API for modules.

9.6 Cellular Certification

Beyond regulatory certification, which is required by all transmitting devices, cellular products also require telecom industry and MNO certification. Cellular certification is both complex and constantly changing, thus this section only attempts to introduce the reader to these certification concepts and rules and should not be considered as a complete set of rules. Cellular certification represents a very large burden to the module vendors and the M2M application developers (or integrators). The industry is continuously trying to minimize this burden by modifying processes and testing procedures, all the while trying not to affect the quality of interoperability. The burden to the integrator is so high that the telecom industry developed the concept of module certification, which most module vendors obtain. But even with module certification, the burden to the integrator is still quite high. The amount of testing required by the integrator is dependent on the integration level of the module. The less integrated the module is (e.g., a solder-down module), the more testing is required by the integrator. For this reason, fully integrated modules are very attractive for integrators with smaller volumes or with a fast TTM requirement.

9.6.1 Telecom Industry Certification

For GSM-based modules, the telecom industry certification is controlled via two associations, that is, GCF (global certification forum) and PTCRB (PCS terminal certification review board). For CDMA-based modules, a new forum called the CDMA certification forum (CCF) has been established. Telecom industry certification typically requires:

- testing against some technical specification, such as TS 51.010, using an accredited laboratory;
- field testing, e.g., drive testing;
- proof of a quality assurance programme, e.g., ISO9001.

As mentioned above, most module vendors acquire "module certification," which greatly minimized the testing that an integrator will have to perform. Even if an integrated module has "module-certification," the integrator can be expected to perform the following types of tests (depending on the module form factor and the integrator's implementation):

- radiated spurious emissions;
- measurement of radiated RF power and receiver performance, e.g., total isotropic sensitivity (TIS) and total radiated power (TRP);
- UICC/SIM electrical test cases;
- man-machine-interface-related test cases.

The module vendor should provide mechanisms for placing the module in a mode of operation to perform the above tests. In addition, the module vendor may also provide an accredited laboratory to perform and expedite the above testing. It should be noted that there are some complex timing restrictions when using "module certification." If the module was tested against a test specification that is too old – that is, the latest test specification has many new/updated test cases – the "module certification" may need to be retested against the latest applicable test specifications.

9.6.2 MNO Certification

Unlike telecom industry certification, MNO certification is not standardized, so the amount of testing, the test limits, and the test cases vary between MNOs. Some MNOs only carry out a small subset of tests where others do more testing and have tighter performance criteria than the telecom industry certification. MNO certification is focused on the equipment deployed in the MNO network, where the field testing and IOT (interoperability testing) is conducted on the MNO's network and not on test equipment, such as Agilent test equipment. MNO certification only deals with complete products (not modules) but a module can be precertified, similar to telecom industry certification, to reduce the burden of testing required by the integrator. Because there are so many MNOs in the world, it is not practical for a module vendor to obtain module precertification from all MNOs.

10

Smart Cards in M2M Communication

François Ennesser
Gemalto S.A., Meudon, France

10.1 Introduction

This chapter investigates how to best address the increasingly sensitive security and privacy issues arising from the M2M context, while minimizing the cost of terminal devices and their management infrastructure.

The security of many modern telecommunication networks already relies on an independent hardware element (e.g., a smart card based on the ETSI UICC platform, as in 3GPP networks) to secure the access credentials provided to terminal devices. This option also offers flexibility for end-device personalization and resolves ownership and liability issues when the service provider is not fully able to trust terminal devices. We explore how this model can address the new challenges arising from the M2M context, while resolving its inherent security needs.

10.2 Security and Privacy Issues in M2M Communication

General security issues related to M2M were considered in Chapter 8 of this book. It appears that while security in the context of M2M communication inherits most principles from human-based communication systems, a number of new constraints add extra security requirements, leading to increased system complexity.

At the network level, a large number of connected devices, and the need to efficiently manage them in an affordable manner, creates the need for enhanced device-management

M2M Communications: A Systems Approach, First Edition.
Edited by David Boswarthick, Omar Elloumi and Olivier Hersent.

infrastructures with group-addressing capabilities, which can use broadcast and multicast technologies.

Furthermore, the tremendous diversity of M2M devices that may connect through a common WAN creates a need to delegate most data-processing and protocol-conversion functions to intermediary gateways.

At the device level, M2M communicating devices may themselves be part of complex systems (such as automotive vehicles) where they may not be easily accessible. In other instances, the device may be unattended by its owner-organization and/or located in hard-to-reach places, with the result that sending maintenance personnel would be too expensive. Therefore, some of the assumptions that apply to regular mobile terminals need to be revisited when considering the wide diversity of foreseen M2M applications:

- Service subscriptions related to existing devices should be prevented from being reused for unintended purposes. This requires appropriate security countermeasures for devices in unguarded locations, such as some smart meters, traffic lights, or vending machines.
- Some regulations, such as the European Universal Service Directive (EUSD) [EUSD], further require that subscription data, such as identity and relevant keys, shall be exchangeable in the field whenever a user decides to change provider. This could take place any number of times during the lifetime of an M2M device, not just once.

In industrial applications involving large numbers of unattended devices, the need for subscription changes between service providers may occur, and remote management solutions are certainly required to address this need, since the costs involved with physically intervening on a potentially wide and geographically spread-out basis would be prohibitive. On the other hand, for applications on personal consumer devices or prepaid models in various businesses where user privacy is at stake, the traditional "plastic roaming" scenario used in SIM-based wireless mobile networks may remain highly pertinent as it offers convenience and cost advantages in comparison to remote management solutions. In this case, the subscription is embedded in a smart card that can easily be swapped from one device to another, according to needs.

This highlights the fact that there is no "one-size-fits-all" solution for addressing the great diversity of M2M applications. Evaluating the most suitable solutions needs to take market specificities into account. Nevertheless, the flexibility offered by secure remote management capabilities should become strategic in M2M networks.

Furthermore, specific device characteristics are often required in M2M environments, for example:

- **device lifetime**: Up to 20 years is desirable in some fields such as smart metering. This is particularly problematic for non-main-powered devices such as certain gas or water meters or some mobile devices. Great flexibility needs to be provided to accommodate the evolution of computing and telecommunication technologies during such time frames, and address new security threats, such as the possible evolution of cryptographic algorithms and change of access network technologies during product lifetime.
- **environmental conditions**: This includes storage and operating temperature or humidity, as well as exposure to chemical corrosion, such as a salty atmosphere, or mechanical constraints, such as shocks and vibrations. These factors combine together to make the evaluation of an expected device lifetime even more challenging.

At the same time, several M2M applications, such as smart power grids or healthcare devices, have an even more pressing need for security (and reliability) than regular telecommunication services, as they may affect more vital human needs.

Finally, the increasing amount of data that may potentially be gathered from those devices surrounding a human user – such as home energy and water meters, cars, and healthcare devices – is likely to compromise individual rights to privacy. Therefore, beyond fulfilling traditional security requirements, devices or gateways need to have (potentially computing-intensive) mechanisms to protect the anonymity and privacy of the user. Failing to address this need may compromise the success of any application, as already illustrated in the Netherlands, where parliament initially rejected laws enforcing the adoption of smart meters or automatic road-pricing technologies due to consumer concerns over privacy. In several applications, however, the use of anonymous prepaid subscriptions offers a well-known way to preserve customer anonymity and should therefore be maintained.

10.3 The Grounds for Hardware-Based Security Solutions

Any software-based security implementation, whether on embedded devices, mobile phones, personal computers, or digital assistants, is open to potential software attacks, such as Trojan horses, worms, or viruses. Furthermore, reverse-engineering techniques, such as extracting program code and disassembly/debugging methods, are greatly simplified in a software environment shared with other applications, leading to potential exposure of sensitive secret components, such as algorithms, private keys, and other information that is assumed to be secure. Therefore, hardware-based mechanisms are always necessary to provide sufficient protection. This includes tamper-resistant memory that is not readable or alterable by simple physical attacks, physical protection mechanisms concealing critical operations, and hardware detectors protecting against classical attacks. Yet the use of dedicated hardware is insufficient to ensure adequate protection of secrets. Protecting the execution environment from potential side-channel attacks, such as simple or differential power analysis, which analyze electromagnetic emissions to infer the data manipulated by a processor, requires special skills, and clever writing of the software when implementing security algorithms in order to fool invasive or non-invasive attacks. Since such implementation issues cannot be addressed by standardization, they have to be assessed by a dedicated industry certification process relying on established methodologies, such as the Common Criteria evaluation assurance level (ISO 15408) or the North American FIPS.

We can further classify the hardware-based security solutions into three categories: removable independent secure element, fixed independent secure element and built-in (non-independent) hardware. A secure element provides a device with security functions, such as storing keys and performing cryptographic functions. It can be removable or soldered onto the device. The SIM card in a mobile device is a good example of the removable independent secure element. The user can remove the SIM card from the device and insert a new one. A secure microcontroller (e.g., a smart card chip) embedded in a security device, such as a USB token, is a good example of fixed independent secure element. Some hardware may not exhibit an independent secure element and still have some security solutions built into the design. These different software- and hardware-based security solutions have different characteristics and capabilities. Table 10.1 provides a comparative assessment of these different security solutions. The next few sections further explain and justify the contents of the table.

10.4 Independent Secure Elements and Trusted Environments

10.4.1 Trusted Environments in M2M Devices

10.4.1.1 Need for Secure "Environment"

Securing the storage, processing, and input and output of sensitive data on an open platform is of critical importance. That is why initiatives have appeared to secure general-purpose computing platforms, such as the trusted computing group (TCG).

As explained above, there is a need to guarantee a tamper-resistant environment to perform sensitive operations and store credentials with sufficient protection. The need to adequately secure a device or its communications module presents several difficulties such as:

- established security standards generally applicable to communication modules or M2M devices are scarce. Currently, the best-known solutions, such as TCG-related work, do not actually address attacks on the underlying hardware. The notion of "secure environment" or "trusted environment" cannot be associated with well-recognized security levels from established methodologies. Furthermore, as noted above, the proper implementation of security functions requires special skills that are not traditionally available in companies without specific security expertise.
- different implementations may not necessarily secure the same functionalities. Even when there is agreement on which parts of a module or device should be secure, the actual security level will depend on the design and the implementations of the functionality. Hardware countermeasures need to be involved, which would be costly if applied to the whole device/module computing environment. In the end, there is no guarantee that all particular implementations will always make proper use of the available protection mechanisms.

When roaming involves different service providers, as in modern telecommunication networks, the weakest link in the security chain – attacks against the most vulnerable device – resulting from one service provider may potentially affect all related service providers. As a result, the need to rely on broadly recognized security-certification schemes, such as Common Criteria, is even more critical for module-/device-integrated security solutions. This is where the advantages of an independent security module become apparent, as detailed later in this chapter.

Nevertheless, when comparing trusted environments built into general-purpose processors with dedicated independent hardware, such as smart cards, besides providing secure storage for sensitive data, such as cryptographic keys, both can be designed to include similar functionalities in terms of support for advanced cryptographic algorithms, random number generators, and key generation, as well as encryption/decryption and

Table 10.1 Comparative assessment of different security solutions

		Removable independent secure element	Fixed independent secure element	Built-in (non-independent) hardware	Software-based security
General security of the solution					
	Physical	Medium	High	High	Low
	Logical	High	High	Medium	Low
Algorithms					
	Secrecy of the algorithm	High	High	Medium[a]	Low[a]
	Possibility to customize the algorithm for the asset owner	High	Medium	Low[a]	Low[a]
	Facility to update the algorithm	High	Medium	Low[a]	Low[a]
Credentials					
	Difficulty to retrieve credentials	High	High	Medium[a]	Low[a]
	Secrecy of service providers data (data not known by device manufacturers)	High	High	Low[a]	Low[a]
Manufacturing/deployment costs		High	Medium	Medium	Low
Personalization					
	Easiness of personalization	High	High	Low[a]	Low[a]
	Trust in the security of personalization process	High	High	Low	Low
Portability – subscription management		High	Medium	Medium	Medium
Anti-theft					
	Trust in the anti-theft solution	Medium	High	High	Low
Human intervention					
	Ease of use without human intervention	Low	High	High	High

[a]The status depends on the way in which the terminal is secured (e.g., standardized or proprietary solution).

signatures methods. Both can be used to provide disk encryption and identification functions for a variety of networks and applications. The main difference is that trusted environments typically include a root of trust, which enables the signature to be computed for software modules loaded into the general-purpose processor on which they are implemented. Those signatures can later be compared with safely stored reference values in order to provide integrity validation. As long as there is a built-in root of trust to provide the trusted signatures, the latter step does not require a built-in secure environment within the trusted environment itself and could just as easily be performed within an independent secure element.

On the other hand, independent secure elements, such as smart cards, today provide a security-dedicated microprocessor (embedded or removable) that can be safely programmed and updated with new applications, even after the deployment of the device in the field, thanks to virtual machine support, such as the JavaCard platform. This capability has laid the ground for the success of the smart card form factors in a variety of applications, such as payment cards, public transportation passes or security modules for telecommunication networks. The benefit of this solution is strongly reinforced when the service provider, whose access credentials are to be protected, does not own the hosting device itself. Indeed, in such business cases, healthy competition between service providers requires the following:

- the remote management of the Service Provider data must be independent from device configuration and applications management.
- the cryptographic algorithms protecting the service provider's assets must remain independent from device manufacturers.

Applications loaded on smart card virtual machines, as well as the corresponding stored data, have the required capability to be remotely managed independently from hosting devices using standard and secure mechanisms that will be detailed later.

10.4.2 Trusting Unknown Devices: The Need for Security Certification

10.4.2.1 Certification for M2M Devices

In the context of M2M, the certification process of a device or its security module faces the following difficulties:

- defining the boundaries of the part(s) to secure.
- defining the specific protection profiles for each particular industry.
- the risk that the scope of the security target may be large:
  - the bigger the security target, the more difficult the certification required, and the higher the resulting cost;
  - the complexity of the certification is directly affected by the impacted environment – e.g., more actors, more process, more risks – which further increases the cost;
  - the need to take maintenance into account has a further impact.

To ensure the integrity of secure elements, it is essential that the supplier-side manufacturing environment is secure. Indeed, beyond certification of the device itself, some initial provisioning of trusted credentials has to be performed, initially during device manufacturing, so that the service provider can trust the device as a safe receptacle capable of protecting its own access credentials.

To prevent such credentials from being exposed, an accreditation is necessary for all the people, facilities, and processes involved in manipulating credentials in order to deliver trusted devices ready to work in the field.

Such processes are well-managed and integrated in internal procedures and implemented in the manufacturing facilities of security specialists, such as smart card manufacturers, as they are required to comply with specific industry regulations, such as the payment card industry data security standards or the GSMA security accreditation scheme of mobile telecommunication. On the other hand, attempting to comply with such processes for an organization that is not already familiar with such issues requires important transformations that generally end up proving very costly.

It is worth mentioning that the GSMA has acknowledged the need to enable remote personalization of M2M security elements, which will require changes or adaptations to their accreditation schemes.

10.4.3 Advantages of the Smart Card Model

Establishing a trusted security chain in M2M device implementation and production cannot bypass the certification and accreditation process. The added cost resulting from such a process is not competitive compared to smart card-based solutions. These rely on well-defined, mass-produced independent secure elements with well-defined boundaries that can be easily integrated in a device's design, and leveraging the huge volumes delivered to other industries, which are more than 1 billion per year. Smart cards have already demonstrated their adaptability to an ever-increasing range of business domains, ranging from banking to mobile telecommunication and digital identity.

The application-agnostic UICC platform that has been specified by ETSI TC SCP in TS 102 221 [TS102221] provides a general-purpose, multiapplication smart card platform that can be used for independent secure elements embedded in M2M devices (Figure 10.1).

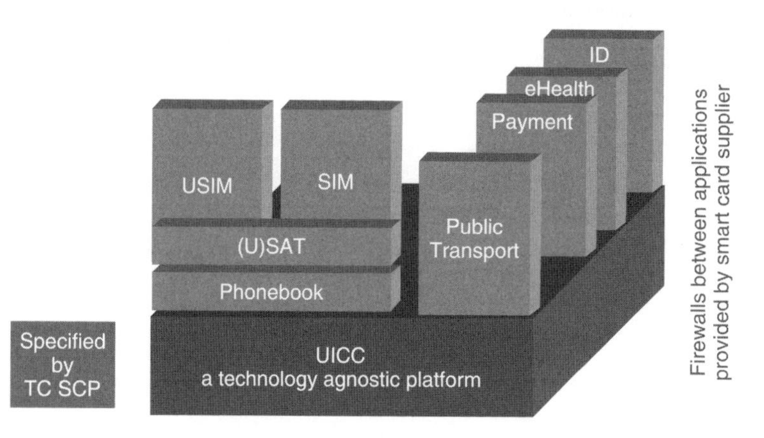

Figure 10.1 The ETSI UICC platform.

This platform provides a highly secure environment that can be remotely managed and will provide security services to locally connected devices; the UICC itself acts as a secure application platform with post-issuance management capabilities. More specifically, it integrates the following features:

- a specific extension for M2M applications (ETSI TS 102 671) [TS102671], defining specific form factors for embedded use and offering specific environmental and lifetime characteristics.
- a virtual machine API (specified for JavaCard in ETSI TS 102 241) [TS102241] and an extensible card application toolkit (CAT) (ETSI TS 102 223) [TS102223] that enables on-card applications to interact with features provided by a hosting device or its telecommunication module, while providing firewalls to preserve confidentiality between different applets. Such virtual machines have reached a Common Criteria evaluation assurance level certification of EAL4+.
- solutions for secure remote management of the files and applications loaded on the card, as per ETSI TS 102 225 [TS102225] and ETSI TS 102 226 [TS102226]. The remote management channel establishes a security end-point in the UICC itself, therefore preventing sensitive information from being compromised at the hosting communication device level, and supports a variety of bearers and protocols to allow flexibility. Furthermore, it relies on features from the GlobalPlatform specifications to provide independent management of specific security domains for each application, thus preserving the confidentiality of their sensitive data.

This platform also provides the following optional interface extensions (Figure 10.2):

- an inter-chip USB interface (ETSI TS 102 600) [TS102600] that provides a modern, higher data rate, and more versatile alternative to the legacy serial smart card interface using the RST, CLK, and I/O contacts;
- a single-wire interface to a contactless module in the hosting device (SWP: ETSI TS 102 613 [TS102613]) with host controller interface (HCI: ETSI TS 102622) [TS102622]);
- an IP extension allowing easy integration in TCP/IP networks, offering client- and server-mode operation for both local and remote connectivity (specified in ETSI TS 102 483 [TS102483]).

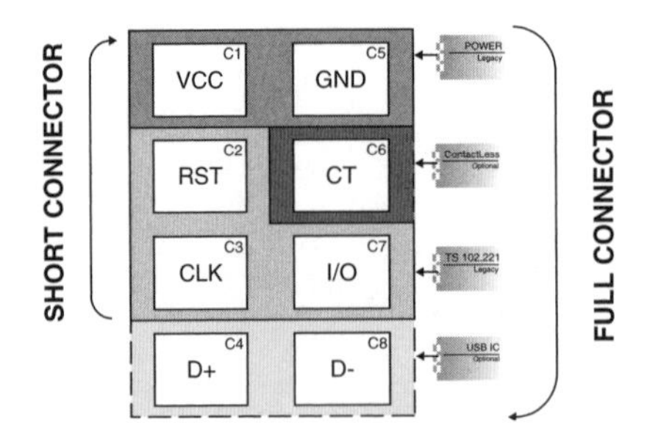

Figure 10.2 UICC interface contacts assignment.

This platform has already been adopted as a placeholder for secure credentials for several telecommunication network technologies, as standardized by UICC-based network access applications specifications from the following organizations:

- 3GPP that uses a 3GPP UICC (3GPP TS 31.101) to hold a USIM application (3GPP TS 31.102) for network access and an ISIM application to secure service on the IMS (3GPP TS 31.103);
- 3GPP2 (CSIM application: 3GPP2 C.S0065);
- ETSI TETRA (TSIM application: EN 300 812); and
- WiMAX forum (WiMAX SIM specification: http://www.wimaxforum.org/sites/wimax forum.org/files/technical_document/2009/09/WMF-T33-114-R015v01_WiMAX-SIM.pdf).

The flexibility provided by smart-card-based provisioning alone has proven to be a great facilitator for the successful deployment of the above-mentioned technologies in many countries around the world.

Of course, the interface between an independent secure element, such as a UICC and its hosting device, has to be standardized. It could therefore, in some situations, be spied upon by an attacker. However, such security elements incorporate potentially large programmable secure-memory and advanced cryptographic capabilities relying on hardware co-processors, providing fast computation despite the low operating frequency resulting in minimized power drain. Combined with the highly flexible application development and management environment of their virtual machine with ready-to-use API capabilities, this enables all sensitive security functions to be contained internally, avoiding the need to exchange any unprotected security information with the hosting device. When this is not deemed as sufficient, ETSI TS 102 484 [TS102484] enables the establishment of a secure communication channel with the hosting device.

10.5 Specific Smart Card Properties for M2M Environments

10.5.1 Removable Smart Cards versus Embedded Secure Elements

A removable smart card, using one of the form factors specified by ETSI TS 102 221 (Figure 10.3) or the miniaturized MFF1 form factor with a connector specified in ETSI TS 102 671 (Figure 10.4), could provide a truly portable subscription solution, since the well-standardized smart card interfaces specified by ETSI for the UICC platform provide universal interoperability with all accepting devices. It also offers all the existing tools secure over-the-air (OTA) management of UICC, whenever subscription-related parameters need to be modified.

For consumer device applications, it is often desirable for the user to be able to use the same service subscription in different devices at different times, or to use a single device with different service providers. The use of physically removable UICCs, preferably in one of the traditional form factors specified in TS 102 221, is the most appropriate approach in such applications. Another advantage of using removable UICCs, especially in devices that have a long lifetime, is to enable upgrade of the security module (cryptographic protections, processor, and algorithms) at an affordable cost to address new threats that

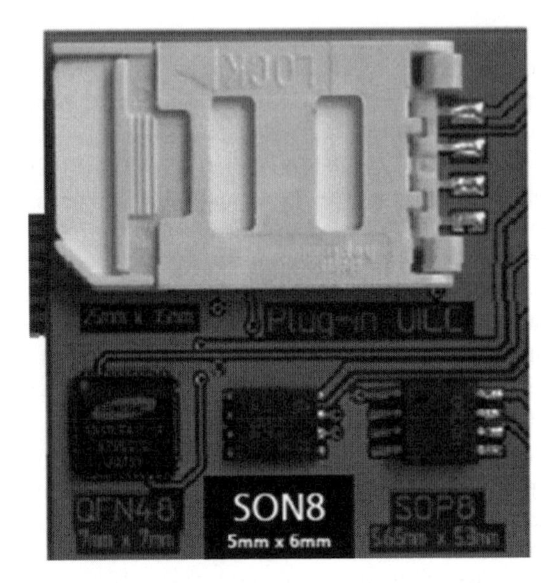

Figure 10.3 Plug-in smart card in its connector next to a soldered MFF2 (aka SON8) form factor.

Figure 10.4 The versatile MFF1 form factor of TS 102 671 next to its connector. The shown MFF component is courtesy of Gemalto and the connector is courtesy of Yamaichi.

may emerge over time. Security is a continuous race and in the space of a decade, new threats and tools are discovered that require upgrades to previously safe implementations.

Though removable UICCs can be stolen, there are several solutions to counter this threat and render them useless for the thief. Such solutions can be of interest when service necessity requires that accessibility to the UICC be preserved.

The accessibility of the UICC, whether it is soldered or removable, depends entirely on the entire device design: where it is placed and in which type of machine (car, meter, vending machine, etc.). In fact, some industrial M2M devices are designed to make the UICC almost impossible to access by users, in order to discourage theft of the UICCs. This is because the "plastic roaming" advantage of removable UICCs is no longer desirable for such applications. In such situations, the use of a mechanically attached UICC, as enabled by soldered MFF1 or MFF2 modules, specified in ETSI TS 102 671 with a remote subscription management solution offers an appropriate alternative.

Non-removable UICCs require remote management solutions to comply with regulations governing free competition between service providers. Though UICC OTA, as standardized today, does not yet support the possibility of remotely creating a new subscription or replacing an existing subscription on a card remotely, it still provides a way to efficiently manage an existing subscription in a device. Some proprietary models that enable the change of subscriptions in the field, such as white-labeled SIMs initially provisioned with multiple subscriptions and managed by a central authority, are already emerging, as in the model proposed by the Brazilian Departmento Nacional de Transito (DENATRAN) for automotive communication modules. A standardization effort to provide an even more flexible and interoperable model is also being initiated in ETSI SCP, as described later in this chapter.

When not combined with a remote management solution, replacing the whole module on which a UICC is soldered may in some instances provide an acceptable alternative, provided the cost of the whole module remains sufficiently low.

10.5.1.1 M2M Secure Element Distribution and Logistics Impact

For non-removable UICCs, logistics and deployment are quite different from traditional removable smart cards. Such secure elements are not generally delivered to the service provider, but to the M2M device manufacturer who will incorporate it into the M2M device. At this stage, the card may not yet be provisioned with service credentials, since for logistical reasons it is generally necessary to only personalize the card at the end of the device manufacturing chain.

More complex distribution models where the smart card is sent to an initial issuer, then to the retail channels, and finally to the end-customer, to be finally inserted in a device, lead to a significant increase in costs and are therefore avoided for M2M. It can be estimated that the network cost (identity and credentials provisioning) for supplying a card to an M2M customer is significant compared to its hardware cost.

The following elements for further cost reduction could be envisaged:

- Reduce smart card cost in the provisioning process by:
  - greater efficiency;
  - improved scalability;
  - reducing distribution costs.
- Reuse the smart card and the security it provides for multiple devices in capillary networks.

- Reduce network side costs by:
  - leveraging typical IT database technologies in the network, e.g., lightweight directory access protocol (LDAP), universal decimal classification (UDC);
  - improving the licensing model so that storage capacity is not the main cost element but processing power becomes the limiting factor instead;
  - reducing processing load by having fewer overall attach/detach events and minimizing the total number of separate device entries in the database, e.g., using device groups or delegating locally.

Whatever security solution is used for device personalization and service bootstrapping, the way in which M2M devices are secured and provisioned becomes a major challenge, especially if the number of devices deployed increases dramatically. In this context, solutions based on independent secure elements enable the leveraging of the already established large-scale logistics know-how developed by the smart card industry to address the constraints of mobile telecommunication networks.

10.5.1.2 Removable UICC with Logical Device Pairing

A removable UICC may be bound to the machine, or to a group of machines, in which it is inserted with a pairing mechanism, such as specified in ETSI TS 102 671, so that there is no need to physically protect the location in which the UICC is inserted. Such UICCs will then fail to work in any device not belonging to the intended group (or not being its originally intended host).

A simple "reverted" lock mechanism, comparable to the SIM-lock mechanism used to bind a GSM terminal with a specific identity (IMEI, IMEISV, MEID, or ESN) to an intended SIM card, but working the other way around, may be used: an "IMEI-lock" instead of an "IMSI-lock" in GSM vocabulary. This requires UICCs with proactive capabilities – that is, supporting the CAT of ETSI TS 102 223 – since smart cards are otherwise designed to only execute commands from their host, for example, to read IMSI, as opposed to requesting actions from them. Even so, such mechanisms, if improperly implemented, may be bypassed in several ways, for example, by using fake terminals or a man-in-the-middle mechanism allowing the desired identity to be presented to the stolen smart card.

10.5.1.3 Secure Channel between the UICC and Its Host

A more secure alternative is to use the locking mechanism specified in ETSI TS 102 671 based on immediate establishment of a secure channel between the card and terminal. This assumes, however, that the hosting device itself provides some sort of trusted environment, allowing for a security key for encrypting communication with the card to be safely stored and utilized, according to the secure channel specification ETSI TS 102 484 [TS102484].

Depending on the security needs of the application, TS 102 671 allows the possibility of initially generating a security association between the UICC and terminal, without actually securing the resulting communication. Otherwise, the platform-to-platform option of ETSI TS 102 484, which forces encryption of all communications between the card and terminal, has to be used for this purpose.

Note that ETSI TS 102 484 also provides the option of using application-to-application secure channels, offering the possibility of selectively encrypting traffic related to a sensitive UICC-based application, independently from the others. Also, the secure channel of TS 102 484 not only addresses legacy traffic over the serial interface, but also traffic over the USB interface of TS 102 600, including Internet-protocol-formatted traffic.

10.5.1.4 UICC Service Restriction

For devices and UICCs that cannot satisfy the above requirements, a combination of business processes and security measures can still assist in preventing unauthorized access to networks and systems, by rapidly detecting suspicious activity, to minimize the risk that credentials from an embedded device may be used for theft of service by:

- restricting services to the minimum levels required by the intended embedded application, e.g., the Call Barring feature of GSM (U)SIM may be used to restrict calling service to intended parties only.
- developing a normal usage profile for predictable embedded applications by
  - limiting traffic volumes to those that might be reasonably required for the application;
  - triggering alarms on high usage or deviations from the normal profile.
- providing securely, where feasible, e.g., through the use of IPSec, dedicated private APNs intended solely for embedded services for use by the embedded applications.

Note that even if pure software-based security is used, the risk of theft still applies to the subscription credentials contained in the communicating application implemented on the M2M device. Although not as visible as a physical theft, the exposure of such data could, however, be more damaging than the theft of a smart card holding a secured subscription.

10.5.2 UICC Resistance to Environmental Constraints

The use of an M2M device can take place in various environments. Some can be extreme in terms of temperature, humidity, chemical, or mechanical exposure. Furthermore, in some applications, it is important to maximize the lifetime of the device as a whole.

The environmental profile generally affects the whole M2M device in the same way as any related (removable or soldered) smart card. All the components of the M2M device are impacted. In some cases, however, removable UICCs may be subject to slightly different conditions from the rest of the device, for example, because part of the device may be better protected from humidity by an additional casing, or because the UICC reader may be placed in a more comfortable environment, such as a car's dashboard instead of the engine's compartment.

ETSI TS 102 671 provides a classification system for environmental conditions that are met by the M2M UICCs based on the following parameters:

- operational and storage temperature;
- moisture and reflow conditions;
- humidity;
- chemical corrosion and possibly fretting corrosion;
- vibration;
- shocks;
- memory data retention time;
- minimum updates requirements.

This environmental classification system allows each industry sector to specify the target profiles that are appropriate for smart card usage in their specific environment.

Supported requirements are generally tested by accelerated methodologies that concentrate, over a reduced time period, the constraints that the product is expected to endure during its full intended lifetime. Applicable test specifications for each categories are referred to in ETSI TS 102 671.

10.5.2.1 Lifetime

Given the wide variety of use cases and market models, the desired lifetime of M2M devices can vary from two (car rental fleet management) to 20 years (smart meters). Lifetimes in the order of 10 or 15 years are often desirable in many other industrial applications.

For the UICC, the achievable lifetime not only depends on the data-retention time and the number of read/write cycles supported by its non-volatile memory (EEPROM or Flash technology), but also on exposure to environmental constraints, such as storage and operational temperature, humidity, and so on.

For removable UICCs, oxidation of the contacts is not a determining factor when appropriate materials are used. The number of removal and insertion cycles is generally more critical, but demanding applications in that respect – such as consumer-attended devices – can normally accommodate replacement every few years.

As regards the read/write cycles, the translation in terms of lifetime greatly depends on each individual use case and internal product implementation. UICC chips today support hardware read/write cycles of 100 000–500 000, but low-level wear-leveling mechanisms can further be used to prevent premature failure of any particular memory cell. Even taking one cycle per hour on a given cell as an example (to give an idea of the range), the hardware lifetime will range (from a writing point of view) from 11 to 57 years. This means that the hardware technology is not the limiting factor, while appropriate wear-leveling techniques have the most drastic impact on product lifetime. Early localized memory failures can easily occur in frequently updated locations when proper countermeasures are not implemented. Efficient wear-leveling techniques must be transparent to the user and therefore implemented in the lower layers of the card-operating system.

ETSI TS 102 671 proposes to assess such capabilities by verifying that UICC information intended for frequent updates support a choice of 100 000, 500 000, or even 1 000 000 update cycles. This can be combined with a choice of memory data-retention requirements of 10, 12, or 15 years.

10.5.2.2 Storage and Operating Temperature Exposure

ETSI TS 102 221 and TS 102 671 provide test methodologies to assess whether UICCs are suitable for operation and storage in any of the following temperature ranges:

- −25 to +85°C;
- −40 to +85°C;
- −40 to +105°C;
- −40 to +125°C (this last set of conditions involves specific design constraints that should only be restricted to the most severe environments).

10.5.2.3 Manufacturing and Lifetime Exposure to Humidity

TS 102 671 provides the following features:

- classification of moisture/reflow conditions applicable during the manufacturing process for soldered UICCs (for removable UICCs, a fretting corrosion requirement with the connector is instead considered);
- operating conditions under humid environments;
- corrosion in salty atmospheres.

10.5.2.4 Shocks, Vibrations, and Other Mechanical Constraints

ETSI TS 102 671 allows for the provision of UICCs that meet the vibration specification [JESD22Vib] and shock-conditions specification [JESD22Shk] defined by JEDEC (Joint Electron Device Engineering Council) for automotive applications.

Both M2M UICC form factors specified in ETSI TS 102 671 have been designed to meet the tight vibration and shock constraints required for use in transportation equipment, such as automobiles. Besides the electrical contact points used for connection, they provide large independent mechanical pads intended to solidly anchor them to their hosting circuit board.

However, it is important to note that mechanical constraints, such as shocks or vibrations, do not prevent the use of removable UICCs, because robust UICC readers are available that ensure reliable connections even in the toughest mechanical environments. Readers for the MFF1 format of TS 102 671 are the toughest because the design of both the MFF1 form factor and its connector was intended to address such constraints. However, even some mini-UICC readers are available that can reasonably address demanding mechanical constraints.

10.5.3 Adapting the Card Application Toolkit to Unattended Devices

UICCs are able to run applications on a virtual machine such as JavaCard and provide APIs (ETSI TS 102 241 [TS102241]) to use their proactive capabilities (CAT, ETSI

TS 102 223 [TS102223]). This has been a key factor behind the success of UICC in mobile telecommunication networks, where telecommunication operators make use of this platform feature to provide value-added services to customers or optimize certain capabilities, such as roaming. In the M2M market, it is expected that the capabilities of these UICC platforms will continue to be used as a competitive differentiator by the (telecommunication or M2M) service providers. However, some differences have to be taken into account when deploying UICC-based applications in an M2M environment.

The main point to consider is that many value-added services in the mobile phone business relied on customer interactions enabled by the user interface of mobile handsets (screen and keyboard, in particular). M2M devices, on their side, are largely unattended, especially in industrial applications, and some of them have no (or only greatly reduced) user interface capabilities. This is why recent versions of the CAT specification (from ETSI SCP and 3GPP Release 9) include specific provisions for terminals lacking certain features.

- no display or other user information capability;
- no keypad or other user input capability;
- no audio-alerting capability;
- no speech-call capability;
- no support for multiple languages.

The result is that proactive toolkit commands that otherwise rely on such features are either inapplicable or only partially supported – that is, with some restrictions – in such terminals. However, the ability to develop useful applications relying on communication capabilities is largely preserved.

10.5.4 Reaching UICC Peripheral Devices with Toolkit Commands

One other factor to take into account is that M2M device design may vary from handset design in several aspects. Though modern handsets often exhibit a bi-processor architecture (separating a digital baseband modem functionality and an application processor), the UICC generally remains connected through its serial interface (defined in ETSI TS 102 221 [TS102221]) only to the modem, which appropriately handles the proactive requests from the card. In an M2M device it could be easy to interface the UICC directly to the application processor in the device – for example, by using the inter-chip USB interface (specified in ETSI TS 102 600 [TS102600]). However, due to the above legacy, it is expected that many M2M devices will retain an architecture where the UICC is only connected to the communication module through the serial interface, with this module being itself connected to the application processor of the device through an interface which generally remains based on AT command specifications (e.g., in 3GPP TS 27.007 [TS27.007]) as shown in Figure 10.5.

In M2M devices, it will be greatly beneficial for proactive UICC applications to be able to interact with the application processor of the device itself. Therefore, in architectures where the M2M UICC (which may be the one that already holds access to a wireless WAN transport network) only has an indirect connection to the application processor through the AT interface of the communication module, there is a need to transport toolkit commands

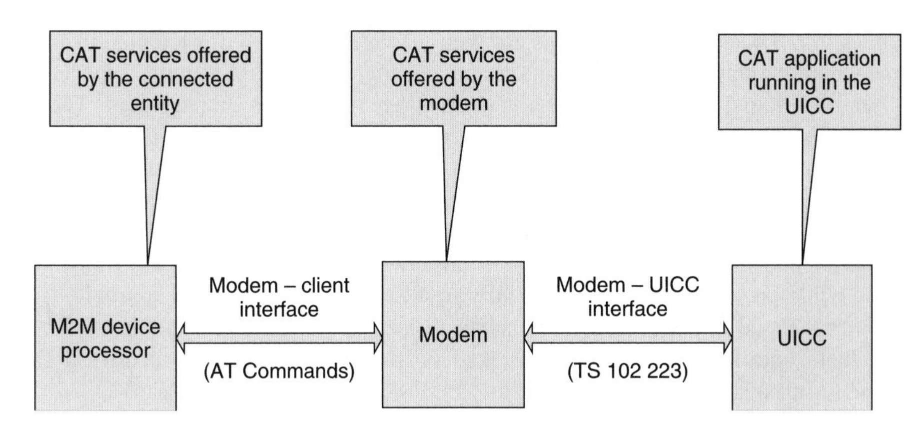

Figure 10.5 UICC applications use of toolkit services in an M2M device.

through the AT interface. This requirement has been addressed by ETSI SCP by means of changes to TS 102 223 Release 9, allowing it to transport CAT traffic over the AT interface with only minimal changes to the AT command set. The connected entity, for example, the application processor of the M2M device, can be identified and even authenticated if necessary, and indicate its supported set of CAT functionalities, which provide a complement to the functionalities that may already be provided by the modem itself.

10.5.5 Confidential Remote Management of Third-Party Applications

Another difference between the M2M and mobile communication world is that several entities involved in an M2M business agreement, such as the M2M service provider and access network operator, typically need to hold their own credentials and subscription on the same M2M device. Fortunately, the UICC platform has been designed to support multiple applications, so that a single secure element in an M2M device can secure all the associated subscriptions. However, only one entity, perhaps the device owner, owns the UICC. There are different warranty and liability implications, depending on whether the secure element warranty is bound to the UICC owning-entity or to the device, such as a car, in which it is integrated. Furthermore, a third party might not be allowed to remotely change its UICC functionality without the permission of the UICC-owning-entity, depending on the roles played by the third party in providing the service. For example, a network operator whose network access application is held on a UICC owned by an M2M service provider may only be allowed to modify the connectivity parameters.

To address such issues, the GlobalPlatform card specifications version 2.2 [GP22] and their amendment A [GP22A] provide the ability to accommodate a variety of models with great flexibility. These specifications, supported by ETSI TS 102 225 [TS102225] and ETSI TS 102 226 [TS102226] in the UICC platform, provide the ability to independently manage several security domains on a smart card, each of them representing a potentially different off-card entity and being protected by its own set of credentials. This enables each application provider to manage its security domain(s) independently of the others by means of a secure (OTA) channel providing integrity, confidentiality, and authentication. These security domains are then managed remotely for loading, installing, and personalizing applications, without compromising confidentiality with respect to the other security domains and the access network used for remote management. The possibility for establishing hierarchy of security domains in the card allows representation of potentially complex business relationships, while preserving the control of the card issuer. The card issuer can choose to manage the card content by itself or to delegate the management of security domains to third parties, so that application providers can fully operate and manage applications that they own on the card through a (third-party) network, while preserving the confidentiality of all related card content (as required, e.g., for payment-related applications). Permission to load applications can be handled by an independent controlling authority represented on the card by its own security domain. This enables all of the following scenarios:

- **issuer-centric**: Only the card issuer is allowed to manage the card content.
- **delegated management**: The issuers retain the control of loading third-party applications, but the associated third party can check the loading and perform card-content management in its security domain.
- **dual management**: Multiple entities are authorized to perform card-content management, each in their respective security domain(s).

Several solutions for the creation and handling of new security domains are also enabled, either based on a pull model with onboard key generation, or on a push model using temporary keys.

Such capabilities are essential for enabling the cohabitation on a single UICC of applications managed by an access network operator, and others managed by an M2M service provider.

10.6 Smart Card Future Evolutions in M2M Environments

10.6.1 UICC-Based M2M Service Identity Module Application

The specification for a UICC-based application (first-level UICC application in the sense of ETSI TS 102 221) managing access to the M2M service layer is still to be developed within ETSI SCP. This "M2M service identity module" (MSIM) application could be operated in conjunction with a network access application, such as USIM or CSIM, on the same UICC in the same way as the ISIM application specified in 3GPP TS 31.103 is used to secure IMS access. Such a specification is necessary to provide an interoperable definition of the related data structures stored on the UICC and the associated operating procedures. This specification should be completed by specific APIs and toolkit command extensions, allowing it to work in conjunction with toolkit applets (second-level applications in the sense of ETSI TS 102 221) to meet the needs of specific M2M vertical markets.

10.6.2 Internet Protocol Integration of the UICC

ETSI TS 102 483 [TS102483] enables easy integration of the UICC platform in TCP/IP infrastructures by standardizing the following mechanisms:

- for allocating an IP address to become integrated into the local IP networks (IPv4 or IPv6) of the hosting terminal without user interaction.
- for publishing and finding available services in the network.
- for addressing devices and applications in devices via TCP/IP.

This allows some limitations of the legacy SIM card protocols inherited by the UICC to be overcome, such as the inability to access devices that are locally connected to the hosting terminal or applications residing in the hosting terminal or in locally connected devices. This is why the IP framework works best in conjunction with the inter-chip USB UICC interface specified in ETSI TS 102 600 [TS102600]. While the legacy UICC interface of ETSI TS 102 221 may only support IP connectivity by using a proxy provided by the bearer independent protocol (BIP) feature of ETSI TS 102 223, the inter-chip USB interface of ETSI TS 102 600 enables UICC supporting a full TCP/IP stack.

To complement this specification, the GlobalPlatform 2.2 specifications referred to above [GP22] have defined a network framework enabling secure remote UICC file and application management over TCP/IP networks based on HTTP. This paradigm is now supported in the latest releases of the ETSI UICC remote management specifications, TS 102 225 and TS 102 226. Furthermore, the Open Mobile Alliance has developed a smart card web server specification [OMASCWS] that provides a powerful and flexible framework for deploying UICC-based services in IP networks by leveraging JavaCard virtual machine capabilities.

Furthermore, ETSI SCP, as well as GlobalPlatform, is working on a complete migration from the old SMS-based toolkit application and remote management framework to the IP world, based on TCP/IP, HTTP/HTTPS, and other relevant protocols from the Internet world. The work in ETSI SCP will support secure communication between devices and their applications with the UICC over IP network and leverage recent IETF developments generally referred as “Web 2.0” technologies. The work in GlobalPlatform will address the integration of the smart card into a generic IP network in a thorough manner, considering all associated technologies in the light of associated benefits and potential risks.

This whole work package will be completed in ETSI SCP by the development of an enhanced application runtime environment based on the new JavaCard 3 engine, which will fully enable UICC applications to act as secure clients or servers in TCP/IP networks.

10.7 Remote Administration of M2M Secure Elements

10.7.1 Overview

As highlighted above, ETSI SCP and GlobalPlatform specifications already include the means for secure remote management of smart cards using either IP connection or wireless network carriers, such as SMS or packet-switched data bearers. One important point to realize is that the secure remote management of a security element requires that the end-point of the security association used for remote management be located in the security element itself, as opposed to being located in the hosting device. This is a key difference with traditional device management protocols, such as OMA DM or broadband forum TR069. Since this difference is fundamental for enabling remote management of security credentials, it is expected that both types of protocol, some for secure element management and some for general device management, will continue to co-exist.

However, as enabled by current standards, the existing UICC remote management protocols have limitations that need to be addressed in the M2M context. Indeed, the smart card approach in telecommunication networks assumed that they would be bound once and for all to a unique service subscription and owned by the service providing entity, while the change of subscription would be handled by changing the SIM card in the terminal equipment (plastic roaming). As previously explained, such assumptions no longer hold true in M2M environments, where secure elements may no longer be removable and several applications require a way of managing connections remotely without having to access the devices physically.

10.7.2 Late Personalization of Subscription

The first constraint that emerged from early M2M deployments was the need for late personalization of the subscription credentials in the secure element once it has left its manufacturing facility. The subscription credentials are downloaded remotely to the secure element once it has been embedded in the hosting device, that is, outside the secure element manufacturer's facility that has been accredited for such purposes. This requires the use of trusted, and therefore duly accredited, remote management solutions. The telecommunication industry has yet to adapt the accreditation schemes needed to support this new requirement, which is necessary to reduce, in particular, inventory management costs. However, such one-shot OTA subscription-provisioning solutions are already available, although not yet standardized. They solve the logistical issues, enabling early secure element integration in the hosting device during manufacturing.

10.7.3 Remote Management of Subscriptions on the Field

There is also a need for methods to securely and remotely provision access credentials on these embedded secure elements, and manage subscription changes between service providers after they have been issued in the field, and potentially more than once during their lifetime.

10.7.3.1 Existing Deployments

Early deployments are already emerging, such as the DENATRAN (the Brazilian governmental transport authority) model being discussed in Brazil for a module to track stolen cars. In this case, the UICCs would be initially provisioned with the service access credentials of multiple operators and an activated DENATRAN profile, while this trusted authority, which manages its own HLR and OTA server, would be in charge of remotely enabling the operator subscription selected by the end-user. This model has limitations, as it provides the ability to service providers to manage their own add-on services on the UICC. It does, however, enable the pre-provision of proprietary algorithms and credentials for up to 30 service providers, giving the end-customer the ability to switch between them, as desired.

10.7.3.2 Standardization Efforts

Acknowledging the need to address the limitations of such models and to harmonize the standards based deployments instead of proprietary schemes, ETSI SCP is now addressing the standardization of a remote personalization and a subscription management solution for embedded UICCs. This new effort will propose solutions in respect of the following issues:

- The solution will involve a new role, that is, the subscription manager, who will own remote management keys and handle the different subscriptions. This role can be taken on by different players in the ecosystem but entails fundamental responsibilities to prevent potential attacks. Several models need to be distinguished, depending on who is acting as the subscription manager. To accommodate all possible business configurations, the client and server processes involved in remote personalization and provisioning, as well as the corresponding protocols and file-exchanged formats, may need to be part of the standardized solution.
- A standardized solution providing the possibility to remotely load proprietary cryptographic algorithms may be difficult to achieve, since the proper writing of the algorithm should be bound to a specific target hardware, and therefore needs to be written in native code. Therefore, a more restrictive solution may need to be considered, requiring the competing service providers to agree either on a common algorithm (and all of its subsequent evolutions) or on a restrictive list that would be preloaded with a standardized selection mechanism.

10.7.3.3 Expected Benefits

The following savings can be expected from the introduction of such embedded secure elements with remote management solutions:

- reduced distribution costs for the secure elements, due to the fact that the secure elements will be distributed with the device before personalization. This eliminates the need for stocks of secure elements.
- as the subscription will only be provisioned/personalized once the device is installed, subscription costs postponed until after installation.
- the type of subscription can be determined at the last minute, or even changed over time, introducing greater flexibility.
- the possibility to efficiently make mass changes to all subscriptions belonging to a common group (possibly using broadcast technology), e.g., for loading secure applications adapted to a specific type of target device.

In summary, such embedded secure elements are expected to allow the cost optimization of subscription management thanks to a more flexible and dynamic account-provisioning process.

10.7.3.4 Open Issues

However, beyond standardization, a number of issues arising from the remote subscription management solutions will need to be addressed by the industry. These include:

- ownership of the embedded secure element before, during and after the allocation of the new subscription.
- liability in the event of a problem occurring during reallocation to a new service provider.
- what warranty terms of the secure element are granted after the change of subscription and how.

References

[EUSD] Directive 2002/22/EC of the European Parliament and the Council of 7 March 2002 on Universal Service and Users' Right Relating to Electronic Communications Network and Services.
[TS102221] ETSI TS 102 221. Smart Cards; UICC-Terminal Interface; Physical and Logical Characteristics (2011).
[TS102671] ETSI RTS/SCP-T090071v910 Smart Cards; Machine to Machine UICC; Physical and logical characteristics (Release 9, 2011).
[TS102241] ETSI TS 102 241. Smart Cards; UICC Application Programming Interface (UICC API) for Java Card (TM) (2011).
[TS102223] ETSI TS 102 223. Smart Cards; Card Application Toolkit (CAT) (2011).
[TS102225] ETSI TS 102 225. Smart Cards; Secured packet Structure for UICC Based Applications (2011).
[TS102226] ETSI TS 102 226. Smart Cards; Remote APDU Structure for UICC Based Applications (2011).
[TS102600] ETSI TS 102 600. Smart Cards; UICC-Terminal Interface; Characteristics of the USB Interface (2010).
[TS102613] ETSI TS 102 613. Smart Cards; UICC-CLF Interface; Physical and Data Link Layer Characteristics (2011).
[TS102622] ETSI TS 102 622. Smart Card; UICC-Contactless Front-end (CLF) Interface, Host Controller Interface (HCI) (2011).
[TS102483] ETSI TS 102 483. Smart Cards; UICC-Terminal Interface; Internet Protocol Connectivity Between the UICC and Terminal (2009).
[TS102484] ETSI TS 102 484. Smart Cards; Secure Channel Between a UICC and an End-Point Terminal (2011).
[JESD22Vib] JEDEC JESD22-B103B. (2002) Variable Frequency Vibration, http://jedec.org/download.
[JESD22Shk] JEDEC JESD22-B104C. (2004) Mechanical Shock, http://jedec.org/download.
[TS27.007] GPP TS 27.007. 3rd Generation Partnership Project; Technical Specification Group Core Network and Terminals; AT command set for User Equipment (UE) (2011).
[GP22] GlobalPlatForm (2006) "GlobalPlatform Card Specification", Version 2.2 Including "Errata and Precision List" Version 0.2, http://www.globalplatform.org/.
[GP22A] GlobalPlatform (2009) GlobalPlatform Card Specification Version 2.2, Amendment A Version 1.0 Including "Errata and Precisions" Version 1.0.
[OMASCWS] OMA (2007) "Smart Card-Web-Server", OMA-TS-Smart Card-Web-Server-V1-0-20070209-C.

Part Three

Book Conclusions and Future Vision

11

Conclusions

David Boswarthick[1] and Omar Elloumi[2]
[1]*ETSI, Sophia Antipolis, France*
[2]*Alcatel-Lucent, Velizy, France*

The lessons learned from early M2M deployments have clearly identified the need to develop a set of operational common practices in order to facilitate the required growth of the M2M market.

In addition, the common capabilities that apply to all major M2M market segments are presently being developed in standards bodies and forums across the globe. The development of these common capabilities should gradually reduce operational deployment timescales, as well as making the cost structures more attractive to potential early adopters. It is important for the market to evolve by considering the lessons learned during the initial deployments of M2M services and for it to actively contribute to the development of a standard and reusable framework for M2M systems. This will ensure the interoperability of devices and services and an M2M solution that can be applied globally.

M2M Communications: A Systems Approach, the first book in the series, examines the subject of M2M using a service-requirements- and system-driven approach. It is intended to provide the reader with an overview of the main issues that should be considered when developing services destined for the M2M market. These issues include network optimization, service architecture, APIs, security and smart cards, M2M devices and modules, and IP protocol suite evolutions, but the list is not exhaustive. These key issues are all considered to be horizontal and globally applicable capabilities for M2M.

However, there is no single solution for all M2M sectors, as the different market segments each have their own specific requirements. This makes it practically impossible

M2M Communications: A Systems Approach, First Edition.
Edited by David Boswarthick, Omar Elloumi and Olivier Hersent.

for wide-scale commercial deployments to be based only upon these horizontal capabilities. This is particularly true when considering the M2M data models and the specific connectivity solutions required in the home or indeed any building environment where sensors and M2M devices are deployed. While multiple and often mature standards for data models and M2M area networks do exist, they are in general permanently evolving due to intensive market competition and the continuing evolution of M2M business. One of the challenges facing the market as it moves forward is the integration of a large number of mostly incompatible sensor technologies within the standardized single horizontal architecture. This is now becoming a hot topic within standards organizations such as ETSI M2M. The answer to this and similar questions are provided in the second book in this series: *The Internet of Things: Key Applications and Protocols*.

This second book further develops the M2M story and allows the readers to familiarize themselves with the individual vertical aspects of M2M. These include smart metering, Smart Grids, home and building automation, and the specific technologies involved in these sectors, such as ZigBee and PLC.

One of the major difficulties encountered by the authors during the writing of these books was the relatively immature level of the different standards within the M2M sector. The target is permanently moving, which is a clear indicator that global standards for M2M are still at a relatively early stage of development and are also inherently linked to the evolving market.

Take, for example, two of the major standards groups working on M2M: ETSI and 3GPP. In both standards groups, literally hundreds of M2M-related contributions are submitted at every meeting, introducing new features, evolved functionality, and alignment. In order to make these books relevant to the reader, and considering the level at which the standards evolve, it became clear that this first book was best written using a systems and architecture approach. Such a high volume of industry contributions and investment from many large ICT companies is clear proof that M2M will have a far-reaching impact, not only on the ICT standards ecosystem and the resulting systems that will be deployed in the coming years, but also on our daily lives. M2M and the future IoT services and applications will certainly have a profound impact on this decade, bringing technological innovation to the home and workplace, and helping to make our planet greener and cleaner. Global standards are just one stepping stone along this road, but an essential one to ensure the smooth rollout of innovative M2M solutions.

Index

M2M Communications: A Systems Approach, First Edition.
Edited by David Boswarthick, Omar Elloumi and Olivier Hersent.